Progress in Inflammation Research

Volume 86

The last few years have seen a revolution in our understanding of how blood and tissue cells interact and of the intracellular mechanisms controlling their activation. This has not only provided multiple targets for potential anti-inflammatory and immunomodulatory therapy, but has also revealed the underlying inflammatory pathology of many diseases.

This series provides up-to-date information on the latest developments in the pathology, mechanisms and therapy of inflammatory disease. Areas covered include: vascular responses, skin inflammation, pain, neuroinflammation, arthritis cartilage and bone, airways inflammation and asthma, allergy, cytokines and inflammatory mediators, cell signalling, and recent advances in drug therapy.

Each volume is edited by acknowledged experts providing succinct overviews on specific topics intended to inform and explain. The series is of interest to academic and industrial biomedical researchers, drug development personnel and rheumatologists, allergists, pathologists, dermatologists and other clinicians requiring regular scientific updates.

More information about this series at http://www.springer.com/series/4983

Tamás Röszer

The M2 Macrophage

Springer

Tamás Röszer
Institute of Neurobiology
Ulm University
Ulm, Baden-Württemberg, Germany

ISSN 1422-7746 ISSN 2296-4525 (electronic)
Progress in Inflammation Research
ISBN 978-3-030-50482-3 ISBN 978-3-030-50480-9 (eBook)
https://doi.org/10.1007/978-3-030-50480-9

This Springer imprint is published by the registered company Springer Nature Switzerland AG.
The registered company address is: Gewerbestrasse 11, 6330 Cham, Switzerland

Preface

היאמר חומר ליוצרו מה־ תעשה

The clay doesn't ask [the potter],
"Why did you make me this way?"
Isa. 45:9

Macrophages are foot soldiers of the innate immune response and probably the evolutionarily most conservative type of immune cells. Their key functions – uptake and neutralization of foreign cells and particles by phagocytosis – can be recognized in the first living cells, and because of the key importance of these functions, they have remained resistant to changes in the course of evolution of the living matter. Today, effector functions of macrophages are conceptualized by the so-called M1/M2 model of polarization, which considers that macrophage activation triggers polarization toward a pathogen killing and inflammatory M1, or an antiinflammatory and tissue-healing M2 state. Of note, some further classifications of macrophage activation states are unfolding as this book is written. Nevertheless, the M2 macrophage is unusual in the sense that it does not fit into the role one may expect from an innate immune cell. Instead of killing pathogens and protecting the host from invader cells, M2 macrophages often tolerate cellular pathogens, parasites, and cancer cells, and have an antiinflammatory, tolerogenic profile.

The roots of M2 macrophage behavior can be traced back to the first unicellular eukaryotes, which are thought to be phagotrophic cells which have overseen invader prokaryote cells, allowing them to evolve into endosymbiosis and develop cell organelles such as mitochondria and the plant-type plastid system. This happening was a "fortunate mistake," opening the avenue for the development of the eukaryote organisms. Similarly, the first multicellular organisms have M2 macrophage-like phagocytes, which may temporally tolerate foreign cells and utilize them as food. The latter is exemplified by social amoeba colonies that establish farming symbiosis with bacteria and archaea. In tissue-forming animals, homeostatic tissue renewal and regeneration of damaged tissues are the major tasks of M2 macrophages, which serve for maintenance of tissue integrity, and secondarily, support the metabolism of the organism. In vertebrates, M2 macrophages can form a network of tissue-resident macrophage niches, and govern functions at the interface of immunity, tissue development, and turnover, metabolism, and endocrine signaling.

It seems that M2 macrophages find a modus vivendi between "self" and "foreign" cells when the energy demands of the organism requires it, and support the immunological tolerance of "self" cells, which ultimately sustains organism growth. M2 macrophage traits are hence necessary for life. Pathogens, however, have adapted to the traits of M2 macrophages, and have found their own niche within these macrophages. Intracellular parasites utilize metabolites of M2 macrophages, hide inside M2 macrophages from other immune cells, and evade immune response by skewing macrophages into the M2 state. Moreover, dysfunction of M2 macrophages, caused for instance by factors of our industrialized and overconsuming lifestyle, disrupts the healthy interplay between the immune system and metabolic processes, and causes diseases such as insulin resistance, metabolic syndrome, and type 1 and 2 diabetes mellitus. It also leads to allergy, autoimmunity, and cancer.

"The clay doesn't ask [the potter]: Why did you make me this way?" However, the researcher raises such questions, with the intent to understand why living matter has adopted the forms and functions in the way one can observe them today. And because the subject of this book is the M2 macrophage, we raise herein the question: "Why have M2 macrophages been designed by the evolution of living matter in that way?" This book aims to answer this question, and summarizes what we have learnt so far, together with the key unresolved questions in the pipeline. I am thankful to Springer for soliciting this book, and for the opportunity to write it. I hope that readers will find it a useful compass navigating through the biology of the M2 macrophage.

Ulm, Baden-Württemberg, Germany — Tamás Röszer

Acknowledgments

I would like to acknowledge image sources. Fig. 1a, c of chapter "What Is an M2 Macrophage? Historical Overview of the Macrophage Polarization Model. The Th1/Th2 and M1/M2 Paradigm, the Arginine Fork" were been acquired at the Spanish National Cardiovascular Research Center (CNIC), Madrid, Spain. Figure 2 of chapter "What Is an M2 Macrophage? Historical Overview of the Macrophage Polarization Model. The Th1/Th2 and M1/M2 Paradigm, the Arginine Fork" displays reprinted illustrations; the original publications are listed in the figure legend. In Fig. 3d of chapter "What Is an M2 Macrophage? Historical Overview of the Macrophage Polarization Model. The Th1/Th2 and M1/M2 Paradigm, the Arginine Fork" the artwork has been licensed from Dreamstime Stock Images (7900767, Stefan Hermans). Figures 4 *left* and 5c of chapter "What Is an M2 Macrophage? Historical Overview of the Macrophage Polarization Model. The Th1/Th2 and M1/M2 Paradigm, the Arginine Fork", 9c *insert* of chapter "Practical Approaches in M2 Macrophage Biology: Analysis, Pharmacology, and Didactical Interpretation of M2 Macrophage Functions", and Graphic Abstract 2 display standing male reference images are licensed from the image repository Human Anatomy for Artists (human-anatomy-for-artist.com), with a royalty-free commercial license and with model release. Figures 4 *right* of chapter "What Is an M2 Macrophage? Historical Overview of the Macrophage Polarization Model. The Th1/Th2 and M1/M2 Paradigm, the Arginine Fork", 4 and 5 of chapter "Practical Approaches in M2 Macrophage Biology: Analysis, Pharmacology, and Didactical Interpretation of M2 Macrophage Functions" display photographs of the artwork *Statue du tombeau de René de Chalon, le "Squelette,"* attributed to Ligier Richier, 1550 A.D., Centre Pompidou, Metz, France; available in Wikimedia Commons, with a CC license. Credit line: H. Zell/CC BY-SA. Figures 1c of chapter "Immune Functions of the M2 Macrophages: Host Defense, Self-tolerance, and Autoimmunity", 5 B1–B2 of chapter "M2 Macrophages in the Integument and in the Musculoskeletal System", 2a, b of chapter "M2 Macrophages in the Metabolic Organs and in the Neuroendocrine System", and 4 chapter "Practical Approaches in M2 Macrophage Biology: Analysis, Pharmacology, and Didactical Interpretation of M2 Macrophage Functions" display images licensed from Dreamstime Stock Images (406619001,

Kirill Makarov; 155424386, Katerynakon; 12702128, Sarah2; 100077094, Zoltan Tarlacz; 75575001, Jirka13, respectively). Figure 7c of chapter "Practical Approaches in M2 Macrophage Biology: Analysis, Pharmacology, and Didactical Interpretation of M2 Macrophage Functions" displays Adam and Eve with apple and serpent MET DP820307.jpg image, which is available from Wikimedia Commons, under a CC license. This file was donated to Wikimedia Commons as part of a project by the Metropolitan Museum of Art. Credit line: Marcantonio Raimondi/CC0. Figure 1c of chapter "Immune Functions of the M2 Macrophages: Host Defense, Self-tolerance, and Autoimmunity" shows a montage of diatom cells, using images of multiple fields. License numbers of Springer-Nature imprints reused in this publication: 4793590901404 (Langhans cells in Fig. 1 of chapter "What Is an M2 Macrophage? Historical Overview of the Macrophage Polarization Model. The Th1/Th2 and M1/M2 Paradigm, the Arginine Fork"), 4793591127332 (chapter "Signal Mechanisms of M2 Macrophage Activation" of Sect. 5), 4793591338463 (mCT images in Fig. 4c of chapter "M2 Macrophages in the Integument and in the Musculoskeletal System"), 4793600021166 (Fig. 2a–e of chapter "M2 Macrophages in the Circulatory, Respiratory, and Excretory Organs"), 4793600191583 (Fig. 4 of chapter "M2 Macrophages in the Integument and in the Musculoskeletal System"). Images displayed in Fig. 2 of chapter "What Is an M2 Macrophage? Historical Overview of the Macrophage Polarization Model. The Th1/Th2 and M1/M2 Paradigm, the Arginine Fork" are, to my knowledge, in the public domain. Sketches prepared for some anatomical drawings were displayed in lectures and lecture notes (Fig. 7a of chapter "Evolutionary Origin of M2 Macrophage Activation: Invertebrates", Fig. 3b of chapter "Evolution of M2 Macrophage Functions in Chordates" were shown as black-and-white sketches in a lecture note, titled Practicum Zooanatomicum, printed by Debrecen University Press, Hungary, in 2002) during my lecturing activities. Macrophage electron micrographs in Figs. 3a, 4, 6, 7 of chapter "What Is an M2 Macrophage? Historical Overview of the Macrophage Polarization Model. The Th1/Th2 and M1/M2 Paradigm, the Arginine Fork", 3, 4, and 5 of chapter "Practical Approaches in M2 Macrophage Biology: Analysis, Pharmacology, and Didactical Interpretation of M2 Macrophage Functions" are from *Mediators of Inflammation* 2015, 16. The mouse embryo image shown in Graphic Abstract 2 is from *Journal of Leukocyte Biology* 102, 845–855. Schemes in the figures were created with the use of Motifolio tools. The author´s portrait image was taken in Ulm, Germany, in May 2020.

Histology specimens that were photographed as illustrations were obtained in research projects and are from histology specimen collections used for university teaching purposes. Unless otherwise indicated, the histology images, electron micrographs, and anatomical drawings are the author's works, and were created at University of Debrecen, Hungary; at the Spanish National Cardiovascular Research Center (CNIC), Madrid, Spain; and at Ulm University, Germany.

Contents

Part II Specific M2 Macrophage Functions

About the Author

Tamás Röszer is a Research Group Leader and Lecturer (Privatdozent) at the Institute of Neurobiology, Ulm University, Germany. His chief research interests are the endocrine control of immunity and metabolism, and macrophage biology in the context of development and physiology. He is the author of several research articles in the field of immunometabolism, diabetes, and comparative molecular endocrinology, and reference books in nitric oxide biology and pathophysiology.

The original version of this book was revised: This book was initially published with incorrect ISSN numbers which has been corrected now. The correction to this book is available at https://doi.org/10.1007/978-3-030-50480-9_12

Part I
General Concepts of M2 Macrophage Biology

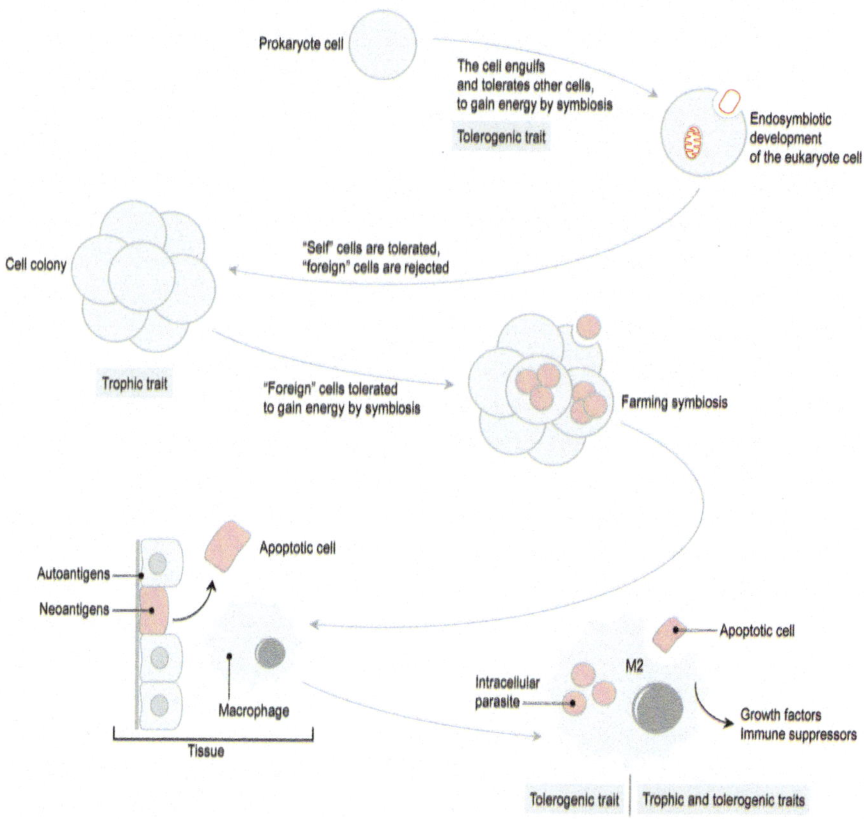

Graphic Abstract of Part I

M2 Macrophage Traits Placed on the Landscape of Evolution

The hypothetical forefather of cellular life is a phagotrophic cell that takes up other cells as food by phagocytosis. Some of the phagocytosed cells have survived and were tolerated by the ancient phagotrophic cell, allowing the development of endosymbiont organelles. This event is a forerunner of the tolerogenic M2 trait, which allows symbiosis or intracellular parasitism. Multicellularity was associated with another M2 trait: trophic features that allow colony expansion and the growth of the organism by supporting metabolism. The tolerogenic M2 trait also serves metabolism by allowing energy gain from symbiont cells. Tolerance of the symbiont cells requires immunosuppression, which is one task of M2 macrophages. Tissue development needs the discrimination of autoantigens from neoantigens that are generated during apoptosis. Apoptosis is key in tissue morphogenesis and turnover. M2 macrophages unify the tolerogenic and trophic traits of their ancestors, and support autoantigen tolerance, neutralization of neoantigens, growth of tissues, and suppression of inflammation that would harm the metabolism of the organism.

What Is an M2 Macrophage? Historical Overview of the Macrophage Polarization Model. The Th1/Th2 and M1/M2 Paradigm, the Arginine Fork

Keywords Macrophage polarization · Alternative macrophage activation · M2 activation · Type 2 macrophage activation · M2 markers · Macrophage · Th2 cytokines · Wound repair · Allergy

1 Introduction

As a method of Socratic teaching, it is logical to begin the understanding of M2 macrophage biology by asserting a definition of the M2 macrophage. As of 2020, when this book is being printed, biologists define macrophages with antiinflammatory, profibrotic, pro-resolving, self-tolerating, regulatory, homeostatic, tolerogenic, neuroprotective, angiogenic, vasculogenic, tumor-tolerating, and tumorigenic roles as M2 macrophages [1–6]. These traits may appear separately or in combination, and vest in M2 macrophages an exceptionally wide repertoire of functions in tissue development, wound healing, immunological self-tolerance, inflammation resolution, immune suppression, Th2 immune response, and tumor development. M2 macrophages are hence often termed as homeostatic, repair, antiinflammatory, pro-resolving, regulatory, tolerogenic, alternatively activated, type 2, deactivated, tumor-associated, or tumorigenic macrophages. The process that results in macrophages displaying these listed features is called, accordingly, M2 macrophage activation, alternative macrophage activation, or type 2 macrophage activation, and sometimes helminth-induced or Th2 cytokine-induced macrophage activation, referring to the immune context in which M2 macrophages appear. M2 macrophages are also defined and classified based on their effector mechanisms and the type of signals that trigger their activation, and based on their specific gene transcription and protein expression patterns. This chapter aims to help readers to understand better some key concepts of M2 macrophage biology and M2 macrophage terminology.

T. Röszer, *The M2 Macrophage*, Progress in Inflammation Research 86,
https://doi.org/10.1007/978-3-030-50480-9_1

2 Classical Macrophage Activation: A Historical Overview

Macrophages are phagocytosing leukocytes and are key elements of the innate immune response [7]. Mammalian macrophages were first identified in the biomedical literature of the nineteenth century, often being described as phagocytes, referring to their prominent phagocytosis ability[1] [7, 8]. For instance, the most abundant macrophages in mammals were described by Karl Wilhelm von Kupffer in 1876 as phagocytosing endothelial cells ("reticuloendothelial cells") of the mammalian liver, and in 1898 these cells were identified as resident tissue macrophages by Tadeusz Browicz [7, 9]. There are even earlier historical accounts of phagocytosing immune cells [7]. Macrophages phagocytose and kill pathogens, and the immunological impact of phagocytosis was explored by Ilya Metchnikoff [Илья Мечников] at the turn of the nineteenth to twentieth century [10, 11].

Key features of a macrophage include substrate adherence and amoeboid movements, phagocytosis of particles, and chemotaxis toward chemical stimuli (Fig. 1). It was recognized early in the history of macrophage biology that macrophages, similarly to other leukocytes, accumulate in inflamed or damaged tissues, and that their phagocytosis capacity increases in response to certain stimuli.[2] Indeed, the first historical accounts of mammalian macrophages observed these cells in inflamed

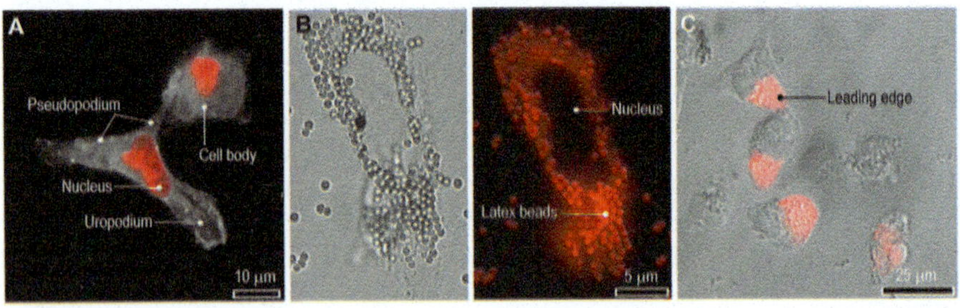

Fig. 1 Morphology and functions of activated macrophages. (**a**) Amoeboid movements and substrate adherence: macrophages cultured in vitro show protrusions of the cell body, which ensure the amoeboid motion of the cell. The fluorescent signal indicates cell adhesion molecules, which anchor the cell body to the substrate. *Red* fluorescence labels nuclear DNA. (**b**) Phagocytosis: a macrophage that is packed with a large amount of phagocytosed red fluorescent latex beads. *Left:* phase contrast image; *right:* fluorescently illumined image. (**c**) Chemotaxis: macrophages migrating toward a chemotactic stimulus in vitro. The leading edge, ensuring forward motion of the macrophage, is rich in cytoskeletal elements, including paxillin. The cells were labeled with a fluorescent antibody against paxillin. (Images by author)

---

[1]The history of macrophage biology has been summarized in the monograph by Ian Carr (*The Macrophage: A Review of Ultrastructure and Function*, Academic Press, 1973).

[2]In the biomedical literature of the late nineteenth century and early twentieth century, macrophages were often denoted with terms now outdated, such as histiocytes, reticuloendothelial cells, adventitial cells, or epithelioid cells. For example, Kupffer cells (hepatic resident macrophages) were described by von Kupffer as hepatic stellate cells ("Sternzellen"). Of note, leukocytes in

tissues [7]. Because phagocytosis is the key weapon of macrophages in the battle against pathogens, increased phagocytosis was considered to be an indicator of the "alarmed," or activated, state of macrophages [7]. Moreover, the ability of macrophages to migrate into damaged or infected tissue by chemotaxis was also considered to be a hallmark of macrophage activation [12]. Macrophage activation was hence primarily used as a term to denote a pathogen-engulfing, pathogen-killing state of macrophages. Activated macrophages show morphological features of their microbicidal activity: they bear several membrane protrusions (filopodia and lamellipodia) that ensure their chemotactic motility and are rich in phagosomes and lysosomes for the uptake and lysis of pathogens (Fig. 1).

However, macrophages, the so-called tissue resident macrophages, have also been identified in noninfected tissues, in which such an activated state seems unnecessary. In this setting, the main macrophage function has been determined as a safe disposal of waste materials [13]. With the development of the understanding of immune cell functions, tissue resident macrophages were considered sentinel cells that present antigens for cells of the adaptive immune response [13]. Macrophages are also present in tumor tissue, and sometimes they seem to be unable to kill pathogens. Metchnikoff himself has described the phenomenon that some pathogens survive phagocytosis, grow within the cytoplasm of infected macrophages, and that the macrophages thus can spread certain infections [14]. Metchnikoff also has noted that some pathogens survive longer within the macrophage than outside of it [14] (Fig. 2). Macrophages hence seem to nurture some pathogens. Today we know that intracellular pathogens such as *Mycobacterium tuberculosis*, *Toxoplasma gondii*, or *Listeria monocytogenes* can skew macrophage immune response, survive within the macrophages, and benefit from metabolites provided by the macrophages. Similarly, tumor cells may be unnoticed by some macrophages (these traits of M2 macrophages are detailed in chapter "Immune Functions of the M2 Macrophages: Host Defense, Self-tolerance and Autoimmunity").

The presence of tumor antigens or pathogens alone hence does not necessarily activate macrophages. For example, the so-called giant Langhans cells,[3] described in tuberculotic granulomatous tissue in the nineteenth century, surround infectious foci [18]. It was already observed in the 1920s that in some granulomatous tissues these cells contain leukocyte remnants and red blood cells [18], suggesting the ability of phagocytosis. In virulent tuberculotic infections, however, such sign was lacking [18]. Decades later the Langhans giant cells were identified as multinucleated macrophages [19–21] (Fig. 2) that develop as a result of the fusion of macrophages.[4]

general were often described as "phagocytes" in the early twentieth-century textbooks, and lymphocytes were described as "microphages" and monocytes as "macrophages."

[3] Named after Theodor Langhans (1839–1915), the Langhans cell is distinct from the antigen-presenting Langerhans cells of the skin and the insulin-producing pancreatic beta cells or Langerhans cells. The Langerhans cells are named after Paul Langerhans (1847–1880).

[4] Langhans cells can develop in response to various pathogens. In addition to the Langhans cells, other forms of fused macrophages are known: e.g., osteoclasts, which are resident macrophages of the bone, also develop as a result of macrophage fusion; crown-like structures in the obese adipose

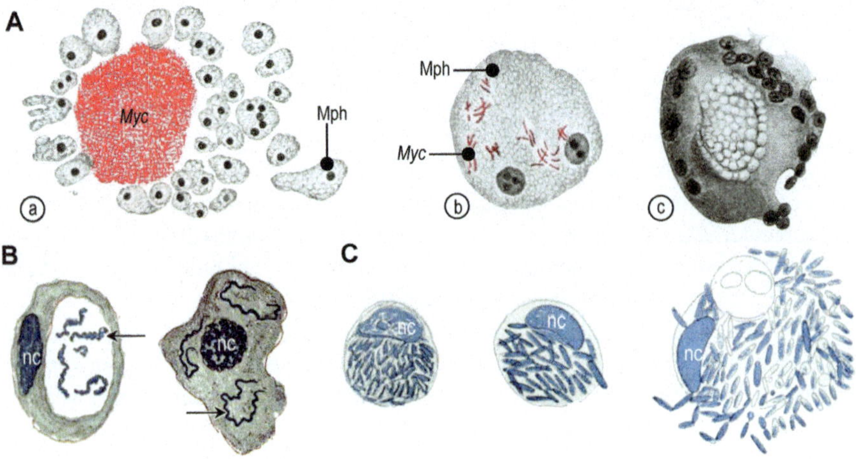

Fig. 2 Intracellular parasitism and tolerance of pathogens by macrophages: accounts from the nineteenth century. It was recognized, early in the history of microbiology, that some bacteria grow extracellularly, whereas others need a living cell host and these grow intracellularly [15]. (**a**) Pathogens such as *Mycobacterium tuberculosis* can be taken up by macrophages without being killed. (a) Drawing of macrophages (*Mph*) gathering around *Mycobacterium* cells *(Myc)*. (b) After being taken up the cells persist in the macrophages. (c) Eventually, the macrophages fuse to form giant cells, called Langhans cells, with reserves of pathogens. (Drawings from the early twentieth century, reprinted from [16]). (**b**) The classical experiments of Metchnikoff. Guinea pig macrophages containing *Spirillum* cells *(left)* and *Spirochaeta* cells *(right)*. The pathogens (labeled with *arrows*) survive within the phagosome *(left)* or in the cytoplasm *(right)* of the macrophages. *Nc*, nucleus. (**c**) Guinea pig macrophage taking up plague bacilli *(Yersinia pestis)*. The bacilli grow inside the macrophage and eventually cause macrophage rupture. (Drawings reprinted from Metchnikoff (1907) [14].) We now know that *Yersinia* species induce a so-called Th2 immune response that favors the development of M2 macrophages [17]

We now know that macrophage fusion is increased by interleukin-4 (IL-4) [22], a typical cytokine that polarizes macrophages toward the M2 state in mammals. The Langhans giant cells may either adopt a pathogen-killing (M1) phenotype with preserved phagocytosis ability or may have diminished ability to phagocytose, thus permitting the growth of *Mycobacterium tuberculosis* [19, 21]. This function is actually an M2 macrophage activation, resulting from the ability of *Mycobacterium* cells to stop M1 activation at its initial stage, and skews macrophage activation toward an M2 state [23, 24] (see details in chapter "Immune Functions of the M2 Macrophages: Host Defense, Self-tolerance and Autoimmunity").

The outcome of an infection, or the emergence of a tumor cell into a malignant lesion, depends on the early action of macrophages. M1 activated macrophages neutralize pathogens and reject tumor cells. M2 macrophages may tolerate pathogens

---

tissue are also fused macrophages; and macrophages in tumors can fuse to form capillaries (so-called vascular mimicry).

and do not reject tumor cells. Hence, it is vital to understand the mechanisms that activate macrophages toward the M1 or M2 state.

3 Alternative Macrophage Activation

3.1 The M1/M2 Polarization Paradigm: Dichotomy of Macrophage Response to Th1 and Th2 Cytokines

It remained elusive, in the 1960s, how *Mycobacterium* cells may evade the immune response of macrophages. It was thought that macrophage activation is determined by a "cell resistance factor," which may be related to the function of lymphocytes [25]. A study published in 1969 states that "antibacterial resistance conferred with immune lymphoid cells is not due to antibacterial antibody; it is mediated indirectly through the macrophages of the recipient. These become activated by a process which appears to depend upon some form of specific interaction between the immune lymphoid cells and the infecting organism." [26]. Observations have suggested that lymphocytes determine whether macrophages are being activated. Further evidence of lymphocyte–macrophage communication was confirmed by the end of the 1970s [27], when it also was shown that T lymphocytes have functionally distinct forms, including the so-called T helper 1 (Th1) and T helper 2 (Th2) cells [28, 29]. Th1 lymphocytes elicit the so-called Th1 immune response by secreting interferon gamma (IFNγ), interleukin-2 (IL-2), or tumor necrosis factor beta (TNFβ), or the Th2 immune response by secreting interleukins IL-4, IL-5, IL-6, IL-10, and IL-13.

Studies by Stein et al., Gordon et al., Fiorentino et al., Munder et al., Mantovani et al., and Mills et al. in the 1990s indicated that macrophages during Th1 and Th2 immune response have distinct immune functions and metabolic features, indicating a functional dichotomy of macrophage activation [30–34]. An IL-4- and IL-10-induced macrophage activation was observed, initially named as alternative macrophage activation [3, 4, 30–34]. In a Th1 immune response, IFNγ activates macrophages to produce inflammatory cytokines and nitric oxide (NO), which allow pathogen killing and cause inflammation (Fig. 3a). In contrast, Th2 immune response diminishes inflammatory cytokine and NO production by macrophages and increases arginase and mannose receptor expression [4, 30, 32] (Fig. 3a).

The dichotomy of Th1- and Th2-induced macrophage functions is well exemplified by the immune response to *Leishmania* infection. *Leishmania* species are intracellular parasites; leishmaniasis is a tropical and subtropical zoonosis, transmitted by the bite of species of sandfly [35]. In some mouse strains, such as B10D2 and SV/129 and the widely used C57/BL6 strain, *Leishmania* elicits a Th1 immune response, allowing effective elimination of the pathogen [36, 37]. The burden and progression of the infection is mouse strain dependent, and when Th1 cytokine production is disrupted, the mice become susceptible to *Leishmania* infection [37, 38].

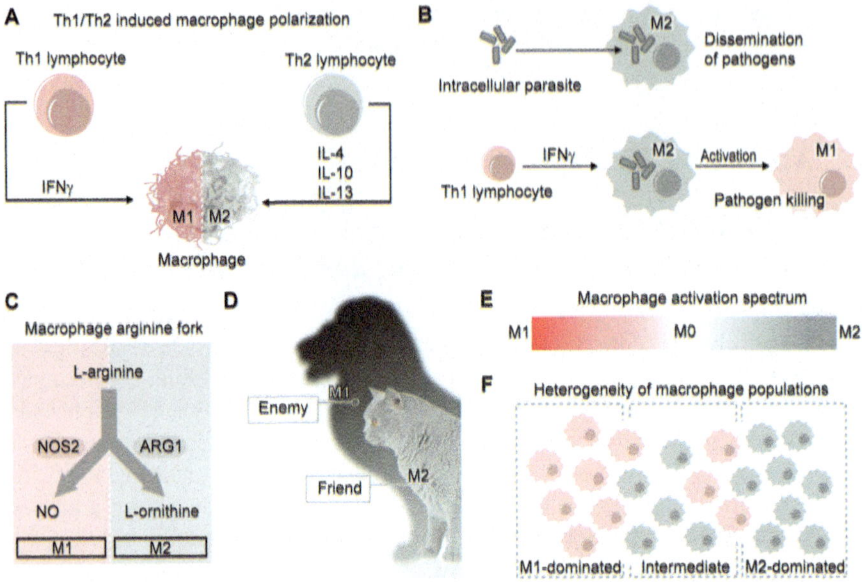

Fig. 3 Development of the M1/M2 macrophage concept from the 1990s to the 2010s. (**a**) Th1 and Th2 cytokines polarize macrophages toward M1 and M2 states. (**b**) Intracellular pathogens may avoid M1 polarization, or maintain M2 polarization. Th1 cytokines serve as triggers of M1 polarization, which eventually eliminates intracellular pathogens. (**c**) Arginine metabolism of M1 and M2 macrophages is distinct in the mouse. M1 macrophages produce cytotoxic amounts of NO from L-arginine, which is an effective pathogen-killing mechanism. M2 macrophages use L-arginine to generate L-ornithine, which can be further used for polyamine and collagen synthesis. (**d**) M2 macrophages support tissue homeostasis and regeneration; M1 macrophages may cause tissue damage. However, M2 macrophages can also cause disease. (**e**) Macrophages may adopt activation states that show hallmarks of both M1 and M2 states. Nonactivated macrophages are often considered as M0 macrophages. (**f**) Tissues usually contain mixed populations of M1 and M2 macrophages

However, some mouse strains, such as BALB/c and DBA/2, respond with Th2 immune response to the same pathogen, which makes these strains more susceptible to *Leishmania* infection [30]. In C57BL/6 mice, acute *Leishmania* infection causes a greater induction of IFNγ gene expression in draining nodes and spleen than in BALB/c mice. In contrast, BALB/c mice express IL-4 during *Leishmania* infection, which response is lacking in C57/BL6 mice [36]. The difference in the cytokine profile of these mouse strains explains the difference in their immune responses and, eventually, their macrophage functions.

The Th1-induced, inflammatory, NO-producing, pathogen-killing macrophage activation has been termed M1, or classical macrophage activation. Macrophage activation caused by Th2 cytokines has been termed M2, or alternative macrophage activation (Fig. 3a) [30]. Certain pathogen-associated molecules, such as lipopolysaccharides (LPS) of the cell wall of gram-negative bacteria, induce M1 activation. However, intracellular pathogens such as *Mycobacterium* cells inhibit M1 macrophage activation, induce or maintain the M2 macrophage activation state (see chapter "Immune Functions of the M2 Macrophages: Host Defense, Self-tolerance and Autoimmunity" for details), and thus survive within the infected macrophages. In this setting, Th1 cytokines are necessary to initiate M1 activation and, eventually, kill pathogens (Fig. 3b). In summary, M2- or alternative macrophage activation was initially considered as a result of Th2 cytokine signaling. The Th2 cytokine-induced macrophage phenotype was considered as an alternative activation state, and today this is what we consider M2 macrophage activation [4].

3.2 The Arginine Fork and the Concept of M1/M2 Macrophage Polarization

It has also been pointed out by early studies that the L-arginine metabolism of M1 and M2 macrophages is distinct [30]. It was already reported in 1977 that arginine metabolism, and specifically, arginase activity, determines the cytotoxic activity of macrophages [39]. In the early 1990s it was shown that arginase activity also determines whether macrophages eliminate or tolerate cancer cells [40]. Today, we know that M1 macrophages use L-arginine to synthesize nitric oxide (NO) through L-arginine oxidation by inducible NO synthase (iNOS or NOS2; EC 1.14.13.39). However, M2 macrophages have compromised or negligible NOS2 expression, and arginine is metabolized by arginase-1 (EC 3.5.3.1) in the urea cycle. Its substrate is the amino acid L-arginine, and it produces L-ornithine and urea. M2 macrophages increase their arginine/ornithine metabolism in response to LPS but M1 macrophages do not. This divergence of arginine metabolism during M1 or M2 activation has led to the "macrophage arginine fork" model (Fig. 3c) [41, 42]. This model suggests that the dominance of NOS2 or arginase-1 determines the outcome of L-arginine metabolism. Although the two enzymes are located in distinct cellular compartments, thus they do not compete for the same L-arginine pool [43]; the relative abundance of NOS2 in M1 macrophages favors NO synthesis. The generated NO kills pathogens. Accordingly, in M2 macrophages the lack of NOS2 and the presence of arginase-1 channels L-arginine metabolism toward ornithine synthesis. Arginase-1 activity generates L-ornithine that may enter polyamine and collagen biosynthesis, eventually promoting fibrosis and tissue healing [44]. Moreover, arginase inhibits other L-arginine-dependent immune functions [45]. L-Arginine depletion suppresses T-cell proliferation [44], which may allow arginase-1-expressing macrophages to dampen the $CD4^{+}$ T-cell effector response [45]. Although this reduces tissue damage in the course of host defense in helminth infections [46], it may worsen immunodeficiency [47].

Because NO kills pathogens, and ornithine is a substrate for polyamine synthesis, which supports collagen synthesis, an additional concept has emerged: a dichotomic shift of macrophage arginine metabolism results in a cytotoxic M1 or a pro-resolving, profibrotic, tissue-regenerating M2 macrophage (Fig. 3c). The concept of a pathogen-killing M1 and a tissue-healing M2 macrophage function was formulated by Charles D. Mills [13, 30, 41, 48].

There are further metabolic differences between M1 and M2 macrophages. For instance, M1 macrophages rely on glycolysis, whereas M2 macrophages have an active fatty acid oxidation and oxidative phosphorylation [49]; M2 macrophages may express mitochondrial uncoupling protein [50]; and M1/M2 macrophages have a distinct repertoire of lipid-metabolizing enzymes [2, 51].

However, the arginine fork does not exist in human macrophages. Upon stimulation with IL-4 and IL-13, or in response to parasitic infection, human monocyte-derived macrophages do not express arginase-1. Instead, some other proteins, such as CLEC10A (human macrophage galactose-type C-type lectin), MRC1 (C type 1 mannose receptor, equivalent of mouse CD206), E-cadherin, PD-L1 (programmed death-ligand 1), and CD86 hallmark the M2 state of human macrophages [52, 53]. Although the arginine fork is lacking from human macrophages, the human M2 macrophages still control Th2 immunity and have an immunosuppressive role. Moreover, human M2 macrophages have reduced expression of MHC-I and MHC-II and of the pathogen-recognizing toll-like receptors TLR3, TLR5, and TLR7 [54].

4 Pros and Cons of M2 Macrophage Functions

Antiinflammatory and tissue-regenerating M2 macrophage functions have been often overemphasized in the biomedical literature of the 2000s (Fig. 3d). In this paradigm, homeostatic tissue functions need M2 macrophages, and to restore tissue lesions caused by M1 macrophages. Depicting M2 macrophages as pious servants of tissues is a likely result of another paradigm of the same era, in which chronic inflammation was considered as a common etiology of several diseases, such as atherosclerosis, cardiovascular diseases, fatty liver disease, insulin resistance, metabolic syndrome, autoimmunity, and neurodegenerative diseases. At various stages of these pathologies, M1 macrophages have certain roles; in most cases they worsen the disease and accelerate its progression. Because M2 macrophages have antiinflammatory effects, the hypothesis was somewhat feasible that reducing M1 macrophage number in favor of M2 macrophages, or restoring a homeostatic balance between M1 and M2 macrophages, could resolve chronic inflammation (this concept is detailed in chapter "Practical Approaches in M2 Macrophage Biology: Analysis, Pharmacology and Didactical Interpretation of M2 Macrophage Functions"). Thus, reprogramming macrophages toward M2 activation has become a therapeutic aim in the recent decade of macrophage research.

Although considerable progress has been achieved in the understanding of M2 macrophage biology, the idea that M2 macrophages antagonize M1 macrophages in chronic inflammatory diseases has been challenged, at least partly because macrophage functions that do not fit into the M1/M2 polarization model were observed. More importantly, it also has been recognized that there are beneficial M1 and unfavorable M2 macrophage functions in specific diseases. For instance, atherosclerotic plaque development is associated with TNFα-expressing, hence M1-like macrophages, and various antiinflammatory (M2, and M2-like, atherosclerotic plaque-specific macrophage subsets) inhibit lesion progression, whereas at least one subset of M2 macrophages is pro-atherogenic. This subset of atherosclerotic plaque macrophages, which express the hemoglobin–haptoglobin scavenger receptor CD163 and are considered to be M2 macrophages, promotes angiogenesis, vessel permeability, and leukocyte infiltration in human and mouse atherosclerotic lesions [55]. Of note, here the angiogenic effect of M2 macrophages is unfavorable. Otherwise, in a different context, for instance, in the setting of tissue development or repair, the same effect would be favorable (Fig. 4).

Another example is the macrophage function in adipose tissue. Adipose tissue harbors macrophages, so-called adipose tissue macrophages (ATMs). ATMs comprise 10% of all cells in the adipose tissue under physiological homeostasis, which increases to 50% in obesity [56]. Being prevalent in obesity, ATMs were first documented in adipose tissue of morbidly obese patients and in extreme obese animal models [57, 58]. In obesity, ATMs gather around lipid-overloaded adipocytes and acquire the M1 activation state [59–68]. ATMs in M1 release inflammatory mediators, which inhibit insulin signaling in the adipose tissue and, by entering the circulation, provoke systemic insulin resistance and exacerbate pancreatic β-cell destruction and type 1 diabetes [60–62, 65, 69] (also see Chapter "M2 Macrophages in the Circulatory, Respiratory and Excretory Organs").

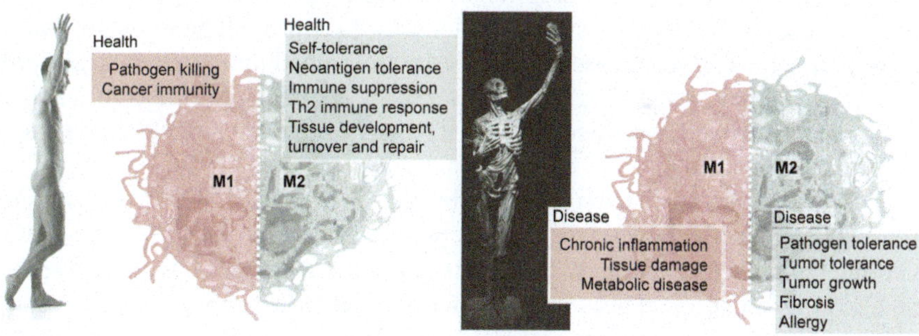

Fig. 4 Functions of M1 and M2 macrophages in health and disease. *Left:* Physiological functions of M1 and M2 macrophages traits. M1 traits are essential for the immune response, elimination of foreign and cancer cells, initialization of tissue healing, and for certain steps of tissue development. M2 traits are required for autoantigen and neoantigen tolerance, immune suppression, resolution of inflammation, so-called Th2 immune response, tissue development, tissue turnover (e.g., apoptotic cell clearance), and tissue repair. *Right:* However, both M1 and M2 macrophage traits can cause disease. M1 macrophages are involved in chronic inflammatory diseases, causing tissue damage and metabolic diseases. M2 macrophages may tolerate certain pathogens and cancer cells, cause fibrosis, and sustain excess Th2 immune response (e.g., allergy). (Image credits: *allegory of health* has been licensed from Human Anatomy References for Artists image repository; *allegory of disease* is a photograph of a sculpture by the late Gothic French sculptor Ligier Richer (ca. 1500–1567). Further license information is provided in the Acknowledgments. Macrophage structures are author's images)

In such critical situations, reprogramming M1 ATMs into an M2 state is certainly the least worst solution. M2 ATMs were hence anticipated to support insulin sensitivity, thus protect from insulin resistance [67, 69]. However, the metabolic role of ATMs is not so simple [67, 70, 71]. For instance, an excess number of M2 ATMs also can have unwanted metabolic effects [72]. M2 macrophages promote fibroblast growth through TGFβ and the chemokine (C-C) motif ligand 18 (CCL18) secretion [73], and they may also undergo a fibrocyte-like phenotype switch and produce collagen [74]. By this pro-fibrotic effect, M2 macrophages can cause adipose tissue fibrosis; it is a condition, when extracellular matrix is produced in excess, encapsulating the fat-storing adipocytes. As a result, the fat-storing capacity of the adipocytes becomes limited [72]. The excess extracellular matrix does not permit the expansion of the fat cells, leading to "spillover" of the intracellular lipids. The unpacked lipids are prone for chemical modifications, can serve as danger signals, and can cause inflammation [65]. Moreover, inflammatory signaling is necessary for normal development of the adipose tissue [75]. Indeed, infant adipose tissue contains ATMs that secrete IL-6, a typical M1 macrophage cytokine, to stimulate heat generation by burning off lipids [51].

M2 activation occurs among tumor-associated macrophages (TAMs) as well [4, 76]. TAMs are major constituents of the tumor microenvironment, and they can adopt an M2 activation state [4, 77]. The macrophage literature is expanding rapidly, and there are many novel views of the classification of macrophages into functional subcategories. TAMs are also denoted as a distinct, M2-like macrophage group [2]. M2 TAMs promote angiogenesis, inhibit T-cell-mediated cytotoxicity against tumor cells, and secrete different factors involved in extracellular matrix remodeling. M2 TAMs hence support tumor growth, tumor cell motility, and metastasis [78, 79]. For this reason, M2 TAMs are sometimes referred to as pro-tumorigenic macrophages (Fig. 4). In this setting, increasing M1 traits of TAMs is the desired therapeutic outcome, because it may allow the immune system to eliminate tumor cells.

These examples are to illustrate that albeit M2 macrophages diminish inflammation: shifting macrophages into the M2 state may lack a favorable outcome in chronic inflammatory diseases. This apparent paradox is caused by the indispensable role of M1 macrophages in tissue homeostasis. Moreover, in certain organs, such as in the lungs, and in the mucosal membranes, M1 macrophages serve as the first line of defense against pathogens. Prevalence of M2 macrophages in the lung indeed causes disease such as asthma, fibrosis, and chronic obstructive pulmonary disease [80, 81].

Altogether, M2 macrophage show critical physiological roles, including pro-resolving, tolerogenic, and tissue-healing functions. These traits however can cause disease, and the outcome of macrophage responses is largely context dependent (Fig. 4).

5 How to Recognize an M2 Macrophage?

The idea that macrophages might be classified into functional subgroups can be traced back to the early 1980s in the literature. Using the technologies available at that time, it was shown that human tissue resident macrophages express distinct proteins, which may hallmark their own, specific activation states [82]. Arginase activity and the expression of the C-type mannose receptor CD206 and the hemoglobin-haptoglobin receptor CD163 was recognized as a hallmark of M2 macrophages by initial studies published in the 1990s [5, 31, 33, 40, 83, 84]. Today M2 macrophages are identified by a set of markers, including transmembrane glycoproteins, scavenger receptors, enzymes, growth factors, hormones, cytokines, and cytokine receptors with diverse and often yet unexplored functions (Fig. 5a) [2]. Key M2 macrophage markers are typically co-expressed, and include arginase-1 (Arg1, EC:3.5.3.1), C-type mannose receptor 1 (MRC1, or CD206), chitinase-like protein 3 (Chil3, Chi3l3, or Ym1), hemoglobin–haptoglobin scavenger receptor (CD163), IL-4-receptor alpha (IL4Rα), IL-10, resistin-like molecule alpha (RELM-α, encoded by *Retnla*, also called as found in inflammatory zone protein 1 (Fizz1)), and others [2]. Gene transcription of these markers is amplified by Th2 cytokines through STAT6 signaling, although they also can be stimulated by STAT6-independent mechanisms (detailed in chapter "Signal Mechanisms of M2 Macrophage Activation"). They are mostly expressed by macrophages (Fig. 5a), although some of them can also be present in other myeloid cells as well, such as granulocytes, dendritic cells, mast cells, and myeloid-derived suppressor cells [45, 85–89].

Many of the M2 marker genes are involved in arginine/ornithine and iron metabolism (Fig. 5a), inhibit cytokine signaling, T-cell activation, and inflammation (Fig. 5b). The impact of arginine metabolism of macrophages has been discussed earlier. Iron metabolism is associated with the uptake of red blood cells, hemoglobin, and haptoglobin by M2 macrophages: this may serve for the safe disposal of aged red blood cells (e.g., by Kupffer cells), or removal of blood clots in damaged tissues. Some intracellular pathogens take advantage of arginase activity and sustained iron support within the M2 macrophages (detailed in chapter "Immune Functions of the M2 Macrophages: Host Defense, Self-tolerance and Autoimmunity"). The main functions of the M2 marker proteins are associated with resolution of inflammation, inhibition of cytokine signaling, and inhibition of Th1 lymphocyte activation [1] (Fig. 5b).

With the advent of single-cell sequencing, it is now possible to measure that macrophages may simultaneously express M1 and M2 markers and may be positioned in a spectrum of activation states (Fig. 3e) [90]. Tissues typically also contain mixed populations of M1 and M2 macrophages [1] (Fig. 3f). Moreover, there are atypical macrophages involved in immune response that display both M1- and M2-associated gene transcription patterns [91, 92] or which do not fit into the prevailing M1/M2 model [51, 93]. Tissue-specific cues, neurohormonal signals [2, 68], and clearing of apoptotic cell remnants during tissue turnover [94] all determine the amount of M1 and M2 macrophages in tissues in addition to signals coming from

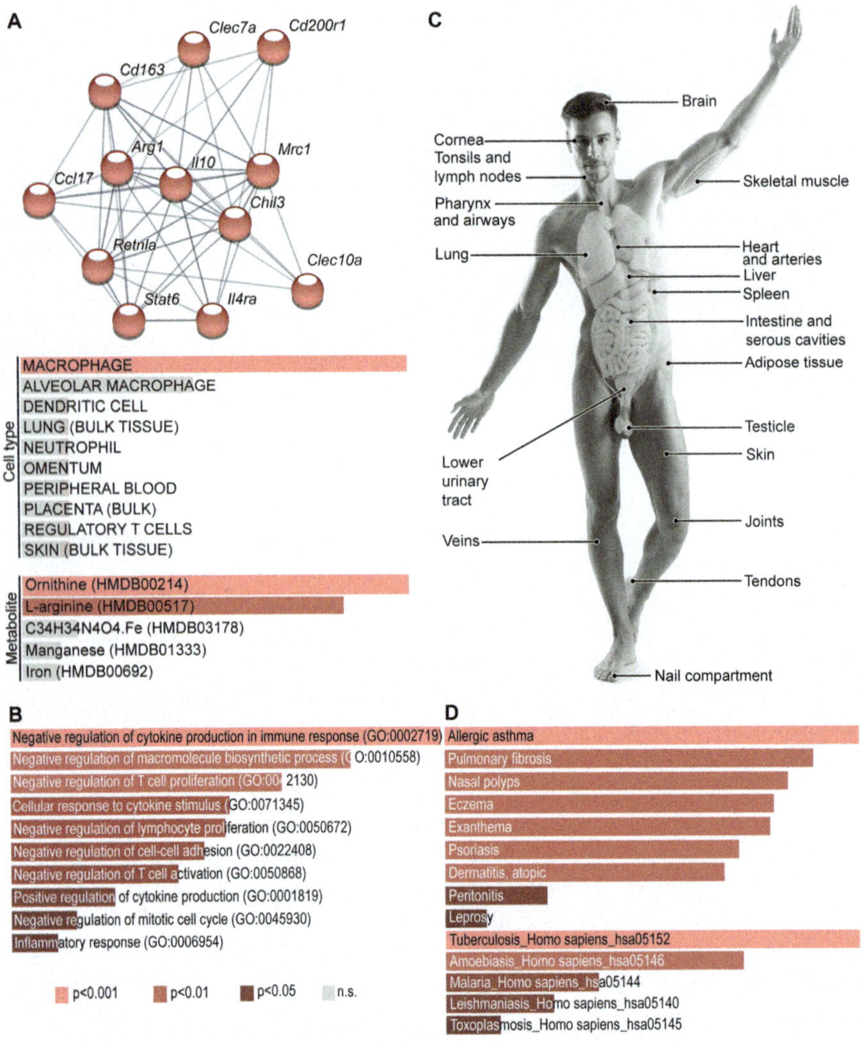

Fig. 5 Markers of M2 macrophages and their functions. (**a**) STRING network of typical M2 macrophage-associated genes. IL-4/STAT-6 signaling and IL-10 are common hubs in the network because they increase gene transcription of a network of genes responsible for M2 macrophage functions. (**b**) Gene ontology terms of M2 activation genes shown in panel **a**. Bar size proportional to relative abundance. (**c**) Overview of organs with M2-activated resident macrophage pools: *right*, organs with physiologically relevant M2 macrophages; *left*, organs in which low M2 content is beneficial. (**d**) Diseases associated with M2 macrophages. Bar size proportional to relative abundance. (Graphs were created with Enrich-R [100, 101])

pathogens and immune cells. Because M2 macrophages occupy specific niches in various organs, certain M2 markers are organ specific (these traits are detailed in chapters "M2 Macrophages in the Integument and in the Musculoskeletal System", "M2 Macrophages in the Circulatory, Respiratory and Excretory Organs" and "M2 Macrophages in the Metabolic Organs and in the Neuroendocrine System").

Some tissues have prevalent M2 macrophage pools under physiological conditions, such as the central nervous system, skin, skeletal muscle, liver, and adipose tissue depots. Immune privileged organs, such as the testicles, are also rich in M2 macrophages (Fig. 5c). However, some tissues that are challenged with pathogens, such as the respiratory tract, harbor more M1 than M2 macrophages, and a pro-fibrogenic effect of M2 macrophages can be unfavorable in many organs [95] (Fig. 4). Certain diseases, such as allergy, asthma, fibrotic tissue lesions, fungal infections, and infections by intracellular parasites are associated with the predominance of M2 tissue macrophages [2] (Fig. 4, Fig. 5d). These pathologies develop from predominant Th2 immune response (e.g., asthma [96]), pro-fibrogenic effect of M2 macrophages (e.g., pulmonary fibrosis [97]), lack of tumor cell recognition (e.g., nasal polyps [98]), and an immunosuppressed state caused by Th2 cytokines and M2 macrophages (e.g., intracellular parasite infections [99]).

6 The M2 Terminology: A Babel-Like Confusion

Classification is an integral part of all sciences, and the best example is the classification of species. How we classify species into groups is constantly changing. Today we live in the era of cladistics, in which species that have evolved from a shared ancestor are sorted into a common group, a so-called clade. A clade should be monophyletic, which means that all species of the clade should be proven descendants of a common ancestor species. Vertebrates are, for instance, considered as a monophyletic group, unlike "invertebrates," which involve phylogenetically diverse polyphyletic groups. Although the term "invertebrate" is scientifically inaccurate, still it is in use and will remain in use in biology. Similarly, the expression "fish" is scientifically inaccurate, because it denotes a phylogenetically diverse group of vertebrates [102]. However, there is still "fish science," or ichthyology; and even an ichthyologist would not change the term "fish and chips" to "elopomorph species and chips" for scientific accuracy. Certainly, some general terms remain in use, and probably this is the case with the M2 macrophage terminology. As the literature of M2 macrophage has grown, it has become necessary to classify M2 macrophages and refine M2 terminology (Fig. 6).

One initiative in the 2010s was to establish subcategories of M2 macrophages. M2 macrophage activation was initially identified as a response to Th2 cytokines. However, several other M2-activating signals have been recognized by further research, as detailed in chapter "Signal Mechanisms of M2 Macrophage Activation". Based on the applied stimuli and the achieved transcriptional changes, the M2

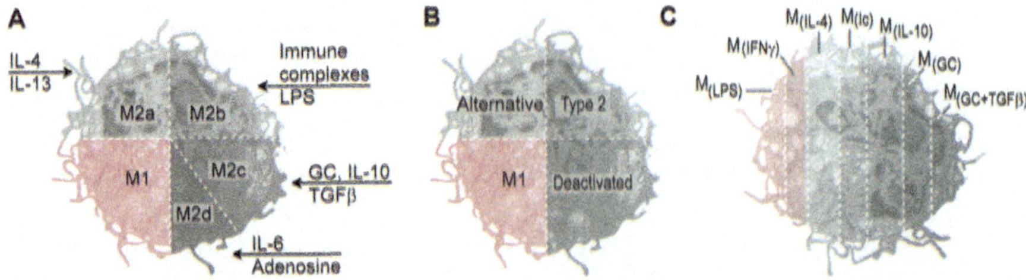

Fig. 6 Models of M2 macrophage subcategories. (**a**) M2a, M2b, M2c, and M2d subtypes of M2 macrophages have been established based on in vitro assays using IL-4/IL-13, immune complexes, LPS, glucocorticoids (GC), IL-10, TGFβ, IL-6, and adenosine. (**b**) Some authors consider the M2a subtype as alternative activated macrophage, the M2b subtype as a type 2 macrophage, and the M2c as a deactivated macrophage. (**c**) Macrophages categories named after their activation stimuli. This model omits the use of M1 or M2; however, it would require the establishment of infinite number of macrophage types. It is more feasible to adhere to the principal M1 and M2 categories, without further specifying macrophage subdivisions

macrophages may be classified into subclasses [1, 2, 103][5], the M2a, M2b, M2c, and M2d macrophages [93, 104]. The M2a activation is a response to IL-4 and IL-13; the M2b to immune complexes and bacterial LPS; the M2c to glucocorticoids, IL-10, and TGFβ, and the M2d activation is a response to IL-6 and adenosines [105, 106] (Fig. 6a). This classification does not cover many further M2 types, and is applicable only if the activating stimuli is known, such as in in vitro experiments.

Moreover, some authors restrict the M2 terminology to Th2 cytokine-elicited macrophages and consider the M2a subtype as the only representative of alternatively activated M2 macrophages. The M2b macrophage can be considered as a type 2 macrophage and the M2c as a deactivated macrophage [1] (Fig. 6b). The deactivated terminology refers to the in vitro ability of macrophage populations to be enriched in M2 activated macrophages following an M1 activation. This phenomenon is considered to be an M1 deactivation, and in vivo it may reflect the replacement of an M1-dominated macrophage population with M2 or naïve macrophages [107]. The M1 macrophages (at least in murine models) have a NO burst and lack the cellular mechanisms that would allow them to survive the cytotoxic effects of NO, which questions whether such deactivation can occur at the single-cell level [43].

Some years ago it was proposed to tag macrophage subdivisions by the name of the applied activation stimuli [1]. In the case of the M2 macrophages these activating stimuli are IL-4, immune complexes (Ic), IL-10, glucocorticoids, and TGFβ (GC + TGFβ), or glucocorticoids (GC). Using this naming convention the M2a group would be termed as $M_{(IL\text{-}4)}$, M2b as $M_{(Ic)}$, and the M2c would be divided into $M_{(IL\text{-}10)}$, $M_{(GC + TGF\beta)}$ and $M_{(GC)}$ macrophages (Fig. 6c). This classification still fails to

[5]An extended version of this section has been published in Röszer, T. (2015) Understanding the mysterious M2 macrophage through activation markers and effector mechanisms. *Mediators of Inflammation* 2015, 16.

cover the wide range of other signals with the ability to induce M2 macrophage activation and M2 macrophage number, and adherence to this terminology would expand M2 subcategories to infinite (these signals are detailed in chapters "Signal Mechanisms of M2 Macrophage Activation" and "Mechanisms Which Control the Size of M2 Macrophage-Dominated Tissue Macrophage Niches"). The aim of the M1/M2 model is to provide an easily understandable framework of macrophage functions, and introducing an infinite number of M2 categories hardly serves this aim.

Another initiative defines further macrophage types, which do not belong to either the M1 or M2 category. Therefore, there are M3 and M4 macrophage types in literature. One example is the use terminology used to describe atherosclerotic plaque-specific macrophage types [93, 108, 109]. Proinflammatory macrophages of the atherosclerotic plaque are considered M1 if they secrete TNFα, IL-6, and IL-12; and as M4 if they secrete IL-6, TNFα, metalloproteinase 7 (MMP-7), and have reduced phagocytosis ability. The M1 macrophages develop in response to TLR4 activation and the similarly proinflammatory M4 macrophages develop in response to CXCL4. The antiinflammatory and antioxidant macrophages of the atherosclerotic plaque are M2 macrophages, $M_{(Hb)}$ macrophages, and $M_{(hem)}$ macrophages. $M_{(Hb)}$ and $M_{(hem)}$ macrophages develop in response to hemoglobin, haptoglobin, and heme, respectively [110]. This terminology is specific to a research field and, in my opinion, does not invalidate a universal framework model of M2 macrophage activation.

Although I believe that the M2 macrophage terminology is useful, there are traits of macrophages, such as their antiviral activities, that do not fit into the current classification of macrophages. It is always the responsibility of scientific authors to communicate their findings in an understandable manner. If better understanding requires it, it is acceptable to adhere to existing classification of M2 macrophages or introduce an alternative terminology: this depends on how well the new terminology is justified. One must also keep in mind that the general meaning of M2 macrophage is widely known today; hence, it is likely that readers will understand the message. Replacing M2 macrophage terminology with new, unknown expressions will most likely be ineffective in reaching readers and colleagues in the research field.

Finally, let me place here an example how macrophages may be classified in a comprehensive way. In my previously proposed system, macrophages are classified based on the combination of the immune context in which they appear (i.e., activating stimuli and signal transducer pathways), and their hallmark traits, such as gene expression profile, effector mechanisms, and functions [2]. In this model there are M1 macrophages that are activated by LPS via Toll-like receptor 4 (TLR4), by IFNγ via IFNγ-receptor (IFNGR), and express inflammatory cytokines and release NO. There are inflammatory, M1-like macrophages that are activated by various pathogen-associated patterns, such as nucleic acids, cell wall components, modified lipids, via Toll-like receptors (TLR3, TLR2, TLR9), cytosolic nucleic acid sensing pathways (e.g., RIG-I/MAVS, cGAS/STING), and express inflammatory mediators, cell adhesion molecules, etc. In both instances nuclear factor-kappa B (NFκB),

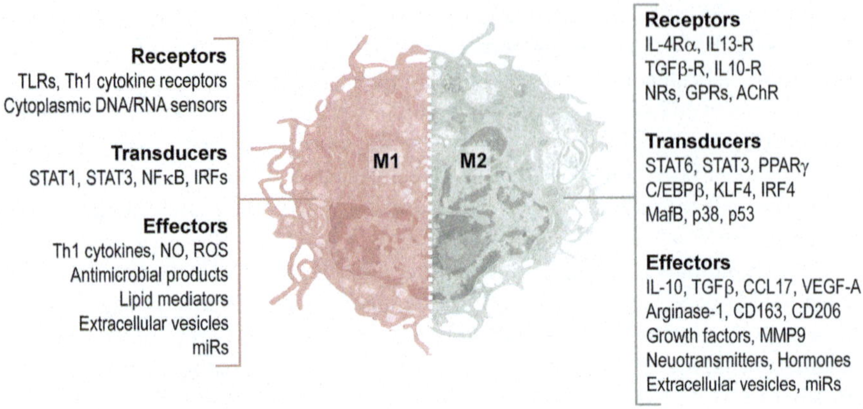

Fig. 7 Receptors and effector mechanisms associated with macrophage activation. Macrophage activation is triggered by signals acting through their respective receptors. Receptors activate specific signal mechanisms, which eventually change gene transcription through transcription factors. In certain situations primary and secondary transcription factors are involved in macrophage activation: STAT6 increases Krüppel-like factor 4 (KLF4) expression, which eventually induces an M2 response. There are synergies between transcription factors, such as STAT6, peroxisome proliferator activator receptor gamma (PPARγ), and interferon regulatory factor 4 (IRF4) can act in concert to elicit the M2 state. STAT3 is involved in both M1 and M2 activation, and by inducing M2-associated gene expression, it is crucial for a self-inhibition of M1 macrophage activation. Effector molecules determine macrophage functions: macrophage-derived cytokines, mediators, micro-RNA (miR) species, and growth factors are hormones that have their own specific functions.

signal transducer and activator of transcription 1 and *3* (STAT1 and STAT3), and interferon regulator factors (IRFs) are involved in the control of gene expression (Fig. 7).

The M2 macrophages are activated by IL-4 and other Th2 cytokines, via the IL-4 receptor and IL-13 receptor (IL-4Ra, IL-13R), IL-10 receptor (IL-10R), TGFβ-receptor (TGFβ-R), apoptotic cargo, hormones, neurotransmitters through G protein-linked receptors (GPRs), and acetylcholine receptor (AChR), and by tumor suppressors and pathogen products (Fig. 7, and chapter "Signal Mechanisms of M2 Macrophage Activation"). The main transcription factors involved are *signal transducer and activator of transcription 6* (STAT6), STAT3, various nuclear receptors (NRs), and other transcriptional regulators, and the expressed genes are immune suppressors and genes required for tissue growth (Fig. 7). For certain reasons macrophages may initiate an M1 response, which is soon inhibited, and the macrophage is shifted toward an M2 state. Examples include macrophages of the intestinal mucosa, which are alarmed by bacterial products of the gut normal flora. However, their M1 activation is blocked by tissue IL-10; this leads to a macrophage population that expresses NOS2 (M1 marker) along with arginase-1 (M2 marker). Similarly, macrophages infected with *Mycobacterium* initiate an M1-like immune response (they increase the expression of the cGAS-STING signal pathway); however, the *Mycobacterium*-derived signals block the eventual expression of the M1 effector genes [23, 24, 111]. Personally, I do not see it necessary to introduce a new macrophage category for the intestinal macrophages or the *Mycobacterium*-infected macrophages. Their features can be described with the existing macrophage terminology.

7 M2 Macrophage: A Valid and Necessary Terminology in Immunology

In summary, M2 macrophages are macrophages with predominantly noninflammatory, antiinflammatory, immune-suppressing, self-tolerating, angiogenic, wound healing, and pro-fibrotic functions. Depending on the context, these functions may be either favorable or unfavorable. For instance, the immune-suppressed environment of M2 macrophages is vital in organs that should tolerate foreign antigens, such as the placenta, where antigens of the developing fetus should be tolerated; or the testis, wherein developing spermatids should be tolerated [112, 113]. M2 macrophages, however, can cause disease, for instance by supporting tumor growth or allowing the spreading of intracellular pathogens. Moreover, one might keep in mind that the original concept of macrophage activation emphasized an antagonistic, divergent, M1 or M2 polarization. In this model one possible misinterpretation is to consider M2 macrophages as antagonists of M1 macrophages. However, M2 macrophages are not the opposite of M1 macrophages. Instead, M2 macrophages involve a complex and functionally diverse group, distinct from M1 macrophages. With these considerations in mind, I hope that the following chapters provide a compass to navigate through M2 macrophage biology.

References

1. Murray PJ, Allen JE, Biswas SK, Fisher EA, Gilroy DW, Goerdt S, Gordon S, Hamilton JA, Ivashkiv LB, Lawrence T, Locati M, Mantovani A, Martinez FO, Mege JL, Mosser DM, Natoli G, Saeij JP, Schultze JL, Shirey KA, Sica A, Suttles J, Udalova I, van Ginderachter JA, Vogel SN, Wynn TA. Macrophage activation and polarization: nomenclature and experimental guidelines. Immunity. 2014;41:14–20.
2. Röszer T. Understanding the mysterious M2 macrophage through activation markers and effector mechanisms. Mediat Inflamm. 2015;2015:16.
3. Gordon S, Plüddemann A. Tissue macrophages: heterogeneity and functions. BMC Biol. 2017;15:53.
4. Mantovani A, Sozzani S, Locati M, Allavena P, Sica A. Macrophage polarization: tumor-associated macrophages as a paradigm for polarized M2 mononuclear phagocytes. Trends Immunol. 2002;23:549–55.
5. Chang CI, Liao JC, Kuo L. Arginase modulates nitric oxide production in activated macrophages. Am J Phys. 1998;274:H342–8.
6. Zorro Manrique S, Zorro Manrique S, Duque MA, Dominguez AL, Mirza NN, Hoelzinger D, Smith SE, Lustgarten J. Identification and characterization of macrophage regulatory cells (Mac-regs) with immunoregulatory properties (89.38). J Immunol. 2009;182:89.38–89.38.

7. Carr I. The macrophage: a review of ultrastructure and function. New York: Academic; 1973.
8. Pujiula P. Manual Completo de Biologia Moderna, Macro- y Microscopica. Barcelona: Tipografia Catolica Casals; 1927.
9. Srodka A, Gryglewski RW, Szczepariski W. Browicz or Kupffer cells? Pol J Pathol Off J Pol Soc Pathol. 2006;57:183–5.
10. Chang ZL. Recent development of the mononuclear phagocyte system: in memory of Metchnikoff and Ehrlich on the 100th anniversary of the 1908 Nobel prize in physiology or medicine. Biol Cell. 2009;101:709–21.
11. Gordon S. Phagocytosis: the legacy of Metchnikoff. Cell. 2016;166:1065–8.
12. Ward PA. Chemotaxis of mononuclear cells. J Exp Med. 1968;128:1201–21.
13. Mills CD. Anatomy of a discovery: m1 and m2 macrophages. Front Immunol. 2015;6:212–212
14. Metchnikoff ÉI. Immunity in infective diseases: Cambridge University Press; 1907.
15. Flügge C. Die Mikroorganismen: Mit besonderer Berücksichtigung der Ätiologie der Infektionskrankheiten. Leipzig: Vogel; 1886.
16. Erdmann R. Praktikum der Gewebepflege oder Explantation Besonders der Gewebezüchtung. Berlin-Wilmersdorf: Springer; 1922.
17. Benoit M, Desnues B, Mege J-L. Macrophage polarization in bacterial infections. J Immunol. 2008;181:3733–9.
18. Masson P. Diagnósticos de laboratorio II. *Tumores. Diagnósticos Histológicos*, vol. XXVII. Barcelona: Editorial Pubul; 1927.
19. Brooks PJ, Glogauer M, McCulloch CA. An overview of the derivation and function of multinucleated Giant cells and their role in pathologic processes. Am J Pathol. 2019;189:1145–58.
20. Hosaka S, Hosaka N. Langhans' giant cell or Langerhans' giant cell? Response to Lin et al. Respir Med. 2006;100:1667.
21. Lay G, Poquet Y, Salek-Peyron P, Puissegur M-P, Botanch C, Bon H, Levillain F, Duteyrat J-L, Emile J-F, Altare F. Langhans giant cells from M. tuberculosis-induced human granulomas cannot mediate mycobacterial uptake. J Pathol. 2007;211:76–85.
22. McNally AK, Anderson JM. Macrophage fusion and multinucleated Giant cells of inflammation. In: Dittmar T, Zänker KS, editors. Cell fusion in health and disease. Dordrecht: Springer; 2011. p. 97–111.
23. Lopes RL, Borges TJ, Zanin RF, Bonorino C. IL-10 is required for polarization of macrophages to M2-like phenotype by mycobacterial DnaK (heat shock protein 70). Cytokine. 2016;85:123–9.
24. Roy S, Schmeier S, Kaczkowski B, Arner E, Alam T, Ozturk M, Tamgue O, Parihar SP, Kawaji H, Itoh M, Lassmann T, Carninci P, Hayashizaki Y, Forrest ARR, Guler R, Bajic VB, Brombacher F, Suzuki H. Transcriptional landscape of Mycobacterium tuberculosis infection in macrophages. Sci Rep. 2018;8:6758.
25. Fong J, Chin D, Elberg SS. Studies on tubercle bacillus-histiocyte relationship. V. Passive transfer of cellular resistance. J Exp Med. 1962;115:475–89.
26. Mackaness GB. The influence of immunologically committed lymphoid cells on macrophage activity in vivo. J Exp Med. 1969;129:973–92.
27. Herscowitz HB, Pennline KJ, Conrad RE, Ullrich SE, Gerber HR. Macrophage-lymphocyte interactions mediated by soluble factors*. Ann N Y Acad Sci. 1979;332:464–81.
28. Evans RL, Breard JM, Lazarus H, Schlossman SF, Chess L. Detection, isolation, and functional characterization of two human T-cell subclasses bearing unique differentiation antigens. J Exp Med. 1977;145:221–33.
29. Reinherz EL, Schlossman SF. Con A-inducible suppression of MLC: evidence for mediation by the TH2+ T cell subset in man. J Immunol. 1979;122:1335–41.
30. Mills CD, Kincaid K, Alt JM, Heilman MJ, Hill AM. M-1/M-2 macrophages and the Th1/Th2 paradigm. J Immunol. 2000;164:6166–73.
31. Munder M, Eichmann K, Moran JM, Centeno F, Soler G, Modolell M. Th1/Th2-regulated expression of arginase isoforms in murine macrophages and dendritic cells. J Immunol. 1999;163:3771–7.

32. Fiorentino DF, Zlotnik A, Mosmann TR, Howard M, O'Garra A. IL-10 inhibits cytokine production by activated macrophages. J Immunol. 1991;147:3815–22.
33. Stein M, Keshav S, Harris N, Gordon S. Interleukin 4 potently enhances murine macrophage mannose receptor activity: a marker of alternative immunologic macrophage activation. J Exp Med. 1992;176:287–92.
34. Bonecchi R, Sozzani S, Stine JT, Luini W, D'Amico G, Allavena P, Chantry D, Mantovani A. Divergent effects of interleukin-4 and interferon-γ on macrophage-derived chemokine production: an amplification circuit of polarized T helper 2 responses. Blood. 1998;92:2668–71.
35. Torres-Guerrero E, Quintanilla-Cedillo MR, Ruiz-Esmenjaud J, Arenas R. Leishmaniasis: a review. F1000Res. 2017;6:750–750.
36. Heinzel FP, Sadick MD, Holaday BJ, Coffman RL, Locksley RM. Reciprocal expression of interferon gamma or interleukin 4 during the resolution or progression of murine leishmaniasis. Evidence for expansion of distinct helper T cell subsets. J Exp Med. 1989;169:59–72.
37. Pérez-Cabezas B, Cecflio P, Gaspar TB, Gärtner F, Vasconcellos R, Cordeiro-da-Silva A. Understanding resistance vs. susceptibility in visceral Leishmaniasis using mouse models of Leishmania infantum infection. Front Cell Infect Microbiol. 2019;9
38. Wilhelm P, Ritter U, Labbow S, Donhauser N, Röllinghoff M, Bogdan C, Körner H. Rapidly fatal leishmaniasis in resistant C57BL/6 mice lacking TNF. J Immunol. 2001;166:4012–9.
39. Kung JT, Brooks SB, Jakway JP, Leonard LL, Talmage DW. Suppression of in vitro cytotoxic response by macrophages due to induced arginase. J Exp Med. 1977;146:665–72.
40. Mills CD, Shearer J, Evans R, Caldwell MD. Macrophage arginine metabolism and the inhibition or stimulation of cancer. J Immunol. 1992;149:2709–14.
41. Mills CD, Lenz LL, Ley K. Macrophages at the fork in the road to health or disease. Front Immunol. 2015;6:59.
42. Rath M, Müller I, Kropf P, Closs EI, Munder M. Metabolism via arginase or nitric oxide synthase: two competing arginine pathways in macrophages. Front Immunol. 2014;5:532–532.
43. Röszer T. Phagosomal and lysosomal NO synthesis. In: The biology of subcellular nitric oxide. Dordrecht: Springer; 2012. p. 145–55.
44. Munder M. Arginase: an emerging key player in the mammalian immune system. Br J Pharmacol. 2009;158:638–51.
45. Munder M, Schneider H, Luckner C, Giese T, Langhans CD, Fuentes JM, Kropf P, Mueller I, Kolb A, Modolell M, Ho AD. Suppression of T-cell functions by human granulocyte arginase. Blood. 2006;108:1627–34.
46. Stempin CC, Dulgerian LR, Garrido VV, Cerban FM. Arginase in parasitic infections: macrophage activation, immunosuppression, and intracellular signals. J Biomed Biotechnol. 2010;2010:683485.
47. Cloke TE, Garvey L, Choi BS, Abebe T, Hailu A, Hancock M, Kadolsky U, Bangham CR, Munder M, Muller I, Taylor GP, Kropf P. Increased level of arginase activity correlates with disease severity in HIV-seropositive patients. J Infect Dis. 2010;202:374–85.
48. Ley K. M1 means kill; M2 means heal. J Immunol. 2017;199:2191–3.
49. Nomura M, Liu J, Rovira II, Gonzalez-Hurtado E, Lee J, Wolfgang MJ, Finkel T. Fatty acid oxidation in macrophage polarization. Nat Immunol. 2016;17:216–7.
50. Gundra UM, Girgis NM, Ruckerl D, Jenkins S, Ward LN, Kurtz ZD, Wiens KE, Tang MS, Basu-Roy U, Mansukhani A, Allen JE, Loke P n. Alternatively activated macrophages derived from monocytes and tissue macrophages are phenotypically and functionally distinct. Blood. 2014;123:e110–22.
51. Yu H, Dilbaz S, Coßmann J, Hoang AC, Diedrich V, Herwig A, Harauma A, Hoshi Y, Moriguchi T, Landgraf K, Körner A, Lucas C, Brodesser S, Balogh L, Thuróczy J, Karemore G, Kuefner MS, Park EA, Rapp C, Travers JB, Röszer T. Breast milk alkylglycerols sustain beige adipocytes through adipose tissue macrophages. J Clin Invest. 2019;129:2485–99.
52. Raes G, Van den Bergh R, De Baetselier P, Ghassabeh GH. Arginase-1 and Ym1 are markers for murine, but not human, alternatively activated myeloid cells. J Immunol. 2005;174:6561–2.

53. Van den Bossche J, Bogaert P, van Hengel J, Guérin CJ, Berx G, Movahedi K, Van den Bergh R, Pereira-Fernandes A, Geuns JMC, Pircher H, Dorny P, Grooten J, De Baetselier P, Van Ginderachter JA. Alternatively activated macrophages engage in homotypic and heterotypic interactions through IL-4 and polyamine-induced E-cadherin/catenin complexes. Blood. 2009;114:4664–74.
54. Semnani R, Mahapatra L, Moore V, Sanprasert V, Nutman T. Functional and phenotypic characteristics of alternative activation induced in human monocytes by interleukin-4 or the parasitic nematode Brugia malayi. Infect Immun. 2011;79:3957–65.
55. Guo L, Akahori H, Harari E, Smith SL, Polavarapu R, Karmali V, Otsuka F, Gannon RL, Braumann RE, Dickinson MH, Gupta A, Jenkins AL, Lipinski MJ, Kim J, Chhour P, de Vries PS, Jinnouchi H, Kutys R, Mori H, Kutyna MD, Torii S, Sakamoto A, Choi CU, Cheng Q, Grove ML, Sawan MA, Zhang Y, Cao Y, Kolodgie FD, Cormode DP, Arking DE, Boerwinkle E, Morrison AC, Erdmann J, Sotoodehnia N, Virmani R, Finn AV. CD163+ macrophages promote angiogenesis and vascular permeability accompanied by inflammation in atherosclerosis. J Clin Invest. 2018;128:1106–24.
56. Odegaard JI, Ganeshan K, Chawla A. Adipose tissue macrophages: "amicus adipem?". Cell Metab. 2013;18:767–8.
57. Lumeng CN, Bodzin JL, Saltiel AR. Obesity induces a phenotypic switch in adipose tissue macrophage polarization. J Clin Invest. 2007;117:175–84.
58. Lumeng CN, DelProposto JB, Westcott DJ, Saltiel AR. Phenotypic switching of adipose tissue macrophages with obesity is generated by spatiotemporal differences in macrophage subtypes. Diabetes. 2008;57:3239–46.
59. Epelman S, Lavine KJ, Randolph GJ. Origin and functions of tissue macrophages. Immunity. 2014;41:21–35.
60. Osborn O, Olefsky JM. The cellular and signaling networks linking the immune system and metabolism in disease. Nat Med. 2012;18:363–74.
61. Lee BC, Lee J. Cellular and molecular players in adipose tissue inflammation in the development of obesity-induced insulin resistance. Biochim Biophys Acta. 2014;1842:446–62.
62. Glass CK, Olefsky JM. Inflammation and lipid signaling in the etiology of insulin resistance. Cell Metab. 2012;15:635–45.
63. Qiu Y, Nguyen KD, Odegaard JI, Cui X, Tian X, Locksley RM, Palmiter RD, Chawla A. Eosinophils and type 2 cytokine signaling in macrophages orchestrate development of functional beige fat. Cell. 2014;157:1292–308.
64. Li P, Spann NJ, Kaikkonen MU, Lu M, Oh DY, Fox JN, Bandyopadhyay G, Talukdar S, Xu J, Lagakos WS, Patsouris D, Armando A, Quehenberger O, Dennis EA, Watkins SM, Auwerx J, Glass CK, Olefsky JM. NCoR repression of LXRs restricts macrophage biosynthesis of insulin-sensitizing omega 3 fatty acids. Cell. 2013;155:200–14.
65. Fuentes L, Röszer T, Ricote M. Inflammatory mediators and insulin resistance in obesity: role of nuclear receptor signaling in macrophages. Mediat Inflamm. 2010;2010:219583.
66. Amano SU, Cohen JL, Vangala P, Tencerova M, Nicoloro SM, Yawe JC, Shen Y, Czech MP, Aouadi M. Local proliferation of macrophages contributes to obesity-associated adipose tissue inflammation. Cell Metab. 2014;19:162–71.
67. Boutens L, Stienstra R. Adipose tissue macrophages: going off track during obesity. Diabetologia. 2016;59:879–94.
68. Waqas SFH, Hoang A, Lin Y, Ampem G, Röszer T, et al. Neuropeptide FF increases M2 activation and self-renewal of adipose tissue macrophages. J Clin Invest. 2017;127:2842–54.
69. Rosen ED, Spiegelman BM. What we talk about when we talk about fat. Cell. 2014;156:20–44.
70. Fischer K, Ruiz HH, Jhun K, Finan B, Oberlin DJ, van der Heide V, Kalinovich AV, Petrovic N, Wolf Y, Clemmensen C, Shin AC, Divanovic S, Brombacher F, Glasmacher E, Keipert S, Jastroch M, Nagler J, Schramm KW, Medrikova D, Collden G, Woods SC, Herzig S, Homann D, Jung S, Nedergaard J, Cannon B, Tschop MH, Muller TD, Buettner C. Alternatively activated macrophages do not synthesize catecholamines or contribute to adipose tissue adaptive thermogenesis. Nat Med. 2017;23:623–30.

71. Nguyen KD, Qiu Y, Cui X, Goh YP, Mwangi J, David T, Mukundan L, Brombacher F, Locksley RM, Chawla A. Alternatively activated macrophages produce catecholamines to sustain adaptive thermogenesis. Nature. 2011;480:104–8.
72. Spencer M, Yao-Borengasser A, Unal R, Rasouli N, Gurley CM, Zhu B, Peterson CA, Kern PA. Adipose tissue macrophages in insulin-resistant subjects are associated with collagen VI and fibrosis and demonstrate alternative activation. Am J Physiol Endocrinol Metab. 2010;299:E1016–27.
73. Bellon T, Martinez V, Lucendo B, del Peso G, Castro MJ, Aroeira LS, Rodriguez-Sanz A, Ossorio M, Sanchez-Villanueva R, Selgas R, Bajo MA. Alternative activation of macrophages in human peritoneum: implications for peritoneal fibrosis. Nephrol Dial Transplant. 2011;26:2995–3005.
74. Medbury HJ, James V, Ngo J, Hitos K, Wang Y, Harris DC, Fletcher JP. Differing association of macrophage subsets with atherosclerotic plaque stability. Int Angiol. 2013;32:74–84.
75. Asterholm IW, Tao C, Morley TS, Wang QA, Delgado-Lopez F, Wang ZV, Scherer PE. Adipocyte inflammation is essential for healthy adipose tissue expansion and remodeling. Cell Metab. 2014;20:103–18.
76. Mantovani A, Sozzani S, Locati M, Schioppa T, Saccani A, Allavena P, Sica A. Infiltration of tumours by macrophages and dendritic cells: tumour-associated macrophages as a paradigm for polarized M2 mononuclear phagocytes. Novartis Found Symp. 2004;256:137–45; discussion 146–138, 259–169
77. Sica A, Mantovani A. Macrophage plasticity and polarization: in vivo veritas. J Clin Invest. 2012;122:787–95.
78. Pathria P, Louis TL, Varner JA. Targeting tumor-associated macrophages in cancer. Trends Immunol. 2019;40:310–27.
79. Laviron M, Boissonnas A. Ontogeny of tumor-associated macrophages. Front Immunol. 2019;10:1799.
80. He S, Xie L, Lu J, Sun S. Characteristics and potential role of M2 macrophages in COPD. Int J Chron Obstruct Pulmon Dis. 2017;12:3029–39.
81. Kaku Y, Imaoka H, Morimatsu Y, Komohara Y, Ohnishi K, Oda H, Takenaka S, Matsuoka M, Kawayama T, Takeya M, Hoshino T. Overexpression of CD163, CD204 and CD206 on alveolar macrophages in the lungs of patients with severe chronic obstructive pulmonary disease. PLoS One. 2014;9:e87400.
82. Radzun HJ, Parwaresch MR. Differential immunohistochemical resolution of the human mononuclear phagocyte system. Cell Immunol. 1983;82:174–83.
83. Hrabak A, Bajor T, Temesi A. Comparison of substrate and inhibitor specificity of arginase and nitric oxide (NO) synthase for arginine analogues and related compounds in murine and rat macrophages. Biochem Biophys Res Commun. 1994;198:206–12.
84. Ritter M, Buechler C, Langmann T, Orso E, Klucken J, Schmitz G. The scavenger receptor CD163: regulation, promoter structure and genomic organization. Pathobiology. 1999;67:257–61.
85. Yang Y, Zhang R, Xia F, Zou T, Huang A, Xiong S, Zhang J. LPS converts Gr-1(+) CD115(+) myeloid-derived suppressor cells from M2 to M1 via P38 MAPK. Exp Cell Res. 2013;319:1774–83.
86. Bhatt S, Qin J, Bennett C, Qian S, Fung JJ, Hamilton TA, Lu L. All-trans retinoic acid induces arginase-1 and inducible nitric oxide synthase-producing dendritic cells with T cell inhibitory function. J Immunol. 2014;192:5098–108.
87. Raes G, De Baetselier P, Noel W, Beschin A, Brombacher F, Hassanzadeh Gh G. Differential expression of FIZZ1 and Ym1 in alternatively versus classically activated macrophages. J Leukoc Biol. 2002;71:597–602.
88. Welch JS, Escoubet-Lozach L, Sykes DB, Liddiard K, Greaves DR, Glass CK. TH2 cytokines and allergic challenge induce Ym1 expression in macrophages by a STAT6-dependent mechanism. J Biol Chem. 2002;277:42821–9.

89. Munder M, Mollinedo F, Calafat J, Canchado J, Gil-Lamaignere C, Fuentes JM, Luckner C, Doschko G, Soler G, Eichmann K, Muller FM, Ho AD, Goerner M, Modolell M. Arginase I is constitutively expressed in human granulocytes and participates in fungicidal activity. Blood. 2005;105:2549–56.
90. Li C, Menoret A, Farragher C, Ouyang Z, Bonin C, Holvoet P, Vella AT, Zhou B. Single cell transcriptomics based-MacSpectrum reveals novel macrophage activation signatures in diseases. JCI Insight. 2019;5:e126453.
91. Joshi AD, Oak SR, Hartigan AJ, Finn WG, Kunkel SL, Duffy KE, Das A, Hogaboam CM. Interleukin-33 contributes to both M1 and M2 chemokine marker expression in human macrophages. BMC Immunol. 2010;11:52.
92. Tatano Y, Shimizu T, Tomioka H. Unique macrophages different from M1/M2 macrophages inhibit T cell mitogenesis while upregulating Th17 polarization. Sci Rep. 2014;4:4146.
93. Colin S, Chinetti-Gbaguidi G, Staels B. Macrophage phenotypes in atherosclerosis. Immunol Rev. 2014;262:153–66.
94. Röszer T. Transcriptional control of apoptotic cell clearance by macrophage nuclear receptors. Apoptosis. 2017;22:284–94.
95. Zhang L, Wang Y, Wu G, Xiong W, Gu W, Wang C-Y. Macrophages: friend or foe in idiopathic pulmonary fibrosis? Respir Res. 2018;19:170–170.
96. Barnes PJ. Th2 cytokines and asthma: an introduction. Respir Res. 2001;2:64–5.
97. Hou J, Shi J, Chen L, Lv Z, Chen X, Cao H, Xiang Z, Han X. M2 macrophages promote myofibroblast differentiation of LR-MSCs and are associated with pulmonary fibrogenesis. Cell Commun Signal. 2018;16:89–89.
98. Wang Z-C. Increased accumulation of alternatively activated macrophages in eosinophilic chronic rhinosinusitis with nasal polyps. J Allergy Clin Immunol. 2017;139:AB64.
99. Muraille E, Leo O, Moser M. TH1/TH2 paradigm extended: macrophage polarization as an unappreciated pathogen-driven escape mechanism? Front Immunol. 2014;5:603.
100. Chen EY, Tan CM, Kou Y, Duan Q, Wang Z, Meirelles GV, Clark NR, Ma'ayan A. Enrichr: interactive and collaborative HTML5 gene list enrichment analysis tool. BMC Bioinform. 2013;14:128.
101. Kuleshov MV, Jones MR, Rouillard AD, Fernandez NF, Duan Q, Wang Z, Koplev S, Jenkins SL, Jagodnik KM, Lachmann A, McDermott MG, Monteiro CD, Gundersen GW, Ma'ayan A. Enrichr: a comprehensive gene set enrichment analysis web server 2016 update. Nucleic Acids Res. 2016;44:W90–7.
102. Betancur-R R, Wiley EO, Arratia G, Acero A, Bailly N, Miya M, Lecointre G, Ortí G. Phylogenetic classification of bony fishes. BMC Evol Biol. 2017;17:162.
103. Martinez FO, Gordon S. The M1 and M2 paradigm of macrophage activation: time for reassessment. F1000Prime Rep. 2014;6:13.
104. Ferrante CJ, Leibovich SJ. Regulation of macrophage polarization and wound healing. Adv Wound Care (New Rochelle). 2012;1:10–6.
105. Wang Q, Ni H, Lan L, Wei X, Xiang R, Wang Y. Fra-1 protooncogene regulates IL-6 expression in macrophages and promotes the generation of M2d macrophages. Cell Res. 2010;20:701–12.
106. Ferrante CJ, Pinhal-Enfield G, Elson G, Cronstein BN, Hasko G, Outram S, Leibovich SJ. The adenosine-dependent angiogenic switch of macrophages to an M2-like phenotype is independent of interleukin-4 receptor alpha (IL-4Ralpha) signaling. Inflammation. 2013;36:921–31.
107. Porcheray F, Viaud S, Rimaniol AC, Leone C, Samah B, Dereuddre-Bosquet N, Dormont D, Gras G. Macrophage activation switching: an asset for the resolution of inflammation. Clin Exp Immunol. 2005;142:481–9.
108. Moore KJ, Sheedy FJ, Fisher EA. Macrophages in atherosclerosis: a dynamic balance. Nat Rev Immunol. 2013;13:709–21.
109. Chinetti-Gbaguidi G, Colin S, Staels B. Macrophage subsets in atherosclerosis. Nat Rev Cardiol. 2014;12:10.

110. Bobryshev YV, Ivanova EA, Chistiakov DA, Nikiforov NG, Orekhov AN. Macrophages and their role in atherosclerosis: pathophysiology and transcriptome analysis. Biomed Res Int. 2016;2016:9582430–9582430.
111. Manzanillo PS, Shiloh MU, Portnoy DA, Cox JS. Mycobacterium Tuberculosis activates the DNA-dependent cytosolic surveillance pathway within macrophages. Cell Host Microbe. 2012;11:469–80.
112. Tang Z, Niven-Fairchild T, Tadesse S, Norwitz ER, Buhimschi CS, Buhimschi IA, Guller S. Glucocorticoids enhance CD163 expression in placental Hofbauer cells. Endocrinology. 2013;154:471–82.
113. Monika F, Andreas M. The testis in immune privilege. Immunol Rev. 2006;213:66–81.

Evolutionary Origin of M2 Macrophage Activation: Invertebrates

Keywords Evolution · Invertebrates · Phagocytes · Immune response · Comparative immunology · Arginase · Regeneration · Development

1 Introduction

Phagocytic macrophage-like cells of invertebrates were the first immune cells ever studied scientifically. These cells are described as *Amoeba*-like ("amoeboid") cells in the book by Thomas Huxley on invertebrate anatomy in 1878 [1]. In the 1880s, Ilya Metchnikoff [Илья Мечников] (1845–1916) carried out experiments on echinoderm, molluscan, and insect hemocytes [2, 3]. He noticed an encapsulation of foreign particles by amoeboid cells of the sea urchin larva, and eventually he conducted extensive studies on phagocytosing cells, which opened the avenue for the development of the theory of innate immunity. He was honored with the Nobel prize in 1908 [4]. Indeed, Metchnikoff postulated that metazoans have evolved from a common ancestor, a phagocytosing cell colony, which he termed as "*Phagocytella*" in 1886 [5, 6]. Another interesting hypothesis from Rudolf Virchow (1821–1902) and Ernst Haeckel (1834–1919) went even further: the first cells were probably formed by the aggregation of simple vesicles, and because *Amoeba* cells are rich in such vesicles, an *Amoeba*-like "Cytode," or "*Protamoeaba*," was the possible forefather of all cells [7, 8]. These theories are interesting pieces of science history today. However, macrophage-like cells are present in the simplest metazoans, and the Toll receptor, for instance, one of the most important pathogen recognition receptor of macrophages, was also initially identified in an invertebrate [9]. This chapter reviews how tolerogenic and growth-supporting macrophage traits, that is, M2 macrophage traits, have evolved in invertebrates.

T. Röszer, *The M2 Macrophage*, Progress in Inflammation Research 86,
https://doi.org/10.1007/978-3-030-50480-9_2

2 Macrophages in Invertebrates

Free-living amoeboid eukaryotes (Amoebozoa, or Rhizopoda) and the phagocytosing amoeboid cells of invertebrates share considerable morphological similarities with the macrophages of vertebrates (Fig. 1). Colony-forming amoeboid cells of the

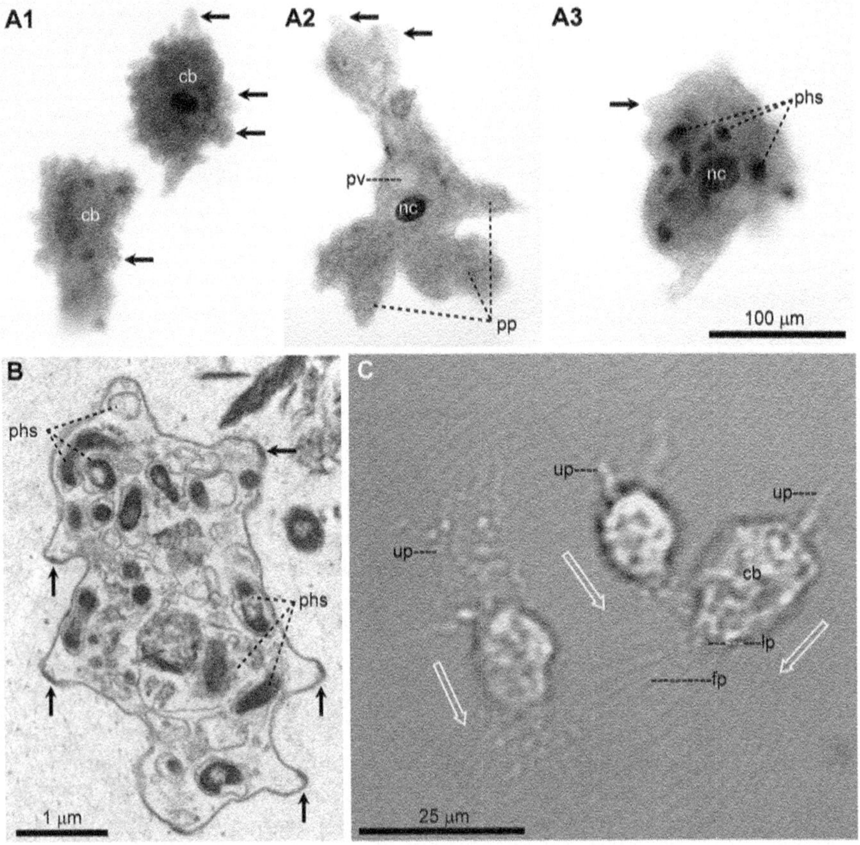

Fig. 1 Phagocyte morphology in invertebrates and in mammals. (**a1–a3**) *Amoeba proteus* cells seen with light microscopy. The amoeboid (amorphous) cell shape is a common feature of free-living phagocytes and of macrophages. *Arrows* label cell-surface protrusions, the lamellipodia, which form pseudopodia (*pp*). Pseudopodia protrude toward the direction of cell movement. Rearrangement of the cell shape allows migration of the cell, and this trait is preserved in all phagocytes. The cells also phagocytose other cells, and hence the cell body (*cb*) contains phagosomes (*phs*). Further visible organelles are the nucleus (*nc*) and the pulsating or contractile vacuole (*pv*), which serve for osmoregulation in freshwater species. (**b**) Transmission electron microscopy image of an amoeba cell. Note the abundance of phagosomes (*phs*). *Arrows* label lamellipodia. The specimen was collected in a flooded ditch, Cuenca Alta del Manzanares, Madrid, Spain. (**c**) Mouse peritoneal macrophages, cultured in vitro. Differential interference contrast image. Note irregular cell body (*cb*) shape and the numerous protrusions of the cell membrane. Direction of the cell motion is labeled with *arrows*. The leading edge forms advancing lamellipodia (*lp*) with numerous forward-projecting filopodia (*fp*). The opposite pole of the cell, called the uropodium (*up*), is retreating. (Images by author)

social amoeba *Dictyostelium discoideum* serve as an important model of macrophage functions and metazoan development [10]. Phagocytosing amoeboid cells, which are considered as macrophages herein, are known in all representative phyla of invertebrates, including sponges [11], worms [12], mollusks [13], arthropods [14–16], ascidians [17], and echinoderms [18, 19]. An extensive literature is available on the possible classification of invertebrate macrophages, and hence there is a diverse nomenclature of these cells [20]. Invertebrate macrophages are often called amoebocytes, phagocytes, phagocytic coelomocytes, wandering cells, phagocytic eleocytes, free hemocytes, plasmatocytes, or granulocytes [20–24]. In sponges these phagocytic cells reside within the cell colony and are embedded in the so-called mesohyl, a collagen-rich extracellular matrix [25–27]. In cnidarians macrophage-like cells, called granulated amoebocytes, are embedded into the extracellular matrix, the mesoglea. These cells are motile, and thus may also be present in the epithelial layer covering the body, and can be enriched in the gonads and accumulate at sites of tissue injury [28]. In animals that have body cavities[1] and a circulatory system, macrophage-like cells circulate in the hemolymph or blood, or settle in tissues. Hence, they resemble vertebrate monocytes and tissue resident macrophages, respectively.

Although invertebrate macrophages have been known since the nineteenth century, our understanding of their phylogeny, ontogeny, and cell and molecular biology is still largely unexplored in the twenty-first century. Their vertebrate counterparts have gained much more attention in the past two decades, most likely for their central role in the human immune system.

3 Macrophage Activation in Invertebrates

3.1 Metabolic Benefit of Macrophage Activation in a Multicellular Organism

Here, we should address the evolutionary aim of a multicellular organisms in developing macrophage-like cells. Let us first be carried away by the concept, that the first living cell was heterotrophic, and we may consider that a free-living, phagocytosing cell, a phagotrophic organism, was the simplest unicellular living matter. This cell did engulf other cells as food and did protect itself from being food for other cells. The phagocytosing cell hence largely resembled an M1 macrophage, which engulfs and degrades other cells. On the other hand, the phagocytosing cell

[1] By definition, only the mesoderm forms body cavities (peritoneum, pleural cavity, pericardial cavity). In some animals, such as in helminths, internal organs are located in a large cavity, which is however outside of the mesodermal, "true" body cavity (coelom). In such instance this body cavity is considered a "false" body cavity (pseudocoelom). This distinction is based on the ontogeny of these cavities. The function of true or false body cavity may be very similar. For instance, macrophages may appear in both body cavities.

also allowed the survival of the engulfed cells for certain reasons (e.g., establishing a symbiosis). That behavior resembles M2 macrophages. Tolerance of foreign cells, which is a key feature of M2 macrophages, thus most likely appeared in the simplest unicellular organisms (Graphic Abstract 1).

The simplest form of multicellularity is colony formation by the daughter cells of an organism. The daughter cells are the "self" cells of the colony (Fig. 2). The colony aims to survive and use the environment's resources to grow. Eventually, the colony size increases to a critical point at which the colony is too large to support its own nutrient demand, so its cells generate a new, separate colony. Because the cells of the colony are built from valuable materials (proteins, carbohydrates, lipids), the colony becomes a potential nutrient resource for of other organisms. In other words, some other organisms may want to feed on the materials of the colony. The foreign cells may grow inside the colony, which may kill the colony because of resource competition between the self- and foreign cells and the digestion of the self-cells by the foreign cells. The self-cells can acquire innate immune defense mechanisms. Free-living cells also have a wide range of immune mechanisms: bacterial colonies can inhibit the growth of other cells, fungal cells produce antibiotics, protozoa emit polymeric nests to protect their cells from chemical insults. Innate immune functions have evolved in epithelial cells of the colony-forming organisms, and in many invertebrates epithelial cells are important in immunity, such as for example the epithelial lining of the intestine in vertebrates. However, when immune functions are concentrated in one specific cell type, it is a wandering phagocyte that assumes these functions (Fig. 2).

This cell is patrolling over the colony, either kills or tolerates the foreign cells, and in all instances tolerates the self-cells (Fig. 2). When the foreign cells are killed, we recognize key traits of the mammalian M1 macrophages. In contrast, in the case of the cell-tolerant invertebrate macrophages, we see functional hallmarks of the mammalian M2 macrophages. The benefit of tolerating a foreign cell may be the establishment of various forms of symbiosis between the self- and foreign cells. One model of the development of multicellularity considers that the first

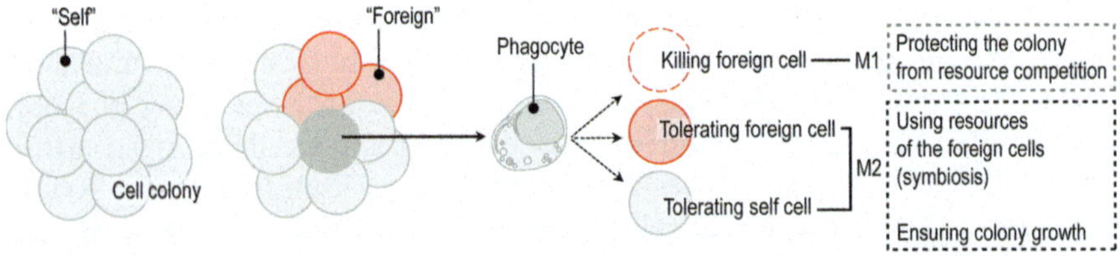

Fig. 2 Metabolic impact of the M1/M2 activation of macrophages. Daughter cells assemble into a loose aggregate and form a colony. Foreign cells can settle within the colony, attempting to consume its materials. One dedicated cell type, a phagocyte, may evolve to acquire immune mechanisms (i.e., phagocytosis, pathogen killing) to eliminate the foreign cells. The phagocytic immune cell always tolerates the self-cells of the colony. The phagocyte may also tolerate invader cells, which leads to temporal or permanent symbiosis between the colony and the foreign cells. The phagocyte eventually increases fitness and survival of the colony, by maximizing its access to nutrients

multicellular organisms were established as a result of symbiosis [29]. In that instance, the wandering phagocyte has been challenged with the same dual role: tolerate foreign cells and eliminate unwanted cells from the symbiotic cell aggregate (Graphic Abstract 1).

One may raise the question whether the development of macrophages is universal in all clades of the living organisms. For instance, multicellular plants lack this cell type, most likely because their cells are encapsulated into cell walls, which impede the mobility of a wandering macrophage-like cell within the plant's body (Fig. 3). Nevertheless, phagocytosis can be evoked in plant cells as well, when their cell wall is removed [31], and several pathogen-killing mechanisms of macrophages are employed by plant cells as well (e.g., nitric oxide synthesis, phenoloxidase activity), without assigning a macrophage-like cell to these functions [32, 33].

3.2 The M1-Like Macrophage Activation in Invertebrates

Signs of macrophage activation in invertebrates include increased rate of engulfing microorganisms, solid particles, or cell debris; increased ability of forming filopodia and spreading on a solid surface; or a respiratory burst and the increment of lysosomal acid phosphatase and peroxidase activity [34, 35]. These features of macrophage activation can be evoked by exposure to pathogens, such as bacterial cells.

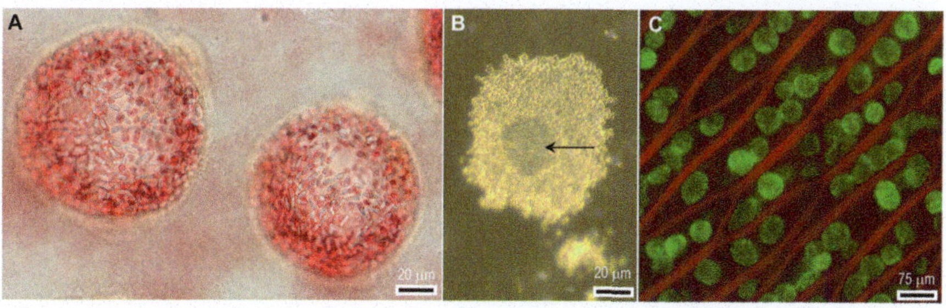

Fig. 3 Representative forms of cell colonies in animals and plants. (**a**) Colonies of *Collozoum inerme*, a colonial radiolarian. The cells are bound in a common gelatinous matrix. The individual cells form pseudopodia and can ingest unicellular dinoflagellate cells, the so-called zooxanthellae. These photosynthesizing cells survive within the host cells, establishing a farming symbiotic relationship between the two organisms [30]. (**b**) Mouse cell colonies in vitro, grown in soft agar. *Arrow* indicates the necrotic core of the colony. A necrotic core develops when the colony reaches a critical volume, and the cells inside the colony die of their limited access to air and nutrients. In nature, colonies can be split into smaller colonies (i.e., by means of asexual reproduction); moreover, apoptosis limits excess cell growth. Innate immune functions are assigned to each member of the colony. (**c**) Moss cells are forming a colony, and the cells are surrounded with a cell wall shown in *red*. The cell wall ensures proper nutrient dissipation within the plant's body; however, this limits the mobility of the individual cells and thus the utility of a wandering phagocyte within the colony. Plant cells have endosymbionts: the chloroplasts, which are seen in *green* fluorescence. (Images by author)

Pathogen-associated patterns known to activate invertebrate macrophages are lipopolysaccharide (LPS), double-stranded RNA polyinosinic:polycytidylic acid (poly I:C), glycoconjugates, and peptidoglycans. These signals elicit a pathogen-killing phenotype in invertebrates that resembles vertebrate-type M1 macrophage activation (Fig. 4). Several gene products have been identified in invertebrates that are involved in pathogen elimination: Toll-like receptors, possible homologue of the vertebrate interleukin-1 receptor-associated kinases (IRAKs), argonaute proteins, proteins increasing bactericidal permeability, possible Toll pathway proteins, proteins associated with H_2O_2 production [36], lysosomal proteins, and many others such as phenoloxidase, nitric oxide (NO), and antimicrobial peptides [37]. Activated invertebrate macrophages also express various homologues of mammalian M1 macrophage cytokines, such as IL-1α, IL-1β, IL-2, IL-6, IL-8, and TNFα [23], although whether they have an equivalent role in invertebrates and vertebrates is still unknown. Inert substances such as latex beads fail to activate invertebrate macrophages [34]. Invertebrates express Toll-like receptors, albeit the number of the possible Toll-like receptors (TLR) is highly variable among species: *Caenorhabditis elegans* has only one TLR, whereas *Strongylocentrotus* has 222 TLRs [37] and a cnidarian species seems to lack TLRs [38]. As a comparison, the mouse has 13 TLRs, and the human has 10 TLRs, each recognizing specific pathogen-associated patterns [37]. This finding suggests that invertebrate macrophages express pattern recognition receptors, which allow them to distinguish self-antigens from danger signals and pathogen-associated patterns.

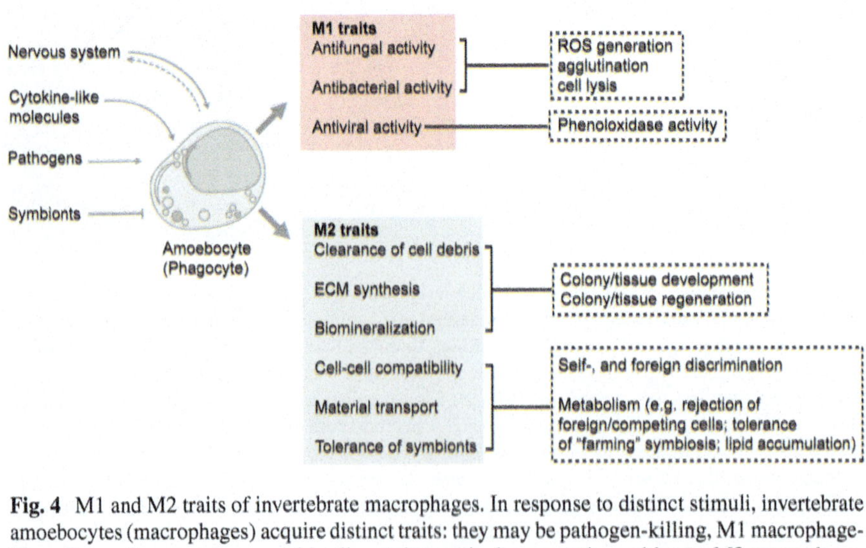

Fig. 4 M1 and M2 traits of invertebrate macrophages. In response to distinct stimuli, invertebrate amoebocytes (macrophages) acquire distinct traits: they may be pathogen-killing, M1 macrophage-like cells, or they may be wound-healing, tolerogenic, homeostatic, and hence M2 macrophage-like cells. Macrophage activation is governed by pathogen-associated molecules, signals of the nervous system, cytokine-like molecules, and is skewed toward a tolerogenic activation state by symbiont cells. *ROS* reactive oxygen species, *ECM* extracellular matrix

The molecular nature of the pathogen-associated pattern recognition molecules in invertebrate macrophages is however largely unexplored. For instance, *Nereis* coelomocytes express various lectins, such as mannose, galactose, and fucose receptors that may allow recognizing cell wall polysaccharides or proteoglycans [39]. Hemocytes of the red swamp crayfish *Procambarus clarkii* also express a mannose receptor-like protein (PcMR): PcMR contains four C-type lectin domains and can bind to a variety of bacteria, as well as glycoconjugates from the bacterial cell surface. PcMR agglutinates bacteria in a calcium-dependent manner [40]. Macrophages also have their own, specific cell-surface glycosylation patterns, which may distinguish them from other immune cells in invertebrates. For instance, fluorescently labeled lectins can distinguish macrophage-like cells from other immune cell types in the echinoderm *Strongylocentrotus purpuratus* [41]. Similarly, the expression of fibrinogen-related proteins distinguishes functional subgroups of macrophages in the snail *Biomphalaria glabrata* [42].

Upon infection, hematopoiesis is accelerated in some invertebrates, such as in gastropods and crustaceans [23, 43]. Increased hematopoiesis replenishes immune cells that are lost in infection or during tissue damage. In snails, a granulin-like snail growth factor increases hematopoiesis and allows combating parasite infections [35]. In crustaceans, the pathogen-associated pattern lipopolysaccharides (LPS) decrease circulating hemocyte numbers to 40% within 3 h. The number of hemocytes is restored in 3–24 h, which implies a high hematopoietic reserve potential [43]. This high degree of adaptability of the hematopoietic tissue requires communication between the pathogen-fighting immune cells and the hematopoietic tissue: this is ensured by invertebrate cytokines, such as astakines [44]. Astakines stimulate the proliferation, differentiation, and survival of hematopoietic tissue cells in crustaceans [44, 45]. In vertebrates, astakine-like cytokines are called prokineticins [46], and these have been recently recognized as inflammation-associated molecules in mammals [44]. The plasma membranes of insect macrophages (granulocytes) form extracellular traps and hair-like nets, which gather other immune cells to increase the elimination of cellular pathogens [16].

In addition to cytokines, neuroendocrine factors also allow a communication between immune cells and hematopoietic cells (Fig. 4). For example, hemocytes in the Chinese mitten crab *Eriocheir sinensis* express a gene encoding platelet-derived growth factor/vascular endothelial growth factor (PDGF/VEGF)-related factor. Its level is increased in response to pathogens and tissue damage, and it also provokes the release of noradrenaline and dopamine in the hemolymph [47]. In turn, immune cells also signal to the nervous system: for instance, lipid mediators that are released during a systemic immune challenge inhibit the establishment of neural circuits in snails [48].

Altogether, pathogen-associated molecular patterns, cytokines, and neuroendocrine signals shape the activation of invertebrate macrophages (Fig. 4). Although this framework seems established, we have much still to explore in the biology of invertebrate cytokine signaling and the neuroendocrine control of immune responses in invertebrates.

Invertebrate macrophages, similarly to their vertebrate counterparts, also employ mechanisms to mitigate their own pathogen-killing mechanisms: these are involved in homeostatic functions of the organism, tolerate symbiont cells, and aid wound healing and tissue remodeling. These features make them similar to M2 macrophages (Fig. 4).

4 M2-Like Features of the Social Amoeba

The mechanisms that mitigate the pathogen-killing activation of invertebrate macrophages is poorly understood, despite the evidence that invertebrate macrophages have the ability to curb their own pathogen-killing mechanisms. For example, the social amoeba *Dictyostelium discoideum* employs mechanisms that allow a temporal tolerance of ingested bacteria [49].

When the environment is rich in bacteria, natural isolates of *D. discoideum* engulf and digest species of *Proteobacteria*, *Actinobacteria*, *Bacteroidetes*, and *Firmicutes*. The ingested bacteria are effectively degraded, so the amoeba cells are rich in multilamellar bodies, the remnants of late phagosomes. Starvation triggers an aggregation of the solitary *D. discoideum* cells into a multicellular, motile, slug-shaped colony. The so-called sentinel cells of the colony are phagocytosing and are involved in pathogen elimination; thus, they are similar to M1 macrophages. The slug-shaped colony develops into a fruiting structure, which is held by a stalk, formed by dead cells. The multicellular form of *D. discoideum* is often called the social stage, in contrast to the solitary (i.e., unicellular) stage [10].

The multicellular form can establish a farming symbiosis with bacteria [49]. This phenomenon develops in about one-third of natural isolates of *D. discoideum*. Establishing symbiosis obviously requires the ability of discriminating pathogenic or useless bacteria from those that are useful to be grown within the amoeba colony [49]. In farming symbiosis, the *D. discoideum* cells do not kill and digest the phagocytosed bacterial cells instantly after ingestion, but instead allow them to grow; this requires mechanisms that can suspend the pathogen-killing mechanisms temporarily. Many examples of farming symbiosis are seen in nature: for example, colonies of *Collozoum inerme*, a colonial radiolarian, engulf unicellular dinoflagellate cells, the zooxanthellae. These photosynthesizing cells survive within the host cells, and the host establishes a farming symbiotic relationship [30]. Symbionts supply metabolites, and eventually energy, to the host (Graphic Abstract 1).

The foregoing examples show that skewing the host immune response toward a tolerogenic state yields metabolic benefits. In macrophages, this means that an M2 state should be adopted to tolerate the symbiont cells. Intracellular pathogens take advantage of this trait of macrophages. For instance, *Mycobacterium* species are able to survive phagocytosis and hence grow in the phagosomes of macrophages, by skewing macrophage activation toward M2. M2 macrophages are prone to tolerate intracellular parasitism (for details, see chapters "Mechanisms Which Control the Size of M2 Macrophage-Dominated Tissue Macrophage Niches" and "Immune

Functions of the M2 Macrophages: Host Defense, Self-Tolerance, and Autoimmunity"). About 25% of natural isolates of *D. discoideum* are infected with *Burkholderia* species [50], which is an intracellular pathogen in mammals. It can evade the phagosomes, and eventually replicates in the cytosol [51]. Moreover, it inhibits M1 activation of vertebrate macrophages, and thus effectively avoids pathogen killing. It is likely that *Burkholderia*, and probably some other bacterial genera, are able to inhibit pathogen killing mechanisms of *D. discoideum* by bacteria-derived signals [52].

One would assume that the multicellular form of *D. discoideum* is hence in an immune-suppressed stage and that this allows the tolerance of the symbiont bacteria. However, *D. discoideum* can still eliminate bacterial pathogens [53], and elimination of pathogenic bacteria is triggered by LPS in both unicellular and multicellular forms [53, 54]. Bactericidal activity and increased phagocytosis are dependent on MAPK signaling pathways as well as on Toll/interleukin-1 receptor domain-containing protein (TirA) in the unicellular form [54]. Pathogen killing of the multicellular *D. discoideum* involves a unique mechanism, the generation of reticulated DNA nets, so-called extracellular traps [53]. Generation of these extracellular traps is triggered by bacteria and LPS, and is dependent on TirA and reactive oxygen species-generating NADPH oxidases [53]. The major component of the extracellular traps is mitochondrial DNA [53]. It is an intriguing question how mitochondrial DNA is released and processed to form extracellular traps. Of note, vertebrate granulocytes also release DNA nets.

Transition from the unicellular form to a multicellular colony is triggered by starvation, which is known to initiate autophagy, a process which eliminates unnecessary organelles and allows recycling their components for cellular metabolism. Autophagy is necessary for the development of the fruiting body of the multicellular *D. discoideum* colony [55]. Depletion of autophagic proteins promotes the accumulation of dysfunctional mitochondria and cytosolic translocation of mitochondrial DNA in response to LPS and ATP in mammalian macrophages [56]. It is likely that, during starvation-induced autophagy of mitochondria, a yet unexplored DNA-processing mechanism allows DNA transport from the mitochondria into the extracellular DNA nets. Interestingly, the second intron of the *D. discoideum* cytochrome oxidase subunit 1/2 fused gene encodes a functional DNA endonuclease [57], which might have a role in processing of DNA during autophagy. Bacteria employ cyclic dinucleotides as messenger molecules, and these nucleotides initiate the innate immune response of vertebrate macrophages (in the so-called cGAS/STING signaling pathway, which serves in recognition of cytoplasmic DNA). The genomic DNA of *D. discoideum* contains sequences that may be targets of cyclic dinucleotide signaling [58]. It is possible that there is a bacteria-to-amoeba signaling in *D. discoideum*, which may be decisive whether macrophages initiate bacterial cell killing or tolerate this.

5 M2-Like Functions of Poriferan Macrophages

Sponges (Porifera) are early diverging colony-forming metazoans that provide insights into conserved mechanisms of host–microbe crosstalk [11]. As sponges are typically filter-feeding organisms, thus the sponge cells are exposed to microbial abundance, and almost all cells of a sponge colony are capable of phagocytosis of microbes. Uptake and digestion of "food" cells is the major task of specialized cells of the colony, the choanocytes and endopinacoytes. These cells line chambers and canals inside the sponge colony, and their major function is the uptake of cells and debris from the water that passes through the colony. In the extracellular matrix of the colony, which is called the mesohyl, dwell various symbiont bacteria [59]. The exact molecular mechanisms that allow the sponge cells to discriminate "food" from pathogenic or symbiont microbiota remains unknown, although this is a key aspect of developing tolerogenic, M2-like macrophages.

It is likely that sponge cells have pattern recognition receptors that allow them to trigger pathogen-killing responses or to tolerate symbiont bacteria. Sponges have a complex chemokine network controlling colony growth and self-tolerance [61]. Pathogen elimination is a major function of macrophage-like cells of sponges, which are often referred as amoebocytes, archaeocytes, or wandering cells. They are embedded in the extracellular matrix, called the mesohyl, of the sponge colony (Fig. 5). They have high phagocytic potential, are able to engulf both bacterial and yeast cells [62], and generate superoxide anions and NO as pathogen-killing agents [63]. There is only one carnivorous sponge genus, the deep-sea *Asbestopluma* (Poecilosclerida) sponges. In these species macrophages migrate toward the prey and concentrate around it. Fragments of the prey are then phagocytosed and digested by macrophages [64]. Symbiotic bacteria can be vertically transferred between generations [65]. There are symbiotic bacteria of sponges that specifically live in sponge colonies only and thus are termed "*Poribacteria*" [66].

Sponge macrophages also retain stem cell-like features: they proliferate and can differentiate into various cell types of the colony (Fig. 5) [27]. Freshwater sponges generate so-called gemmules, aggregates of archeocytes, which are amoeboid cells and resemble the macrophage-like amoebocytes of the sponge colonies. These cells are produced by the dying colony in autumn; they overwinter and generate new colonies in spring.

In addition to their pathogen-killing and particle-digesting properties [11], sponge macrophages also have a role in metabolism [26] and building the extracellular matrix (ECM) [67, 68] of the sponge colony (Fig. 5). Their metabolic role is the result of their tolerance of various symbiont bacteria, which affects the nitrogen balance of the sponge colony. Macrophages also secrete a galectin-like protein, which is a component of the mesohyl and is necessary for the proper development of the colony [67].

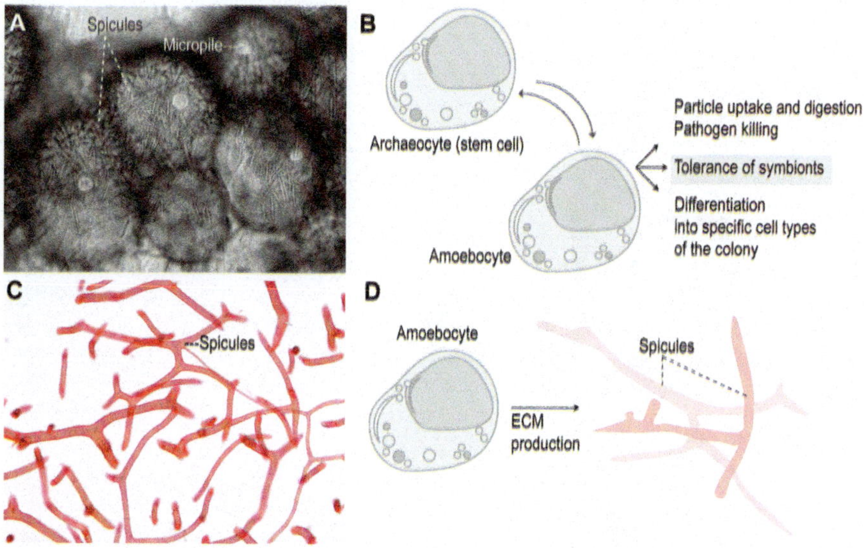

Fig. 5 Macrophage development and functions in sponges. (**a**) Freshwater sponges form stem cell aggregates, called gemmules, in clusters during winter. The gemmules of the freshwater sponge *Spongilla fragilis* are filled with amoeboid archaeocytes (stem cells), bordered by epithelium-like cells, and collated by mesohyl and spicules. The micropile is a still-closed orifice of the gemmule that releases the archaeocytes in spring, when the gemmule is hatching and the stem cells initiate the development of new colonies. Generation of gemmules is under the control of environmental factors [60]. (**b**) Development and function of sponge macrophages (amoebocytes), which develop from amoeboid archaeocytes. The amoebocytes preserve stem cell capacity and can differentiate into various cell types of the colony. The amoebocytes engulf particles, and eliminate pathogens, hence resembling mammalian M1 macrophages. They are also capable of tolerating symbiont cells of the colony, which trait resembles a tolerogenic M2 macrophage. (**c**) Spicules from the mesohyl of *Spongilla lacustris*. Spicules are skeletal elements of the sponge colony. (**d**) M2-like amoebocytes of the mesohyl may produce extracellular matrix (*ECM*) and assist the assembly of spicules. (Images by the author)

6 A Complex Body Plan Requires Tissue Homeostasis: Further Development of M2 Traits

In contrast to sponges, cnidarians build tissues rather than colonies of cells (Fig. 6), which is considered a key evolutionary advance of cnidarians, leading the further development of the metazoans living today. The body plan of cnidarians is more complex than of the colony-forming metazoans; they have a body axis, germ layers, and the so-called endoderm and ectoderm that have fixed positions and functions in the body. The body plan is generated in the process of morphogenesis, which needs the expansion and differentiation of stem cells, along with the programmed removal of unwanted cells. Moreover, cnidarian tissues are rich in stem cells, which constantly replenish the body's cells, leading to the apparent lack of aging and to a high tissue turnover rate and regeneration capacity [69].

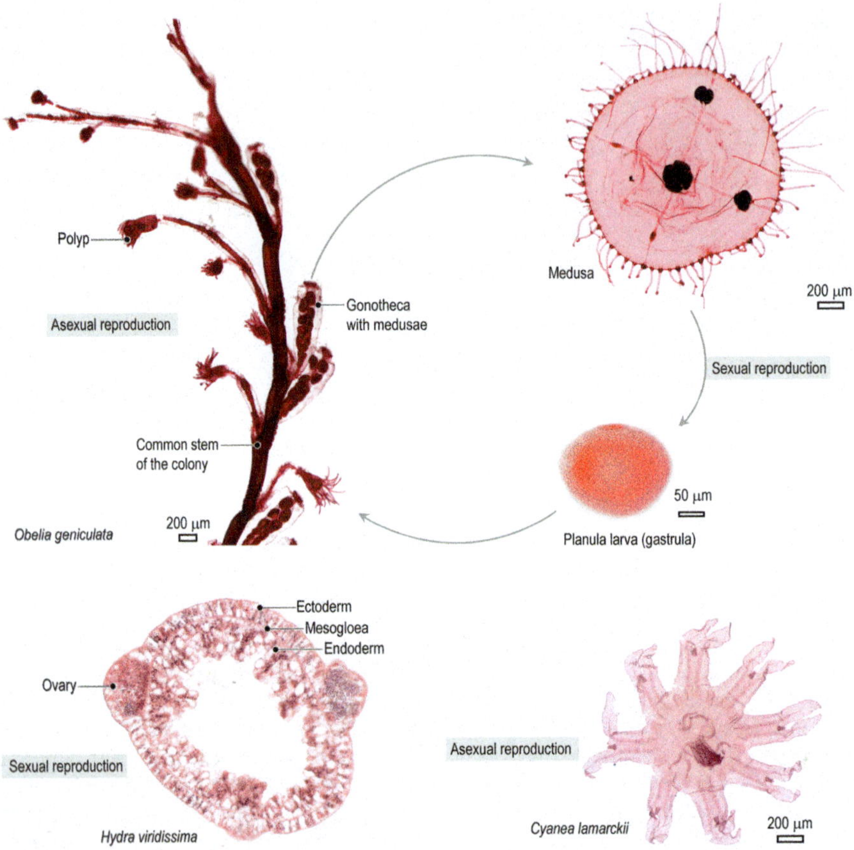

Fig. 6 A complex body plan increases the demand of autoantigen and neoantigen recognition. Cnidarians develop a complex body plan with radial symmetry and with the oral–aboral axis, along with germ layers and specialized tissues. The reproductive cycle is also complex in many cnidarians, which challenges "self" and "foreign" discrimination by generating neoantigen-rich gametes. *Top:* life cycle of *Obelia geniculata* with asexual budding and with sexual reproduction. Gametes are generated in the medusa form. *Bottom left:* cross section of the green hydra, *Hydra viridissima*, a cnidarian which has photosynthesizing endosymbionts. The section shows the two germ layers, the interconnecting extracellular matrix, and the neoantigen-rich gonads. *Bottom right:* ephyra larva of the blue jellyfish, which is generated asexually. (Images by the author)

The ectoderm and the endoderm form specific tissues, and tissue identity also needs a firm discrimination of "self" and "foreign" cells. The complex mechanism that ensures that "self" cells are allowed to grow whereas "foreign" cells of the tissues are rejected is called allorecognition, and ensures histocompatibility [70] (Graphic Abstract 1). Rejection of the "foreign" cells is called allograft rejection, that is, the rejection of genetically nonidentical cells by the host organism. As discussed earlier, it is plausible that allorecognition and histocompatibility evolved to ensure the optimal resource utilization by cell colonies that have given rise to the multicellular organisms. Invader cells, which would compete with the cells of the organism for the same resources (nutrients, oxygen, etc.) and space, are rejected. In the setting of tissue development, "foreign" is not only an invader cell, it can also mean cells that do not belong to the specific tissue. Cnidarian macrophages are hence challenged with two distinct physiological aims: they should protect the body from foreign cells and pathogens and also tolerate self-antigens [71, 72].

Pathogen elimination is caused by M1-like macrophage traits in cnidarians. The majority of these mechanisms eliminate fungal pathogens, because in their natural habitat fungal parasitism is the main threat of some cnidarians, for example, corals [73, 74]. It is still unclear how macrophages discriminate self- and foreign antigens in cnidarians. Cell adhesion molecules bind together "self" cells, putative immunoglobulin-like domain-containing proteins may serve as allorecognition receptors (ALRs), and one possible, and highly variable, gene (*adr2*) may be responsible for the development of the specific antigen pattern of individuals in cnidarians [38, 75]. Cnidarian immune recognition receptors are possible homologues of mammalian nucleotide-binding oligomerization domain (NOD)-like receptors, C-type lectins, rhamnose binding lectins, scavenger receptors, and LPS-binding proteins [38]. Toll-like receptor homologues may be lacking in cnidarians; however, a homologue of nuclear factor kappa B (NFκB), the central hub for inflammatory gene expression in mammals, has been identified in cnidarians [38].

Cnidarians have complex sexual reproduction cycles in addition to asexual cloning of their cells. Sexual reproduction raises a novel physiological setting: the newly generated gametes are genetically foreign cells and are arranged in specific organs, the gonads. The gonads are hence rich in so-called neoantigens, which are distinct from the "self" antigens of the rest of the body. There is hence a risk of rejecting the gametes by the surrounding cells. Amoebocytes are enriched in the cnidarian gonads [71]; however, gametes are tolerated and not rejected by them. Possibly the gonadal macrophages adopt a tolerogenic phenotype. An equivalent mechanism is seen in mammals, where testicular macrophages tolerate the progenitors of male gametes. Moreover, in some cnidarian species, asexually generated offspring can dwell on their parent's body. Although these offspring are asexually generated from the parental cells, there is a probability of genetic diversity between the parent and the offspring (e.g., distinct epigenetics) [76]. Although this may be the source of an immune response against the offspring, such rejection is not triggered; this is an evolutionary forerunner of the embryo developing in the womb of a mammal. The placenta contains M2 macrophages, which maintain an immunological tolerance of the embryo (chapter "Immune Functions of the M2 Macrophages: Host Defense, Self-Tolerance, and Autoimmunity").

Moreover, there are symbiotic associations of cnidarians and photosynthesizing cells in which the cnidarian tissues do not reject symbiotic cells. There is a metabolic co-dependence between the cnidarian host cells and the photosynthesizing endosymbionts, and the endosymbiont cells are inherited between generations of the host [77]. Importantly, cnidarian cells suppress the expression of pathogen-recognizing gene products, and upregulate a transcript encoding a mannose receptor homologue when the endosymbiont is present [78], leading to a tolerogenic phenotype of the host cells. In certain stress situations the host can initiate immune response against the endosymbiont. This process is called bleaching, because the cnidarian host loses its green color by the loss of the photosynthesizing (hence, green) endosymbionts [79]. These tolerogenic immune traits are associated with epithelial cells in cnidarians. Epithelial cells have key importance in invertebrate innate immunity, and they may retain macrophage-like functions, such as in mollusks and in cephalochordates [13].

7 M2-Like Features of Macrophages in Invertebrate Wound Healing

Lesions in cell colonies or tissues are rapidly invaded by macrophages in invertebrates. This role of macrophages is already apparent in sponges, where macrophages remove cell debris during regeneration of a damaged colony [25]. A similar role of macrophages is known in all taxa of metazoan invertebrates [71, 80, 81].

The role of macrophages in tissue regeneration is well studied in leeches. The wounded body wall is infiltrated by macrophage-like cells, which form clots and close the wound [81]. This response is rapid, and detectable within 10 min following the tissue lesion [81]. The responsible macrophages are classified as phagocytosing microglia cells [82]. Their first scientific records date back to the late nineteenth century, when they were detected by various silver impregnation techniques [83]. Microglia cells express complement component 1q (C1q) binding protein, and are recruited to the site of injury by *Hirudo medicinalis* complement component 1q (HmC1q) [84]. Some microglia cells also express *H. medicinalis* calreticulin (HmCalR), which is a secondary binding partner for HmC1q [85]. Binding of microglia to HmC1q allows the chemoattraction of resident microglia to the site of tissue repair. Some further signals, such as ATP and NO, increases microglia movement to nerve injury [86]. The time course of NO release in the injured leech nervous system is partially under the control of endocannabinoids, namely, *N*-arachidonyl-ethanolamide and 2-arachidonyl glycerol [87]. Roles of microglia include the phagocytosis of cellular debris at the lesion site, and secretion of laminin [88]. Laminin is a component of the extracellular matrix of the nervous system that supports axon outgrowth in the regenerating nervous system [88]. Leech microglia express equivalent of calcium-binding adapter molecule 1 (Iba1) [89]. Of note, Iba1 is a known marker of mammalian microglia and is expressed by certain tissue resident macrophages of primates [90, 91]. In the mouse, Iba1$^+$ microglia also express CD200R, which is known to inhibit TNFα and NO release, and hence is associated with M2 macrophages [90].

Tissue injury requires defense from infections, so macrophages at the site of injury express antibacterial agents, resembling M1 macrophages. In crabs, for instance, the eye peduncle is often exposed to injury, and it has a relatively high degree of regeneration potential. In the crab *Carcinus maenas*, the regenerating eye peduncle expresses arcinin, an antibacterial crustin. Crustin is mostly expressed by hemocytes [92] and initiates a unique pathogen-eliminating mechanism by activating pro-phenoloxidases in the hemolymph [93]. Phenoloxidase activity is an evolutionarily ancient, nonspecific immune mechanism that is present in vascular plants, fungi, and also in most invertebrates. This function can be seen when an apple or a mushroom is cut into halves and the damaged cells are exposed to air, causing a tan color of the tissues to develop. In the invertebrate hemolymph the mechanism leads to the scavenging of pathogens, the so-called sclerotization or melanization process [13]. Phenoloxidase activity generates melanin, which has antibacterial and antiviral activity. Also, it contributes to encapsulation of foreign particles [94] (this resembles the M2 macrophage-rich granuloma formation around helminth eggs; see chapter "Immune Functions of the M2 Macrophages: Host Defense, Self-Tolerance, and Autoimmunity"). Secondarily, it also contributes to exoskeleton sclerotization and tanning. Activity of phenoloxidase should hence be well controlled to avoid unwanted tissue damage. In *Tenebrio molitor*, for instance, the so-called Spätzle-processing enzyme converts the inactive pro-phenoloxidase and the so-called clip-domain SPH1 zymogen to an active melanization complex. Pro-phenoloxidase is a competitive inhibitor in the complex formation and thus limits melanin synthesis [95]. Other insects have phenoloxidase inhibitors in their hemolymph. Hemocyanin, the main component of the hemolymph in most invertebrates, also has antiviral activity and its enzymatic cleavage yields antimicrobial and antifungal peptides, serving as a coagulant in insects [13, 96].

8 A Unique Trait of Invertebrate Macrophages: Apoptotic Macrophages Augment Pathogen Elimination

The main task of macrophages is to remove cell debris at the site of tissue injury, which resembles the homeostatic role of macrophages in mammals. In mammals, M2 macrophage activation is apparent when macrophages engulf apoptotic cells or their fragments. In contrast, tissue damage causing necrosis, or infections that cause pyroptosis, induce M1 macrophage activation in mammals [97]. In mammals, when macrophages undergo apoptosis, they are engulfed by other macrophages; this evokes in the engulfing macrophages an M2-like activation state and contributes to the resolution of inflammation at the site of infection (detailed in chapters "Signal Mechanisms of M2 Macrophage Activation" and "Immune Functions of the M2 Macrophages: Host Defense, Self-Tolerance, and Autoimmunity"). Unlike their vertebrate counterparts, invertebrate macrophages do induce pathogen killing when they undergo apoptosis [13]. In *Limulus polyphemus*, phagocytosing hemocytes (i.e., macrophages) engulf cellular pathogens, they eventually undergo apoptosis, which is associated with the enrichment of phosphatidylserine in their cell surface. Phosphatidylserine has an asymmetrical distribution across the inner and outer phospholipid bilayer of the cell membrane, and in living cells the outer layer lacks phosphatidylserine. Early in apoptosis, however, phosphatidylserine appears in the outer cell surface, which serves as a signal for macrophages to initiate the engulfment of apoptotic debris. In *L. polyphemus*, contact with phosphatidylserine induces a conformational change of hemocyanin, a main constituent of the hemolymph. This conformation change gives a phenoloxidase-like activity to the molecule [98, 99].

9 M2-Like Macrophage Functions in Molluscan Shell Building

Mollusks have a specific trait, the ability to form a shell from calcium carbonate crystals. The shell may form symmetrical halves, such as in the majority of bivalves, may appear as a spirally coiled shell, such as in the majority of the gastropods and in some cephalopods; or may be embedded into muscle bundles, such as in slugs and many cephalopods. The shell material is the conchiolin matrix and calcium carbonate ($CaCO_3$) prisms and crystals, in the form of the stable calcite and the metastable aragonite [100–102]. As calcite is the more stable form of calcium carbonate, aragonite is the dominant form in the shell of mollusks, resulting from the effects of the conchiolin matrix [102] and the magnesium content of the mantle tissue [103]. These materials are synthesized and secreted by gland cells of the pallium, or mantle, a specialized region of the skin (Fig. 7). Conchiolin and calcium are released along with mucus from the gland cells, forming a thick layer around the

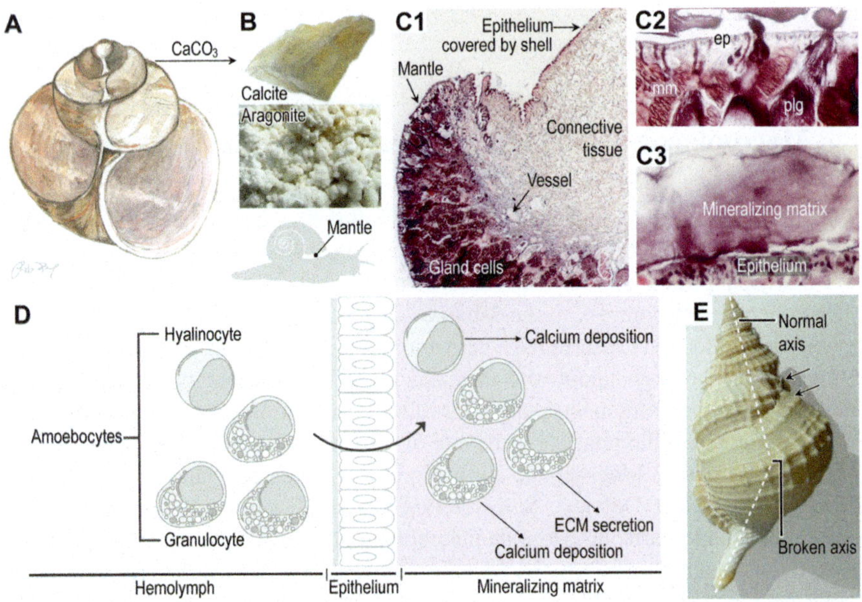

Fig. 7 Shell building in mollusks: role of macrophages. (**a**) Longitudinal section of a snail shell (*Helix pomatia*). (**b**) Crystals of $CaCO_3$ form the mineral component of the shell. Calcite has greater endurance than aragonite crystals of $CaCO_3$. The mineralization process takes place at the mantle (also called the pallium). (**c1–c3**) Histology of the mantle. Note abundant gland cells at the edge of the pallium. The mineralizing matrix and the gland cells show strong basophilia from the presence of calcium. Hematoxylin and eosin. (**d**) Scheme of macrophage functions in the synthesis and mineralization of the shell matrix. (**e**) Example of shell regeneration. The specimen shows severe damage of the shell, which was successfully healed (*arrows*). The axis of the shell healed in a deviated orientation (*dotted lines*). (Drawings and images by author)

skin epithelium. Eventually, as water evaporates from the matrix, the concentration of calcium allows the nucleation of calcium crystals. The shell grows proportionally with the developing body, and when it is damaged, a regenerative process is initiated to restore the shell material. In bivalves, when a foreign body is captured between the shell and the mantle, a so-called pearl is formed that is built up from the same material as the shell and encapsulates the foreign body [104].

In mollusks, hemocytes are often called amoebocytes, or are further classified based on the absence or presence of cytoplasmic granules and termed hyalinocytes and granulocytes, respectively. Both cell types show phagocytosis, form lamellipodia, and are similar to macrophages. The mantle contains both cell types, although granulocytes are more prevalent, and their number increases rapidly in response to shell damage in bivalves and in snails [105–108]. Macrophage contribution to the generation of the pearl has been shown by many studies [104, 109]. In vitro both cell types form extracellular matrix, and the granulocytes also contain calcium crystals [110] (Fig. 7). Similarly, granulocytes contain intracellular calcium carbonate crystals in vivo, which they deposit in the wounded shell matrix [106–108]. It is possible that the calcium is released to the circulation from the midintestinal gland or hepatopancreas, which stores calcium and is responsible for the intermediated metabolism of the body [13, 111]. The midintestinal gland is also an immune organ, and is the possible source of hemocytes [13]. However, it is also plausible that hemocytes are generated locally in the mantle [104].

It is not only calcium that is released by hemocytes in the mantle. Indeed, specific subsets of macrophage-like cells are associated with the mantle tissue in oysters, and they produce various proteins required for mineralization of the matrix, including nacrein and fibronectin [112]. Variation in the expression patterns of biomineralization-related genes combined with differences in the motility and adhesion between different hemocyte fractions demonstrate that different types of hemocytes are predominantly engaged in shell production or the immune response [112]. Macrophages of the regenerating shell express a soluble protein of hemocytes from diseased shell (HDS) of the oyster, *Crassostrea gigas*, which shows enzymatic activity and forms $CaCO_3$ crystals [113, 114].

Although macrophages contribute to the process, the shell matrix is mainly formed by gland epithelia of the mantle, and the geometry of the mantle determines the shape of the shell [115]. Macrophages may also produce matrix components or facilitate the mineralization. For instance, *Haliotis* macrophages express lustrin A, and two carbonic anhydrase enzymes; both are involved in biomineralization [116] (Fig. 7).

Oysters are filtrating organisms that are constantly exposed to bacteria, viruses, and debris from sea sediment; hence, a shell wound opens a potential entry site for pathogens. It is not surprising that shell damage evokes a rise of macrophages in the circulation of the pearl oyster [106, 108]. Moreover, in the oyster *Crassostrea gigas* macrophages express tissue inhibitor of metalloproteinase (Cg-TIMP), and Cg-TIMP mRNA accumulates during shell damage and bacterial challenge [117]. Mammalian metalloproteinases are expressed by M1 macrophages [118], and TIMPs inhibit the activity of tissue metalloproteinases. Thus, these have functions in extracellular matrix biogenesis, and inactivation of certain chemical transmitters, such as cytokines. In the context of macrophage biology, TIMPs are required for proper M2 macrophage activation, and the lack of TIMPs leads to the transcription M1 macrophage activation genes [119].

Further possible homologues of mammalian M2 macrophage-associated genes have been identified in the course of shell regeneration. For instance, the transcription of a molluscan chitinase (PfChi1) is increased rapidly after shell damage in the oyster *Pinctada fucata*. It is plausible that PfChi1 has its main function in mantle cells by allowing mineralization of the newly formed shell matrix. As it is also expressed in the so-called trochophore larva, a free-swimming larval form of bivalves, in which the larval primordium of the shell is formed, this molluscan chitinase may have a physiological function in shell matrix mineralization [120]. Also, vascular endothelial growth factor (VEGF)-like protein has been identified in macrophages involved in shell mineralization [112]. In mammals, VEGF is associated with the angiogenic effect of M2 macrophages [121].

10 Immune Signaling, Macrophages, and Wing Pattern in Insects

Wing morphology and pattern are species-specific traits of many insects, especially in Lepidoptera (Fig. 8). The wings are flat extensions of the chitinous cuticle, covered with colored scales and scaffolded with chitinous tunnels, called veins. The veins contain air channels, called tracheal channels, and hemolymph vessels that perfuse the wing [122]. Each species has a specific venation pattern, and linked clades of insects share similarities in the arborization pattern of their wing veins. The wings serve the flight of the animals; the scales of the wing improve the wing aerodynamics, the pigmented surface absorbs sunlight (i.e., increases body temperature), and the specific and flamboyant colors and patterns ensure mating and can serve as camouflage or mimicry (Fig. 8). Wing development hence has a central role in the life cycle of Lepidoptera. During metamorphosis, the wingless larva develops the primordia of the wings, which will gain their final form at the time of hatching. This morphogenesis process requires programmed elimination of cells, removal of the apoptotic remnants, and the building of a new tissue matrix. Free-moving hemocytes, considered analogous to macrophages, appear in the early wing primordia, around tracheal branches [123]. The hemocytes remove the basal lamina of the wing primordium epithelia, clear apoptotic cells, and eventually help the synthesis of the new basal lamina for the rearranged epithelial cells of the developing wing in *Manduca sexta* [124]. Insect hemocytes express mRNA and protein of type IV collagen, a key component of basal lamina [125]. Similarly, hemocytes are necessary for the wing development in *Drosophila melanogaster*, and they develop locally within the wing [126]. These hemocytes remove epithelial cells and matrix molecules and secrete a matrix that can bond the two surfaces of the wing together

[126]. The role of hemocytes, that is, phagocytosing, macrophage-like cells, is conserved among two diverse insect groups: Diptera (flies) and Lepidoptera (butterflies).

The lepidopteran wings are often richly pigmented and bear species- and gender-specific patterns [128] (Fig. 8). The dorsal and ventral patterns of the wings are markedly different, which offers a valuable model for understanding molecular mechanisms of complex body pattern development. Interestingly, in insects, the dorsal–ventral pattern of the body plan is controlled by Toll receptor signaling, which is also responsible for the initiation of innate immune response [129]. Toll receptor is activated by its ligand, so-called Spätzle, which is produced by distinct enzymes. In the setting of infection, Spätzle is produced by the Spätzle-processing enzyme in response to gram-positive bacteria or fungal pathogens [129]. This enzyme also increases melanin production by activating melanin synthesis. In arthropods, melanin is produced by the activity of phenoloxidase, which is

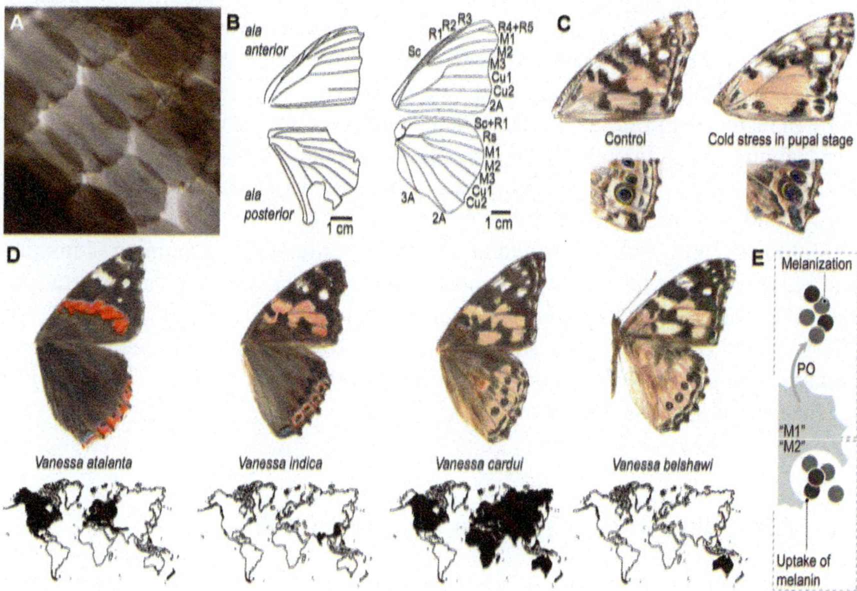

Fig. 8 Nymphalid wing anatomy and pattern. (**a**) Scales on the surface of the wing. (**b**) Veins on the surface of nymphalid wings. *Left:* extinct nymphalid wing anatomy, exemplified by the wing vein pattern of *Vanessa amerindica* [127]. *Right:* recent nymphalid wing anatomy. *Sc* subcostal vein, *R* radial vein, *M* medial vein, *Cu* cubital vein, *A* anal vein. Ala anterior: first wing, ala posterior: second wing. (**c**) Cold stress at the wing morphogenesis changes wing pattern in *Vanessa cardui*. (**d**) Recent *Vanessa* species share common ancestry and have similar wing vein anatomy and largely similar pattern. However, they have colonized distinct geographic niches, and have evolved into distinct species. Wing pattern is a distinctive feature of these species, and is necessary for the recognition of mates. Eventually, alterations in wing pattern, for example, in response to cold stress, can result in unsuccessful mating. (**e**) Macrophage-like cells contribute to the development of the wing pattern: M1-like cells promote melanization and color development, and M2 cells remove pigments. (Photograph (**a**) and paintings (**c, d**) by author)

associated with hemocytes [130, 131]. Hence, when the Toll pathway is activated, the melanization reaction of the hemocytes is increased.

Toll signaling is involved in the butterfly wing pattern development, for example, in the development of mimicry. Toll expression is increased in red pigment-bearing spots in female *Papilio polytes*, which causes them to resemble a nonedible species, *Pachliopta aristolochiae* [132]. The degree of macrophage-like hemocyte contribution to the pattern formation of the lepidopteran wings is yet to be elucidated.[2] It is likely that they may remove unwanted pigments and degrade unwanted tissues in wing morphogenesis [123]. Hemocyte number and phagocytosis activity peak at the pupal stage in lepidopterans, reflecting the role of phagocytosing cells in morphogenesis, which takes place at this time [133]. For instance, macrophage-like hemocytes may function in the development of sexual dimorphism of wing pattern. In response to ecdysteroid, phagocytosing hemocytes are abundant in the female wings [134]. Also, cold stress reduces hemocyte count and impairs hemocyte activities, along with changes in melanin and dopamine biosynthesis. The wing color and pattern can change if the pupa is exposed to cold stress during wing morphogenesis (Fig. 8). The underlying mechanisms involve dopamine release from the local tracheal system, probably from endocrine cells that surround the tracheal channels [135]. In the lepidopterans, hemocytes express dopamine receptors, produce dopamine in response to LPS, and dopamine increases their phagocytosis and increases survival of a fungal infection [136]. In *Drosophila*, inhibition of dopamine biosynthesis or lack of dopamine receptors causes flies to prefer temperatures colder than normal and they lose cold avoidance [137].

Cold stress reduces hemocyte number and causes hemocyte clumping in insects, and reduces macrophage activation in mammals; sharks also have a temperature-sensitive macrophage activation mechanism. It is intriguing to ask whether cold stress affects pigment distribution by interference with immune functions, and, whether dopaminergic stimulation of macrophage activation would be a rescue mechanism to increase immune vigilance in cold.

11 Are There M2 Markers in Invertebrates?

All these examples show that the macrophages of invertebrates have M2-like, that is, tolerogenic and tissue-regenerating, features, but we still lack specific cell-surface markers that would allow discriminating these macrophage populations from pathogen-killing macrophages. Similarly, lineage tracing experiments, flow cytometry analysis, and immunohistochemical detection are still limited in invertebrates. Most studies of invertebrate macrophage development have been focused on *Drosophila melanogaster*.

---

[2] Main pigments of the wing scales are melanin, 3-OH-kynurenine, and ommochrome. Moreover, the scales also can have structural color from light reflectance.

Homologues of mammalian M2 activation genes are present in invertebrates (Fig. 9). Indeed, chitinase- and arginase-like proteins have a high degree of conservation in the course of evolution, and they are present in bacteria, archaea, fungi, and metazoans (Fig. 9). In invertebrates they may have a conserved function as well in host defense and tissue extracellular matrix synthesis [13, 121]. For example, arginase-1 expression has been shown in parasite-infected hemocytes in *Biomphalaria glabrata*, is associated with survival of engulfed parasites [138]. Although it is tempting to assume that arginase-1-expressing invertebrate macrophages represent an M2-equivalent macrophage population, it is more likely that

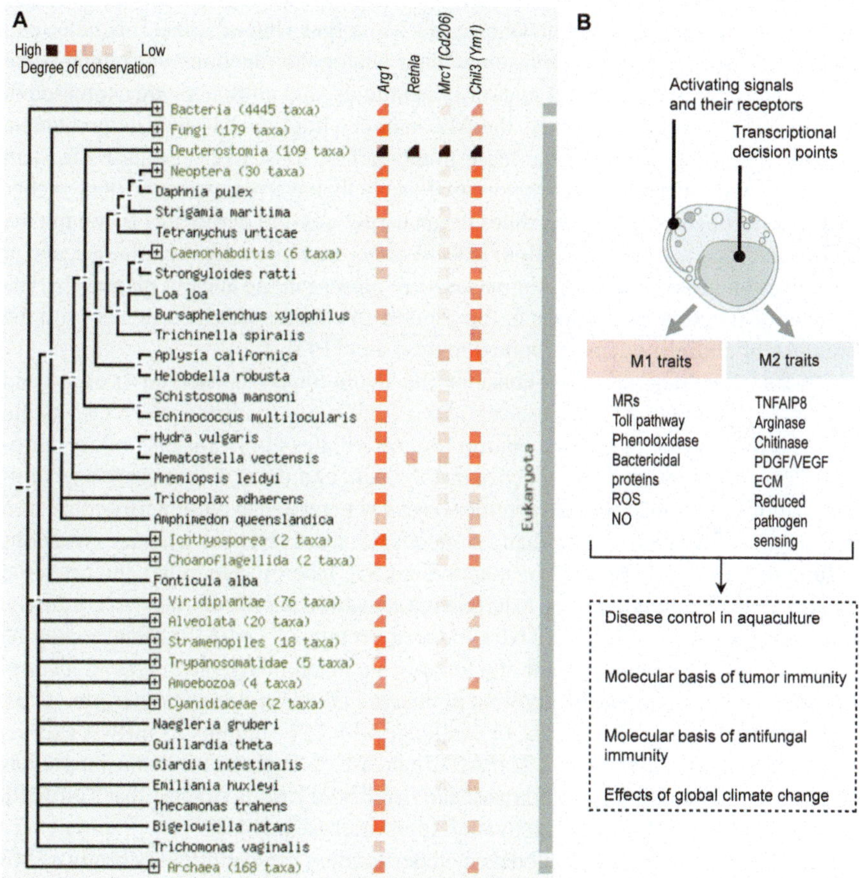

Fig. 9 Phylogenetic homologies of M2 marker genes. Future of M2 macrophage biology in invertebrates. (**a**) STRING analysis of mammalian M2 marker genes showing the degree of their phylogenic conservation. *Arg1* arginase-1, *Retnla* resistin-like molecule alpha, *Mrc1* C-type mannose receptor 1, *Chil3* chitinase. (**b**) Open questions and their impact in the study of invertebrate macrophages. We lack a comprehensive understanding of the signals and receptors that evoke M1 or M2 features; the transcriptional control of macrophage activation; and the markers of M1 or M2 states in invertebrates. *MRs* mannose receptors, *ROS* reactive oxygen species

invertebrate macrophages have their own unique effector mechanisms and hallmarks. Phenoloxidase activity is a hallmark of M1-like activation in invertebrates. The immune evasion mechanism of certain insect parasites curbs phenoloxidase activity and inhibits the melanization process [139]. Study of this immune mechanism can elucidate how pathogens can skew invertebrate macrophages into an M2-like, tolerogenic state. C-type lectins, galectin, mannose receptors, C1q domain-containing proteins, and chitinases are all expressed at various stages of the invertebrate immune response [140–142].

Despite the existence and the macrophage expression of M2 marker homologues in invertebrates, it is challenging to determine if these are specific to the tolerogenic and wound-healing functions of invertebrate macrophages. There are many examples in comparative anatomy in which equivalent organs have distinct functions, and similarities of organs can be a result of their analogous functions. A chitinase-like protein is, for example, an M2 marker in mammals, and chitinases are expressed in invertebrates as well. However, the M2 marker chitinase has lost its enzymatic activity, and mammals have chitinase pseudogenes, most likely because the stem placentals were generally insectivorous. Placentals have radiated in dietary niches that were left behind by extinct dinosaurian carnivores and herbivores following the end-Cretaceous massive extinction. Insectivorous animals express chitinases, and at least five functional copies of chitinases were present in the ancient placentals. The number of functional chitinases in the genome of extant species correlates with the presence of chitin (i.e., invertebrates) in their diet [143].

As an example, one may consider the evolutionary conserved insulin- and insulin-like peptides. Albeit these peptides are present in many invertebrates, insulin does not have a role in glucose control in invertebrates [144]. Instead, other neuropeptides control invertebrate glucose metabolism, and these do not share any similarity with insulin. Insulin-like peptides have a general and evolutionarily conserved role in tissue growth rather than in glucose control. Hence, it is an erroneous assumption that the presence of mammalian M2 macrophage genes indicates the existence of a mammalian-type macrophage activation in invertebrates. Accordingly, a recent study shows that the TNFα-induced protein 8 (TNFAIP8) family may be responsible for the development of a tolerogenic macrophage phenotype in invertebrates: it is a newly identified protein in the sea cucumber *Apostichopus japonicus*. Infection with *Vibrio splendidus*, or challenge with LPS, inhibits its mRNA expression. Importantly, silencing of TNFAIP8 increases NO production and suppresses agmatinase transcription and arginase activity. These results suggest that TNFAIP8 may be decisive in M1 or M2 activation in invertebrates [145].

Invertebrates are valuable models in understanding the evolutionary origin of M2 macrophages. Further, some traits of invertebrate M2 macrophages have timely importance. For instance, macrophage activation has an impact on disease control and prevention of disease outbreaks in the aquaculture of economically important invertebrates. It is not surprising that most of our knowledge about invertebrate immunity and gene expression control is coming from studies using species with economic value: various crabs, oysters, and blue mussels. Moreover, understanding the molecular biology of the switch from symbiont tolerance to elimination of unwanted symbionts can catalyze understanding how macrophages tolerate unwanted cancer cells. Tumor immunity thus can benefit from studies in invertebrate organisms. Last, invertebrate macrophages are involved in the shell development and shell mineralization of mollusks. Global warming leads to ocean acidification, which makes vulnerable those species that build an aragonite-dominated shell. Aragonite is less resistant to an acidic environment; hence, these species become the prey of other animals or become more vulnerable for damage and, eventually, for infections [103]. Climate change affects the annual dynamics of diseases (viral and fungal infections) in the natural habitats of marine invertebrates. Corals, for instance, are responding to increasing sea temperature with increased macrophage numbers [74]. Climate change thus impacts the macrophage-controlled physiological traits of invertebrates, having a significant effect on the future of the marine ecosystem.

References

1. Huxley TH. A manual of the anatomy of invertebrate animals. Churchill, London: J. & A; 1878.
2. Gordon S. Phagocytosis: the legacy of Metchnikoff. Cell. 2016;166:1065–8.
3. Pandey J, Tiwari R. An overview of insect hemocyte science and its future application in applied and biomedical fields. Am J Biochem Mol Biol. 2012;2:82–105.
4. Chang ZL. Recent development of the mononuclear phagocyte system: in memory of Metchnikoff and Ehrlich on the 100th anniversary of the 1908 Nobel prize in physiology or medicine. Biol Cell. 2009;101:709–21.
5. Tauber AI. Metchnikoff and the phagocytosis theory. Nat Rev Mol Cell Biol. 2003;4:897–901.
6. Metchnikoff ÉI. Embryologische studien an Medusen: Ein Beitrag Zur Genealogie Der Primitiv-Organe. Wien: Alfred Hölder; 1886.
7. Tyson J. The cell doctrine. Philadelphia: Lindsay & Blakiston; 1878.
8. Frey H. Handbuch der histologie und histochemie des Menschen. Leipzig: Wilhelm Engelmann; 1876.
9. Anderson KV, Bokla L, Nüsslein-Volhard C. Establishment of dorsal-ventral polarity in the Drosophila embryo: the induction of polarity by the *Toll* gene product. Cell. 1985;42:791–8.
10. Loomis WF. Chapter 2: Amoeboid stage. In: Loomis WF, editor. *Dictyostelium discoideum*. New York: Academic Press; 1975. p. 26–12.
11. Pita L, Hoeppner MP, Ribes M, Hentschel U. Differential expression of immune receptors in two marine sponges upon exposure to microbial-associated molecular patterns. Sci Rep. 2018;8:16081.
12. Kauschke E, Mohrig W. Cytotoxic activity in the coelomic fluid of the annelid *Eisenia foetida* Sav. J Comp Physiol B Biochem Syst Environ Physiol. 1987;157:77–83.
13. Röszer T. The invertebrate midintestinal gland ("hepatopancreas") is an evolutionary forerunner in the integration of immunity and metabolism. Cell Tissue Res. 2014;358:685–95.
14. Lee SY, Wang R, Söderhäll K. A lipopolysaccharide- and β-1,3-glucan-binding protein from hemocytes of the freshwater crayfish *Pacifastacus leniusculus*: purification, characterization, and cDNA cloning. J Biol Chem. 2000;275:1337–43.
15. Kwon H, Bang K, Cho S. Characterization of the hemocytes in larvae of *Protaetia brevitarsis seulensis*: involvement of granulocyte-mediated phagocytosis. PLoS One. 2014;9:e103620.
16. Cho Y, Cho S. Hemocyte-hemocyte adhesion by granulocytes is associated with cellular immunity in the cricket, *Gryllus bimaculatus*. Sci Rep. 2019;9:18066.

17. Donaghy L, Hong H-K, Park K-I, Nobuhisa K, Youn S-H, Kang C-K, Choi K-S. Flow cytometric characterization of hemocytes of the solitary ascidian, *Halocynthia roretzi*. Fish Shellfish Immunol. 2017;66:289–99.
18. Beck G, Ellis T, Zhang H, Lin W, Beauregard K, Habicht GS, Truong N. Nitric oxide production by coelomocytes of *Asterias forbesi*. Dev Comp Immunol. 2001;25:1–10.
19. Beck G, Ellis TW, Habicht GS, Schluter SF, Marchalonis JJ. Evolution of the acute phase response: iron release by echinoderm (*Asterias forbesi*) coelomocytes, and cloning of an echinoderm ferritin molecule. Dev Comp Immunol. 2002;26:11–26.
20. Luchtel D, Martin A, Deyrup-Olsen I, Boer H. Gastropoda: Pulmonata. In: Harrison F, Kohn A, editors. Microscopic anatomy of invertebrates, vol. 6B. New York.: Wiley-Liss; 1997. p. 459–718.
21. Kuchel RP, Raftos DA, Birch D, Vella N. Haemocyte morphology and function in the Akoya pearl oyster, *Pinctada imbricata*. J Invertebr Pathol. 2010;105:36–48.
22. Carr I. The macrophage: a review of ultrastructure and function. New York: Academic Press; 1973.
23. Pila EA, Sullivan JT, Wu XZ, Fang J, Rudko SP, Gordy MA, Hanington PC. Haematopoiesis in molluscs: A review of haemocyte development and function in gastropods, cephalopods and bivalves. Dev Comp Immunol. 2016;58:119–28.
24. Gold KS, Brückner K. Macrophages and cellular immunity in *Drosophila melanogaster*. Semin Immunol. 2015;27:357–68.
25. Smith LC, Hildemann WH. Allograft rejection, autograft fusion and inflammatory responses to injury in *Callyspongia diffusa* (Porifera; Demospongia). Proc R Soc Lond B Biol Sci. 1986;226:445–64.
26. Leys SP, Kahn AS, Fang JKH, Kutti T, Bannister RJ. Phagocytosis of microbial symbionts balances the carbon and nitrogen budget for the deep-water boreal sponge *Geodia barretti*. Limnol Oceanogr. 2018;63:187–202.
27. Sun L, Song Y, Qu Y, Yu X, Zhang W. Purification and in vitro cultivation of archaeocytes (stem cells) of the marine sponge *Hymeniacidon perleve* (Demospongiae). Cell Tissue Res. 2007;328:223–37.
28. Larkman AU. The fine structure of granular amoebocytes from the gonads of the sea anemone *Actinia fragacea* (Cnidaria: Anthozoa). Protoplasma. 1984;122:203–21.
29. Barnes RSK, Calow P, Olive PJW. The invertebrates: a new synthesis. Oxford: Blackwell; 1993.
30. Anderson OR. Ultrastructure of a colonial radiolarian *Collozoum inerme* and a cytochemical determination of the role of its zooxanthellae. Tissue Cell. 1976;8:195–208.
31. Ueda K, Tan Y, Sato F, Yamada Y. Phagocytosis in plant protoplasts. Cell Struct Funct. 1978;3:25–30.
32. Röszer T. Biosynthesis of nitric oxide in plants. In: Khan MN, Mobin F, Corpas F, editors. Nitric oxide in plants: metabolism and role in stress physiology. Heidelberg: Springer; 2014. p. 17–32.
33. Sullivan ML. Beyond brown: polyphenol oxidases as enzymes of plant specialized metabolism. Front Plant Sci. 2015;5:783.
34. Urbański A, Adamski Z, Rosiński G. Developmental changes in haemocyte morphology in response to *Staphylococcus aureus* and latex beads in the beetle *Tenebrio molitor* L. Micron. 2018;104:8–20.
35. Pila EA, Gordy MA, Phillips VK, Kabore AL, Rudko SP, Hanington PC. Endogenous growth factor stimulation of hemocyte proliferation induces resistance to *Schistosoma mansoni* challenge in the snail host. Proc Natl Acad Sci USA. 2016;113:201521239.
36. Niu S, Yang L, Zuo H, Zheng J, Weng S, He J, Xu X. A chitinase from Pacific white shrimp *Litopenaeus vannamei* involved in immune regulation. Dev Comp Immunol. 2018;85:161–9.
37. Nie L, Cai S-Y, Shao J-Z, Chen J. Toll-like receptors, associated biological roles, and signaling networks in non-mammals. Front Immunol. 2018;9:1523.

38. Zárate-Potes A, Ocampo ID, Cadavid LF. The putative immune recognition repertoire of the model cnidarian *Hydractinia symbiolongicarpus* is large and diverse. Gene. 2019;(Amst) 684:104–17.
39. M'Beri M, Debray H, Dhainaut A, Porchet-Hennere E. Distribution and nature of membrane receptors for different plant lectins in the coelomocyte subpopulations of the Annelida *Nereis diversicolor*. Dev Comp Immunol. 1988;12:1–15.
40. Man X, Pan X-T, Zhang H-W, Wang Y, Li X-C, Zhang X-W. A mannose receptor is involved in the anti-*Vibrio* defense of red swamp crayfish. Fish Shellfish Immunol. 2018;82:258–66.
41. Liao W-Y, Fugmann SD. Lectins identify distinct populations of coelomocytes in *Strongylocentrotus purpuratus*. PLoS One. 2017;12:e0187987.
42. Gordy MA, Pila EA, Hanington PC. The role of fibrinogen-related proteins in the gastropod immune response. Fish Shellfish Immunol. 2015;46:39–49.
43. Lorenzon S, de Guarrini S, Smith VJ, Ferrero EA. Effects of LPS injection on circulating haemocytes in crustaceans in vivo. Fish Shellfish Immunol. 1999;9:31–50.
44. Jia Z, Wang M, Wang X, Xu J, Wang L, Zhang H, Song L. A prokineticin (PK)-like cytokine from Chinese mitten crab *Eriocheir sinensis* promotes the production of hemocytes via reactive oxygen species. Fish Shellfish Immunol. 2018;77:419–28.
45. Lin X, Söderhäll I. Crustacean hematopoiesis and the astakine cytokines. Blood. 2011;117:6417–24.
46. Zhou Q-Y. The prokineticins: a novel pair of regulatory peptides. Mol Interv. 2006;6:330–8.
47. Li F, Xu L, Gai X, Zhou Z, Wang L, Zhang H, Gai Y, Song L, Yu J, Liang C. The involvement of PDGF/VEGF related factor in regulation of immune and neuroendocrine in Chinese mitten crab *Eriocheir sinensis*. Fish Shellfish Immunol. 2013;35:1240–8.
48. Hermann PM, Park D, Beaulieu E, Wildering WC. Evidence for inflammation-mediated memory dysfunction in gastropods: putative PLA2 and COX inhibitors abolish long-term memory failure induced by systemic immune challenges. BMC Neurosci. 2013;14:83.
49. Brock DA, Douglas TE, Queller DC, Strassmann JE. Primitive agriculture in a social amoeba. Nature. 2011;469:393.
50. Haselkorn TS, DiSalvo S, Miller JW, Bashir U, Brock DA, Queller DC, Strassmann JE. The specificity of *Burkholderia* symbionts in the social amoeba farming symbiosis: prevalence, species, genetic and phenotypic diversity. Mol Ecol. 2019;28:847–62.
51. Adler B. Strategies for intracellular survival of *Burkholderia pseudomallei*. Front Microbiol. 2011;2:170.
52. Brock DA, Haselkorn TS, Garcia JR, Bashir U, Douglas TE, Galloway J, Brodie F, Queller DC, Strassmann JE. Diversity of free-living environmental bacteria and their interactions with a bactivorous amoeba. Front Cell Infect Microbiol. 2018;8:411.
53. Zhang X, Zhuchenko O, Kuspa A, Soldati T. Social amoebae trap and kill bacteria by casting DNA nets. Nat Commun. 2016;7:10938.
54. Walk A, Callahan J, Srisawangvong P, Leuschner J, Samaroo D, Cassilly D, Snyder MLD. Lipopolysaccharide enhances bactericidal activity in *Dictyostelium discoideum* cells. Dev Comp Immunol. 2011;35:850–6.
55. Otto GP, Wu MY, Kazgan N, Anderson OR, Kessin RH. Macroautophagy is required for multicellular development of the social amoeba *Dictyostelium discoideum*. J Biol Chem. 2003;278:17636–45.
56. Nakahira K, Haspel JA, Rathinam VAK, Lee S-J, Dolinay T, Lam HC, Englert JA, Rabinovitch M, Cernadas M, Kim HP, Fitzgerald KA, Ryter SW, Choi AMK. Autophagy proteins regulate innate immune responses by inhibiting the release of mitochondrial DNA mediated by the NALP3 inflammasome. Nat Immunol. 2011;12:222–30.
57. Ogawa S, Naito K, Angata K, Morio T, Urushihara H, Tanaka Y. A site-specific DNA endonuclease specified by one of two ORFs encoded by a group I intron in *Dictyostelium discoideum* mitochondrial DNA. Gene (Amst). 1997;191:115–21.
58. Schaap P. Cyclic di-nucleotide signaling enters the eukaryote domain. IUBMB Life. 2013;65:897–903.

59. Wehrl M, Steinert M, Hentschel U. Bacterial uptake by the marine sponge *Aplysina aerophoba*. Microb Ecol. 2007;53:355–65.
60. Simpson TL, Vaccaro CA, Sha'afi RI. The role of intragemmular osmotic pressure in cell division and hatching of gemmules of the fresh-water sponge *Spongilla lacustris* (Porifera). Zeitschrift für Morphologie der Tiere. 1973;76:339–57.
61. Müller WEG, Wiens M, Müller IM, Schröder HC. The chemokine networks in sponges: potential roles in morphogenesis, immunity and stem cell formation. In: Beschin A, Müller WEG, editors. Invertebrate cytokines and the phylogeny of immunity: facts and paradoxes. Berlin: Springer; 2004. p. 103–43.
62. Maldonado M, Zhang X, Cao X, Xue L, Cao H, Zhang W. Selective feeding by sponges on pathogenic microbes: a reassessment of potential for abatement of microbial pollution. Mar Ecol Prog Ser. 2010;403:75–89.
63. Mukherjee S, Ray M, Ray S. Phagocytic efficiency and cytotoxic responses of Indian freshwater sponge (*Eunapius carteri*) cells isolated by density gradient centrifugation and flow cytometry: a morphofunctional analysis. Zoology. 2015;118:8–18.
64. Vacelet J, Duport E. Prey capture and digestion in the carnivorous sponge *Asbestopluma hypogea* (Porifera: Demospongiae). Zoomorphology. 2004;123:179–90.
65. Schmitt S, Weisz JB, Lindquist N, Hentschel U. Vertical transmission of a phylogenetically complex microbial consortium in the viviparous sponge *Ircinia felix*. Appl Environ Microbiol. 2007;73:2067–78.
66. Fieseler L, Horn M, Wagner M, Hentschel U. Discovery of the novel candidate phylum "Poribacteria" in marine sponges. Appl Environ Microbiol. 2004;70:3724–32.
67. Wagner-Hülsmann C, Bachinski N, Diehl-Seifert B, Blumbach B, Steffen R, Pancer Z, Müller W. A galectin links the aggregation factor to cells in the sponge (*Geodia cydonium*) system. Glycobiology. 1997;6:785–93.
68. Kuhns WJ, Popescu O, Burger MM, Misevic G. Sulfate restriction induces hyposecretion of the adhesion proteoglycan and cell hypomotility associated with increased $^{35}SO_4^{2}$: uptake and expression of a band 3 like protein in the marine sponge, *Microciona prolifera*. J Cell Biochem. 1995;57:71–89.
69. Holstein TW, Hobmayer E, Technau U. Cnidarians: an evolutionarily conserved model system for regeneration? Dev Dyn. 2003;226:257–67.
70. Rinkevich B. Neglected biological features in cnidarians self–nonself recognition. In: López-Larrea C, editor. Self and nonself. New York: Springer US; 2012. p. 46–59.
71. Patterson MJ, Landolt ML. Cellular reaction to injury in the anthozoan *Anthopleura elegantissima*. J Invertebr Pathol. 1979;33:189–96.
72. Meszaros A, Bigger C. Qualitative and quantitative study of wound healing processes in the coelenterate, *Plexaurella fusifera*: spatial, temporal, and environmental (light attenuation) influences. J Invertebr Pathol. 1999;73:321–31.
73. Hutton D, Smith VJ. Antibacterial properties of isolated amoebocytes from the sea anemone *Actinia equina*. Biol Bull. 1996;191:441–51.
74. Mydlarz LD, Holthouse SF, Peters EC, Harvell CD. Cellular responses in sea fan corals: granular amoebocytes react to pathogen and climate stressors. PLoS One. 2008;3:e1811.
75. Nicotra ML, Powell AE, Rosengarten RD, Moreno M, Grimwood J, Lakkis FG, Dellaporta SL, Buss LW. A hypervariable invertebrate allodeterminant. Curr Biol. 2009;19:583–9.
76. Török A, Schiffer PH, Schnitzler CE, Ford K, Mullikin JC, Baxevanis AD, Bacic A, Frank U, Gornik SG. The cnidarian *Hydractinia echinata* employs canonical and highly adapted histones to pack its DNA. Epigenetics Chromatin. 2016;9:36.
77. Poland D, Coffroth MA. Trans-generational specificity within a cnidarian–algal symbiosis. Coral Reefs. 2016;36:405–15.
78. Hamada M, Schröder K, Bathia J, Kürn U, Fraune S, Khalturina M, Khalturin K, Shinzato C, Satoh N, Bosch TC. Metabolic co-dependence drives the evolutionarily ancient Hydra–Chlorella symbiosis. elife. 2018;7:e35122.

79. Weis VM. Cellular mechanisms of cnidarian bleaching: stress causes the collapse of symbiosis. J Exp Biol. 2008;211:3059–66.
80. Franchini A, Ottaviani E. Repair of molluscan tissue injury: role of PDGF and TGF-beta1. Tissue Cell. 2000;32:312–21.
81. de Eguileor M, Tettamanti G, Grimaldi A, Boselli A, Scarì G, Valvassori R, Cooper EL, Lanzavecchia G. Histopathological changes after induced injury in leeches. J Invertebr Pathol. 1999;74:14–28.
82. Le Marrec-Croq F, Drago F, Vizioli J, Sautière P-E, Lefebvre C. The leech nervous system: a valuable model to study the microglia involvement in regenerative processes. Clin Dev Immunol. 2013;2013:274019.
83. Boidin-Wichlacz C, Vergote D, Slomianny C, Jouy N, Salzet M, Tasiemski A. Morphological and functional characterization of leech circulating blood cells: role in immunity and neural repair. Cell Mol Life Sci. 2012;69:1717–31.
84. Tahtouh M, Garçon-Bocquet A, Croq F, Vizioli J, Sautière P-E, Van Camp C, Salzet M, Nagnan-le Meillour P, Pestel J, Lefebvre C. Interaction of HmC1q with leech microglial cells: involvement of C1qBP-related molecule in the induction of cell chemotaxis. J Neuroinflammation. 2012;9:37.
85. Le Marrec-Croq F, Bocquet-Garcon A, Vizioli J, Vancamp C, Drago F, Franck J, Wisztorski M, Salzet M, Sautiere PE, Lefebvre C. Calreticulin contributes to C1q-dependent recruitment of microglia in the leech *Hirudo medicinalis* following a CNS injury. Med Sci Monit Int Med J Exp Clin Res. 2014;20:644–53.
86. Samuels SE, Lipitz JB, Wang J, Dahl G, Muller KJ. Arachidonic acid closes innexin/pannexin channels and thereby inhibits microglia cell movement to a nerve injury. Dev Neurobiol. 2013;73:621–31.
87. Arafah K, Croix D, Vizioli J, Desmons A, Fournier I, Salzet M. Involvement of nitric oxide through endocannabinoids release in microglia activation during the course of CNS regeneration in the medicinal leech. Glia. 2013;61:636–49.
88. Masuda-Nakagawa LM, Muller KJ, Nicholls JG. Axonal sprouting and laminin appearance after destruction of glial sheaths. Proc Natl Acad Sci USA. 1993;90:4966–70.
89. Drago F, Sautière P-E, Le Marrec-Croq F, Accorsi A, Van Camp C, Salzet M, Lefebvre C, Vizioli J. Microglia of medicinal leech (*Hirudo medicinalis*) express a specific activation marker homologous to vertebrate ionized calcium-binding adapter molecule 1 (Iba1/alias aif-1). Dev Neurobiol. 2014;74:987–1001.
90. Shrivastava K, Gonzalez P, Acarin L. The immune inhibitory complex CD200/CD200R is developmentally regulated in the mouse brain. J Comp Neurol. 2012;520:2657–75.
91. Yu H, Dilbaz S, Coßmann J, Hoang AC, Diedrich V, Herwig A, Harauma A, Hoshi Y, Moriguchi T, Landgraf K, Körner A, Lucas C, Brodesser S, Balogh L, Thuróczy J, Karemore G, Kuefner MS, Park EA, Rapp C, Travers JB, Röszer T. Breast milk alkylglycerols sustain beige adipocytes through adipose tissue macrophages. J Clin Invest. 2019;129:2485–99.
92. Suleiman S, Smith VJ, Dyrynda EA. Unusual tissue distribution of carcinin, an antibacterial crustin, in the crab, *Carcinus maenas*, reveals its multi-functionality. Dev Comp Immunol. 2017;76:274–84.
93. Jang I-K, Pang Z, Yu J, Kim S-K, Seo H-C, Cho Y-R. Selectively enhanced expression of prophenoloxidase activating enzyme 1 (PPAE1) at a bacteria clearance site in the white shrimp, *Litopenaeus vannamei*. BMC Immunol. 2011;12:70.
94. Márkus R, Laurinyecz B, Kurucz É, Honti V, Bajusz I, Sipos B, Somogyi K, Kronhamn J, Hultmark D, Andó I. Sessile hemocytes as a hematopoietic compartment in *Drosophila melanogaster*. Proc Natl Acad Sci USA. 2009;106:4805–9.
95. Kan H, Kim C-H, Kwon H-M, Park J-W, Roh K-B, Lee H, Park B-J, Zhang R, Zhang J, Söderhäll K, Ha N-C, Lee BL. Molecular control of phenoloxidase-induced melanin synthesis in an insect. J Biol Chem. 2008;283:25316–23.
96. Cerenius L, Soderhall K. The prophenoloxidase-activating system in invertebrates. Immunol Rev. 2004;198:116–26.

97. Robinson N, Ganesan R, Hegedűs C, Kovács K, Kufer TA, Virág L. Programmed necrotic cell death of macrophages: focus on pyroptosis, necroptosis, and parthanatos. Redox Biol. 2019;26:101239.
98. Coates CJ, Kelly SM, Nairn J. Possible role of phosphatidylserine-hemocyanin interaction in the innate immune response of *Limulus polyphemus*. Dev Comp Immunol. 2011;35:155–63.
99. Coates CJ, Whalley T, Wyman M, Nairn J. A putative link between phagocytosis-induced apoptosis and hemocyanin-derived phenoloxidase activation. Apoptosis. 2013;18:1319–31.
100. Grégoire C. Structure of the conchiolin cases of the prisms in *Mytilus edulis* Linne. J Biophys Biochem Cytol. 1961;9:395–400.
101. Gregoire CH. Conchiolin remnants in mother-of-pearl from fossil Cephalopoda. Nature. 1959;184:1157–8.
102. Bowen CE, Tang H. Conchiolin-protein in aragonite shells of mollusks. Comp Biochem Physiol A Physiol. 1996;115:269–75.
103. Sun W, Jayaraman S, Chen W, Persson KA, Ceder G. Nucleation of metastable aragonite $CaCO_3$ in seawater. Proc Natl Acad Sci USA. 2015;112:3199–204.
104. Fisher W. Relationship of amebocytes and terrestrial elements to adult shell deposition in eastern oysters. J Shellfish Res. 2004;23:353–67.
105. Liu Q, Zhao LL, Yang S, Zhang JE, Zhao NQ, Wu H, He Z, Yan TM, Guo J. Regeneration of excised shell by the invasive apple snail *Pomacea canaliculata*. Mar Freshw Behav Physiol. 2017;50:17–29.
106. Li S, Liu Y, Liu C, Huang J, Zheng G, Xie L, Zhang R. Hemocytes participate in calcium carbonate crystal formation, transportation and shell regeneration in the pearl oyster *Pinctada fucata*. Fish Shellfish Immunol. 2016;51:263–70.
107. Huang J, Li S, Liu Y, Liu C, Xie L, Zhang R. Hemocytes in the extrapallial space of *Pinctada fucata* are involved in immunity and biomineralization. Sci Rep. 2018;8:4657.
108. Mount AS, Wheeler AP, Paradkar RP, Snider D. Hemocyte-mediated shell mineralization in the eastern oyster. Science. 2004;304:297–300.
109. Beedham GE. Repair of the shell in species of Anodonta. Proc Zool Soc London. 1965;145:107–23.
110. Suja CP, Dharmaraj S. In vitro culture of mantle tissue of the abalone *Haliotis varia* Linnaeus. Tissue Cell. 2005;37:1–10.
111. Rőszer T, Kappelmayer J, Nagy GG, Szentmiklosi AJ, Basnakian AG, Banfalvi G. The neuropeptide FMRF amide can protect cells against apoptosis in the snail digestive gland. Apoptosis. 2006;11:173–82.
112. Ivanina AV, Falfushynska HI, Beniash E, Piontkivska H, Sokolova IM. Biomineralization-related specialization of hemocytes and mantle tissues of the Pacific oyster *Crassostrea gigas*. J Exp Biol. 2017;220:3209–21.
113. Lee SW, Park SB, Choi CS. On self-organized shell formation by bovine carbonic anhydrase II, and soluble protein extracted from regenerated shell. Micron. 2008;39:1228–34.
114. Lee SW, Choi CS. High-rate growth of calcium carbonate crystal using soluble protein from diseased oyster shell. Cryst Growth Des. 2007;7:1463–8.
115. Fleury C, Marin F, Marie B, Luquet G, Thomas J, Josse C, Serpentini A, Lebel JM. Shell repair process in the green ormer *Haliotis tuberculata*: a histological and microstructural study. Tissue Cell. 2008;40:207–18.
116. O'Neill M, Gaume B, Denis F, Auzoux-Bordenave S. Expression of biomineralisation genes in tissues and cultured cells of the abalone *Haliotis tuberculata*. Cytotechnology. 2013;65:737–47.
117. Montagnani C, Le Roux F, Berthe F, Escoubas J-M. Cg-TIMP, an inducible tissue inhibitor of metalloproteinase from the Pacific oyster *Crassostrea gigas* with a potential role in wound healing and defense mechanisms1. FEBS Lett. 2001;500:64–70.
118. Huang W-C, Sala-Newby GB, Susana A, Johnson JL, Newby AC. Classical macrophage activation up-regulates several matrix metalloproteinases through mitogen activated protein kinases and nuclear factor-κB. PLoS One. 2012;7:e42507.

119. Gill SE, Gharib SA, Bench EM, Sussman SW, Wang RT, Rims C, Birkland TP, Wang Y, Manicone AM, McGuire JK, Parks WC. Tissue inhibitor of metalloproteinases-3 moderates the proinflammatory status of macrophages. Am J Respir Cell Mol Biol. 2013;49:768–77.
120. Li H, Wang D, Deng Z, Huang G, Fan S, Zhou D, Liu B, Zhang B, Yu D. Molecular characterization and expression analysis of chitinase from the pearl oyster *Pinctada fucata*. Comp Biochem Physiol B: Biochem Mol Biol. 2017;203:141–8.
121. Röszer T. Understanding the mysterious M2 macrophage through activation markers and effector mechanisms. Mediat Inflamm. 2015;2015:16.
122. Chintapalli RTV, Hillyer JF. Hemolymph circulation in insect flight appendages: physiology of the wing heart and circulatory flow in the wings of the mosquito *Anopheles gambiae*. J Exp Biol. 2016;219:3945–51.
123. Iwata M, Ohno Y, Otaki JM. Real-time in vivo imaging of butterfly wing development: revealing the cellular dynamics of the pupal wing tissue. PLoS One. 2014;9:e89500.
124. Nardi JB, Miklasz SD. Hemocytes contribute to both the formation and breakdown of the basal lamina in developing wings of *Manduca sexta*. Tissue Cell. 1989;21:559–67.
125. Harvey SJ, Thorner PS. Type IV collagen: a network for development, differentiation, and disease. In: Advances in developmental biology, vol. 15. Amsterdam: Elsevier; 2005. p. 1–64.
126. Kiger JA, Natzle JE, Kimbrell DA, Paddy MR, Kleinhesselink K, Green MM. Tissue remodeling during maturation of the *Drosophila* wing. Dev Biol. 2007;301:178–91.
127. Miller JY, Brown FM, Allyn Museum of Entomology. A new Oligocene fossil butterfly, *Vanessa amerindica* (Lepidoptera:Nymphalidae), from the Florissant formation. Sarasota: Allyn Museum of Entomology; 1989.
128. Stavenga DG, Leertouwer HL, Wilts BD. Coloration principles of nymphaline butterflies – thin films, melanin, ommochromes and wing scale stacking. J Exp Biol. 2014;217:2171–80.
129. Mulinari S, Häcker U, Castillejo-López C. Expression and regulation of Spätzle-processing enzyme in *Drosophila*. FEBS Lett. 2006;580:5406–10.
130. Taleh M, Saadati M, Farshbaf R, Khakvar R. Partial characterization of phenoloxidase enzyme in the hemocytes of *Helicoverpa armigera* Hübner (Lepidoptera: Noctuidae). J King Saud Univ Sci. 2014;26:285–9.
131. Delkash-Roudsari S, Zibaee A, Bigham Z. Purification and characterization of a phenoloxidase in the hemocytes of *Ephestia kuehniella* Zeller (Lepidoptera: Pyralidae): effects of insect growth regulators and endogenous inhibitors. J Enzyme Inhib Med Chem. 2015;30:569–74.
132. Nishikawa H, Iga M, Yamaguchi J, Saito K, Kataoka H, Suzuki Y, Sugano S, Fujiwara H. Molecular basis of wing coloration in a Batesian mimic butterfly, *Papilio polytes*. Sci Rep. 2013;3:3184.
133. Kurtz J. Phagocytosis by invertebrate hemocytes: causes of individual variation in *Panorpa vulgaris* scorpionflies. Microsc Res Tech. 2002;57:456–68.
134. Lobbia S, Niitsu S, Fujiwara H. Female-specific wing degeneration caused by ecdysteroid in the tussock moth, *Orgyia recens*: hormonal and developmental regulation of sexual dimorphism. J Insect Sci. 2003;3:11.
135. Mahdi SHA, Gima S, Tomita Y, Yamasaki H, Otaki JM. Physiological characterization of the cold-shock-induced humoral factor for wing color-pattern changes in butterflies. J Insect Physiol. 2010;56:1022–31.
136. Wu S-F, Xu G, Stanley D, Huang J, Ye G-Y. Dopamine modulates hemocyte phagocytosis via a D1-like receptor in the rice stem borer, *Chilo suppressalis*. Sci Rep. 2015;5:12247.
137. Bang S, Hyun S, Sung-Tae H, Kang J, Jeong K, Park J-J, Choe J, Chung J. Dopamine signalling in mushroom bodies regulates temperature-preference behaviour in *Drosophila*. PLoS Genet. 2011;7:e1001346.
138. Dinguirard N, Cavalcanti MGS, Wu X-J, Bickham-Wright U, Sabat G, Yoshino TP. Proteomic analysis of *Biomphalaria glabrata* hemocytes during in vitro encapsulation of *Schistosoma mansoni* sporocysts. Front Immunol. 2018;9:2773.

139. Meng E, Tang B, Hou Y, Chen X, Chen J, Yu X-Q. Altered immune function of *Octodonta nipae* (Maulik) to its pupal endoparasitoid, *Tetrastichus brontispae* Ferrière. Comp Biochem Physiol B Biochem Mol Biol. 2016;198:100–9.
140. Wang L, Zhang H, Wang M, Zhou Z, Wang W, Liu R, Huang M, Yang C, Qiu L, Song L. The transcriptomic expression of pattern recognition receptors: insight into molecular recognition of various invading pathogens in oyster *Crassostrea gigas*. Dev Comp Immunol. 2019;91:1–7.
141. Bai Y, Niu D, Li Y, Bai Y, Lan T, Peng M, Dong Z, Sun F, Li J. Identification and characterisation of a novel small galectin in razor clam (*Sinonovacula constricta*) with multiple innate immune functions. Dev Comp Immunol. 2019;93:11–7.
142. Sricharoen S, Kim JJ, Tunkijjanukij S, Soderhall I. Exocytosis and proteomic analysis of the vesicle content of granular hemocytes from a crayfish. Dev Comp Immunol. 2005;29:1017–31.
143. Emerling CA, Delsuc F, Nachman MW. Chitinase genes (CHIAs) provide genomic footprints of a post-Cretaceous dietary radiation in placental mammals. Sci Adv. 2018;4:eaar6478.
144. Röszer T, Kiss-Toth ED. FMRF-amide is a glucose-lowering hormone in the snail *Helix aspersa*. Cell Tissue Res. 2014;358:371–83.
145. Shao Y, Chen H, Lv M, Li C, Zhang W, Li Y, Zhao X, Bao Y. A novel TNFAIP8 gene mediates L-arginine metabolism in *Apostichopus japonicus*. Fish Shellfish Immunol. 2017;69:26–34.

Evolution of M2 Macrophage Functions in Chordates

Keywords Comparative immunology · Myeloid cells · Evolution · M2 macrophage · Chordates · Cephalochordates · Fishes · Amphibia · Reptiles · Phagocyte · Macrophage

1 Introduction

Extant chordates are divided into three subphyla: tunicates (urochordates), cephalochordates, and vertebrates. Urochordates and vertebrates have macrophages, but cephalochordates are unique in their apparent lack of this cell type. Urochordate and nonmammalian vertebrate macrophages have M2-like functions, such as inhibition of macrophage-dependent cytotoxicity, wound healing, endocrine signaling, and clearance of apoptotic cell debris. However, signals of an M2-like activation in nonmammalian chordates are poorly understood, and we have major gaps in the principles of macrophage biology of some chordate phyla. For instance, studies are scarce in elasmobranch fishes, or in certain "living fossil" species, such as the axolotl, lungfish, or platypus. Moreover, the available data show that macrophage activation has unique features in some chordates that are distinct from the mammalian macrophage activation. For instance, macrophage activation is temperature dependent in sharks. In bony fishes, cAMP activates M2 macrophages, instead of an IL-4-equivalent protein; and there is a sexual dimorphism of macrophage functions in reptiles. This chapter provides a summary of these and other traits relevant for the understanding of M2 macrophage biology in chordates. Future studies of nonmammalian vertebrate macrophages can help us understand antitumor immunity, temperature- and light-sensitive changes in innate immune response, endocrine control of macrophage functions, disease control in aquaculture, and the role of macrophage activation in scar formation or regeneration.

T. Röszer, *The M2 Macrophage*, Progress in Inflammation Research 86,
https://doi.org/10.1007/978-3-030-50480-9_3

2 Macrophages with M2 Functions in Tunicates

Adult tunicates (also called urochordates) are typically sessile filter-feeding animals. Traditionally, they are considered as representing the earliest chordate lineage whereas cephalochordates are the closest living relatives of vertebrates. Although cephalochordates and vertebrates share many morphological similarities, tunicates may be the closest extant relatives of vertebrates [1].

Tunicates share a common ancestry with vertebrates, and it is logical to assume that the evolutionary roots of M2 macrophage activation can also be found in tunicates. Indeed, the genome of the common vase tunicate *Ciona intestinalis* contains orthologues of various mammalian genes of M2 macrophage activation, such as chitinase (*cht*: *Ci-GH18*; gene ID: 100135776), C-type mannose receptor 2 (gene ID: 100176419), protein inhibitor of activated STAT (*PIAS*; gene ID: 778731), and peroxisome proliferator-activated receptor (*Ci-PPAR*; gene ID: 778733). Surprisingly, however, most studies on tunicate macrophage functions are focused on the pathogen-killing (M1-like) macrophage functions, without defining M2-like features of macrophages.

Macrophage-like cells are enriched in specific regions of the body in tunicates, such as the oral siphon and pharynx, and in the so-called tunic (Fig. 1a). The oral siphon and the pharynx are proximal parts of the alimentary canal and function as immune organs. Macrophages associated with these regions are phagocytic sentinel

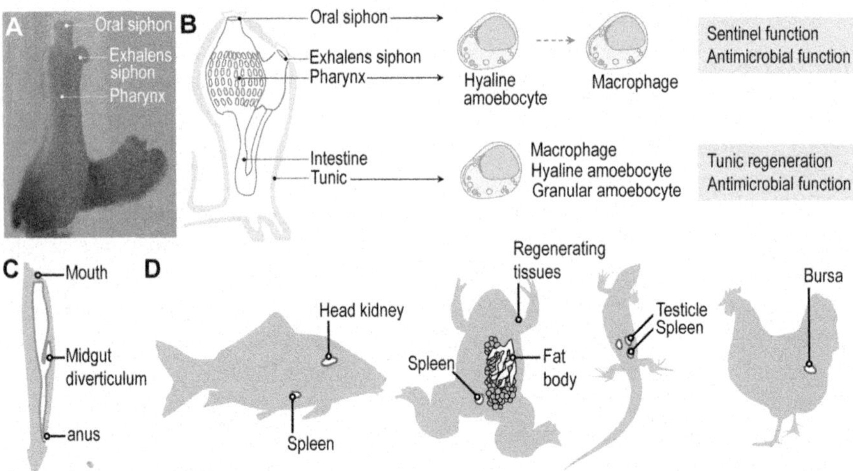

Fig. 1 Macrophage distribution in tissues and organs of chordates. (**a**) Main body parts of the vase tunicate *Ciona intestinalis*. (**b**) Diagram of the major organs of tunicates, indicating regions rich in macrophages (fixed phagocytes). Main functions of the macrophages are summarized in the scheme; tunic regeneration may be an evolutionary forerunner of a tissue-regenerating M2 trait of macrophages. (**c**) The midgut diverticulum in cephalochordates harbors phagocytosing epithelial cells and putative macrophages. (**d**) Major organs and tissues that are rich in macrophages in non-mammalian vertebrates. (Images by author)

cells (Fig. 1b). The tunic is a first line of defense around the body and is built up from a cellulose-like polymer. The main function of the tunic macrophages is to patrol in the tunic and ingest foreign material (Fig. 1b). These macrophage-like cells resemble the fixed, tissue-resident macrophages of mammals. Macrophages in tunics are round large cells 10–15 μm in diameter; they are re rich in lysosomes and phagosomes, and show acid phosphatase and nonspecific esterase activity. The macrophage cell surface has affinity for the alpha-mannose specific agglutinin of *Narcissus pseudonarcissus*, and is recognized by an antibody raised against mammalian CD39 (ectonucleoside triphosphate diphosphohydrolase-1) [2, 3].

The macrophage-like cells of tunicates are also termed phagocytes, hyaline amebocytes, spreading phagocytes, or round phagocytes. However, immune cell classification is mainly based on morphological features, and macrophage traits are attributed to more than only one cell type in tunicates: hyaline amoebocytes and macrophage-like cells are major phagocytic cells; granular amoebocytes and morula cells are cytotoxic cells with phenoloxidase activity; and spherule cells that may be needed for fibrogenesis during tunic repair [4]. It is possible that these cells share common progenitors. For instance, macrophages of the oral siphon develop from hyaline amebocytes. However, further studies are needed to define macrophage ontogeny and functional heterogeneity in tunicates.

Pathogen elimination mechanisms of tunicates involves both invertebrate-type and vertebrate-type features. Phenoloxidase activity is an invertebrate heritage of the tunicate immune system: immune cells express phenoloxidase, which generates cytotoxic quinones that eventually polymerize to melanin and aid pathogen elimination [5]. Tunicate immune response also involves nitric oxide (NO) and reactive oxygen species [5, 6]. As vertebrate traits, tunicate macrophages secrete cytokine-like proteins, which show some degree of similarity to the mammalian M1 cytokines, such as TNFα and IL-1β [7]. Moreover, complement activation and opsonization of pathogens are also apparent in infected tunicates [5].

Apoptotic cell clearance is an M2-like feature of tunicate macrophages that is readily observable during the morphogenesis of colonial ascidians, such as in *Botryllus schlosseri*. This species has a unique body plan and life cycle. The *Botryllus* colonies are formed by large numbers of genetically identical individuals, called zooids, which have a cyclical developmental program. The adult zooids form young zooids by budding. After the buds develop into mature zooids, the parental zooids undergo apoptosis, and their place is taken by the budding offspring. The dying zooids are cleared by macrophage-like cells [8]. In mammals, uptake of apoptotic cells leads to an M2 macrophage phenotype. Clearance of the apoptotic zooids, however, increases oxygen consumption and the production of reactive oxygen species (ROS) by the engulfing macrophages, which eventually may cause apoptosis of the macrophages as well. However, the ascidian macrophages express certain Cu/Zn superoxide dismutase (SOD), γ-glutamyl-cysteine ligase modulatory subunit (GCLM), glutathione synthase (GS), and glutathione peroxidase enzymes, which protect the cells from ROS [9]. If clearance of the apoptotic zooids is inhibited chemically, the entire colony dies. Hence, macrophages are vital for the life cycle of the colony. It is still unknown, however, how the apoptotic zooids are recognized by the macrophages. It is possible that macrophage-like cells recognize modified lipids in the cell membrane of the dying zooids, and this triggers the engulfing process [8]. Another M2-like feature is the aid of tissue regeneration. Regeneration starts with a rapid healing response characterized by hemocyte aggregation and infiltration of immunocytes, followed by recruitment of macrophage-like cells for clearing debris [10].

Interestingly, tunicates synthesize molecules that have antioxidant and antiinflammatory effects in mammalian macrophages. For instance, lipid metabolites such as prenylated hydroquinones [11], chromene 2, and hydroquinones 4 and 5 inhibit lipid peroxidation in mammalian cells, probably by inhibiting lipoxygenase activity [12]. Furthermore, unique antiinflammatory amino acids, such as herdmanines (isolated from *Herdmania momus*), and lipid mediators such as eicosanoids are known in tunicates [13].

3 Do Cephalochordates Have Macrophages?

The European lancelet *Branchiostoma lanceolatum* (previously named *Amphioxus lanceolatus*) is a representative species of the extant cephalochordates. It is an evolutionarily important species because it is a link in the phylogenetic route toward vertebrates. Predicted orthologues of mammalian M2 genes are indeed present in the *Branchiostoma belcheri* genome, such as mannose receptor (NCBI reference sequence: XM_019790581.1) and arginase-1 (NCBI reference sequence: XM_019786923.1). Expression of macrophage inhibitory factor (MIF) has also been shown in *Branchiostoma* [14]. In the mouse, macrophage-derived MIF participates in M2 activation and sustains immune tolerance toward tumor cells [15].

However, we still lack definite evidence whether cephalochordates have macrophages. It was shown in 1982 that fixed and mobile macrophage-like cells were present in *Branchiostoma* [16]. However, the existence of these cells has not yet been confirmed [17]; instead, some fixed phagocytosing cells have been found in the gut epithelium [18]. Phagocytosis is an important trait of the intestinal epithelia in many invertebrates, especially in those species that have a midintestinal gland (often called the hepatopancreas or "liver"). This organ engulfs solid food particles for further intracellular digestion, and also ingests cellular pathogens [19]. This dual function is especially important in filtrating animals, such as bivalves. *Branchiostoma* is similarly a filter-feeding animal and has a hepatopancreas-equivalent midgut diverticulum that is often termed as a liver [20] (Fig. 1c). This organ is rich in phagocytosing epithelial cells and probably also contain fixed macrophage-like cells [18].

Overall, the existence of macrophages in these species is still to be confirmed. A dual immune-metabolic function of the gut epithelia is, however, not without precedent: it is a known trait of the midintestinal gland in mollusks and crustaceans, and also in the so-called fat body of insects [19, 21]. In these instances, phagocytic cells have their primary function in metabolism, such as nutrient uptake or nutrient storage, and are involved in host defense secondarily. In vertebrates, similarly to tunicates, macrophages are key cells of the innate immunity and tissue homeostasis (Fig. 1d). However, vertebrates also retain the ability of phagocytosis by somatic cells, so-called nonprofessional phagocytes. These cells take up particles by mean of phagocytosis but are not related to macrophages [22].

4 Macrophages with M2 Traits in Fishes

In sharks, cytotoxic macrophages have been identified that resemble the immune functions of mammalian tumor-killing M1 macrophages (Fig. 2a), and, at least in part, show functional similarity to natural killer cells [23]. These macrophages constitute 1% of the circulating blood, are adherent cells that lack F_c receptors, and hence have an antibody-independent cytotoxic activity; their activation is triggered by lipopolysaccharides (LPS) [24]. Cytotoxicity is measurable after 2 h of

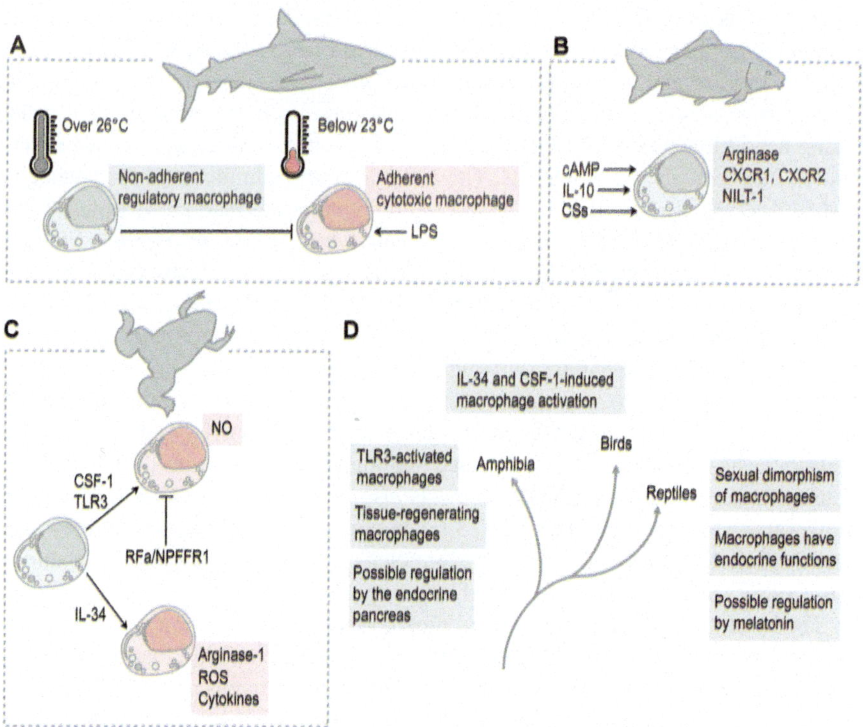

Fig. 2 Macrophage activation in nonmammalian vertebrates. (**a**) Temperature-dependent macrophage activation in sharks. (**b**) Signals and markers of the teleost M2-like macrophage activation. (**c**) Macrophage activation in amphibia. (**d**) Features and unique traits of macrophages in amphibians, reptiles, and birds

stimulation with LPS and is complete within 6 h. Cytotoxic activity of shark macrophages does not require phagocytosis [25]. It is plausible that the cytotoxic activity of these cells is similar to the antitumor function of mammalian M1 macrophages. Of note, the anticancer immunity of sharks seems stronger than of any other vertebrates [26].

Macrophage activation is temperature sensitive in sharks (Fig. 2a). Cytotoxic activation dominates below 23 °C. Above 26 °C, a regulatory, inhibitory (possibly M2-like) macrophage phenotype is dominant and inhibits cytotoxicity [24, 27]. These regulatory macrophage-like cells are nonadherent (Fig. 2a). The temperature-sensitive macrophage activation may compensate the inhibitory effect of low temperature on lymphocyte responses. When the temperature falls, lymphocyte responses are compromised and replaced by macrophage-mediated cytotoxicity [28].

We still lack understanding how macrophage number is controlled in fishes. About two-third of the circulating blood cells in sharks are in mitosis [28]. Hence, it is plausible that signals acting in the circulation can control the expansion and differentiation of macrophage populations.

The so-called head kidney is the remnant of the embryonic kidney (pronephros), and functions as a hematopoietic organ in teleost fishes. It generates myeloid cell lineages (Fig. 1d). The pluripotent myeloid stem cells of the head kidney can be cultured in vitro and differentiated into macrophages, similarly to the mammalian in vitro macrophage differentiation from bone marrow hematopoietic stem cells. Using head kidney-derived macrophages as an in vitro model, macrophage polarization has been shown in teleosts. The M2-polarizing stimulus in this instance is cAMP, and not IL-4 or IL-13. This activation state is hallmarked by upregulated gene expression of *CXCR1* and *CXCR2* and *novel Ig-like transcript* (*NILT*)*-1* (Fig. 2b). Although putative homologues of IL-4, IL-13, and their receptors have been found in teleost fishes, it has not been proven that an IL-4/IL-13-induced macrophage activation exists in fishes. Teleost fishes may lack a homologue of the mammalian IL-4/IL-13 signaling, although the equivalent of cAMP-induced M2 activation is known in mammals. Beta-adrenergic receptor stimulation, which increases cAMP level, increases macrophage arginase-1 activity [29] and inhibits NO synthesis in *Mycobacterium*-infected mammalian macrophages [30]. Similarly, neuropeptides that induce cAMP synthesis also elicit Th2 immune response [31]. However, cAMP also stabilizes STAT3 in mammalian macrophages, so it may also curb excessive M2 activation [32]. Nevertheless, cAMP seems to be an evolutionarily conserved signal transducer in M2 macrophages.

The head kidney contains so-called melano-macrophages, arranged in clusters. These macrophages accumulate pigments such as melanin and lipofuscin. They also capture and store iron, and moreover trap and present antigens and sequester toxic metabolites [33, 34]. Also, they are involved in melanin biosynthesis. Aged erythrocytes are degraded by splenic macrophages in teleost fishes. These macrophages appear in clusters in the spleen, phagocytose erythrocytes, and hence are rich in iron and hemosiderin [35]. Removal of aged erythrocytes by macrophages, called erythrophagocytosis, is crucial for hemoglobin metabolism, serving for the disposal of heme and iron recycling. Aged erythrocytes may undergo lysis, and hence release cytotoxic amounts of heme and iron; macrophages prevent this by taking up and degrading aged erythrocytes. Recycling heme and iron also sustains hematopoiesis and the replenishment of erythrocytes. The ability of macrophages to scavenge iron is increased during infections with extracellular pathogens, which limits the iron supply of the pathogen [36]. M2 macrophages express CD163, which is a hemoglobin-haptoglobin receptor and allows heme uptake, and erythrocyte uptake is controlled by nuclear receptors linked to M2 activation. However, iron-recycling macrophages are prone to be M1 activated and be involved in pathogen elimination (reviewed by [37]).

Macrophages in teleost fishes may allow the growth of intracellular bacteria, and thus resemble intracellular pathogen-infected M2 macrophages. Environmental stress increases melano-macrophage content, and they serve as biomarkers for water quality [33, 34].

The arginine fork exists in fish macrophages, similarly to mammals (for details see chapter "What Is an M2 Macrophage? Historical Overview of the Macrophage Polarization Model. The Th1/Th2 and M1/M2 Paradigm, the Arginine Fork").[1] M1-like macrophages generate NO in an oxidative metabolism of L-arginine, whereas M2-like macrophages show arginase activity and lack NO synthesis (Fig. 2b). Further signals, such as IL-10 or various steroids such as corticoids, may also induce an M2-like activation state in fish macrophages [38]. Given the evolutionary distance between fishes and mammals, however, caution is warranted in interpreting similarities as equivalent mechanisms in the course of macrophage activation in fishes and mammals.

5 M2 Macrophage Functions in Amphibia

Traits of M2 macrophages such as tissue healing and efferocytosis (engulfment of apoptotic cells) can be recognized in amphibia. For instance, salamanders have a high capacity of tissue regeneration, which is absent when macrophages are systemically depleted. Lacking macrophages, the limb is unable to regenerate, and the wound shows extensive fibrosis and disturbed gene expression of extracellular matrix components [39]. Replenishing macrophages restores limb regeneration, showing that macrophages have a key role in tissue healing in amphibia. Similarly, optic nerve regeneration in the frog *Rana pipiens* requires macrophages. The growth factors ciliary neurotrophic factor (CNTF) and fibroblast growth factor 2 (FGF-2) increase macrophage accumulation in the damaged nerve, which benefits the regenerating axons [40].

[1] Human macrophages lack the arginine fork.

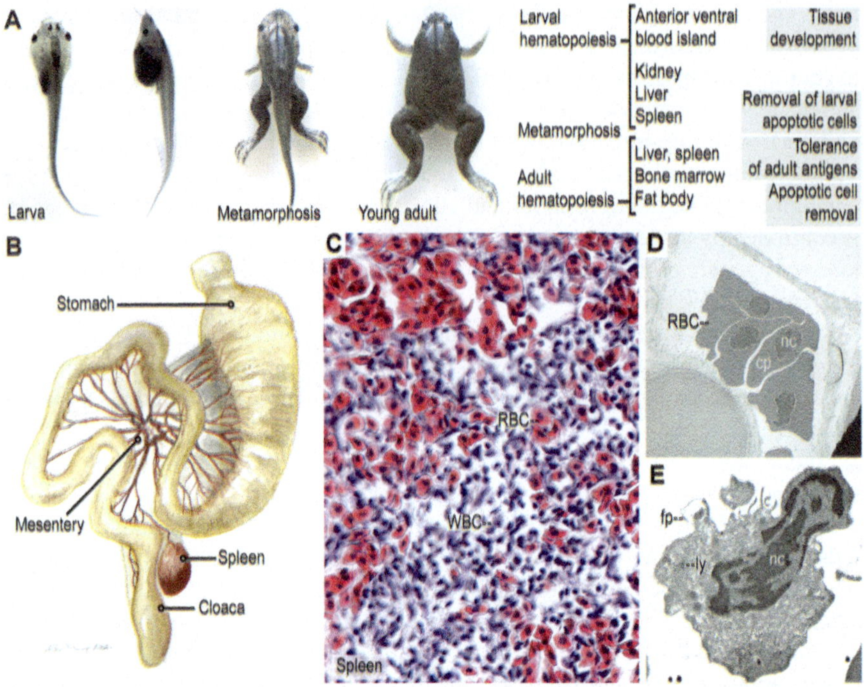

Fig. 3 Amphibian metamorphosis and the adult spleen in amphibia. (**a**) Larva, metamorphosis, and young adult life stages of *Xenopus laevis*. Macrophages develop distinctly in the larvae, during metamorphosis, and in adult frogs. The function of the macrophages is also dependent on the life stage. M2-like homeostatic macrophages support organ development in the larva, whereas at the climax of metamorphosis the macrophages are essential for safe disposal of apoptotic remnants and the self-tolerance of the newly synthesized adult-type antigens. (**b**) Anatomy of the spleen in the adult frog (*Rana ridibunda* or *Pelophylax ridibundus*). (**c**) Histology of the adult spleen (from [42]). *RBC* red blood cell, *WBC* white blood cell. (**d**) Transmission electron microscopy image of RBCs in a capillary. *nc*, nucleus, *cp* cytoplasm. (**e**) Transmission electron microscopy of a macrophage in *Xenopus laevis*. *fp* filopodia, *ly* lysosome, *nc* nucleus. (Author's drawings and images)

Most amphibian species undergo metamorphosis during their life cycle (Fig. 3a).[2] The metamorphosis is associated with massive apoptosis of larval tissues, affecting major organs such as the limbs, the gills, the gut, the nervous system, the kidneys, leukocytes, and red blood cells [41]. Removal of the apoptotic debris, and the suppression of immune response against the newly generated adult-type antigens, require the orchestrated work of macrophages. Macrophages infiltrate the apoptotic tissues. For instance, the skin or the gut epithelial layer allows macrophage infiltration by derangement of the basal membrane of the epithelia, resulting in a loosely arranged basal membrane that allows the migration of macrophages. Macrophages engulf apoptotic remnants that are extruded into the gut lumen [41].

[2] For instance, the axolotl *Ambystoma mexicanum* has a permanent larval stage, without undergoing metamorphosis.

We still lack understanding of the molecular signals that orchestrate this process and sustain macrophage tolerance toward adult antigens. Importantly, macrophage development also undergoes rearrangement during metamorphosis. The first wave of macrophages develops in the anterior ventral blood island, and the macrophages migrate toward the developing tissues. This stage is followed by macrophage generation in the kidney, spleen, liver, fat bodies, and in some species in the bone marrow (Fig. 3a) [42–45]; they colonize the tissues, and have homeostatic functions necessary for tissue development [46, 47]. It is intriguing to explore how these macrophage waves are controlled and which activation state they adopt.

In adult amphibians, macrophages continue phagocytosing apoptotic cells, for instance, red blood cells. The red blood cells retain their nuclei in amphibia, unlike mammals, and thus are considerably larger than the macrophages (Fig. 3b–e). It is intriguing that despite this difference between the macrophage and the phagocytic cargo, the engulfment of red blood cells is possible. In contrast to mammals, aged red blood cells are not accumulated in the spleen of adult amphibia. Instead, they are phagocytosed in the blood, and the spleen serves as a hematopoietic organ. However, the larval red blood cells, but not coexisting adult red blood cells, are selectively sequestered from the systemic circulation by macrophages in the liver and spleen during metamorphosis [41]. How this selective removal is possible is another intriguing trait of macrophages still to be explored.

The adult spleen also contains melano-macrophages, also called splenic pigmented cells [34], which may have a similar function as the melano-macrophages of teleost fishes and synthesize melanin [48]. Macrophages of the amphibian liver, often considered as equivalents of mammalian Kupffer cells, also synthesize and degrade melanin [49, 50].

These examples suggest that M2-like macrophage functions are present in amphibia. However, macrophage activation in amphibia is still to be explored. In amphibia the TLR signaling is distinct from its mammalian counterpart. TLR4 is truncated in amphibia, hence amphibian macrophages have a diminished sensitivity to LPS [42]. Thus we lack the homology of the LPS/TLR4 signaling, which is the prototypical M1 macrophage activation route in mammals. The pathogen-eliminating macrophage activation is achieved by TLR3 signaling and through colony-stimulating factor-1 receptor (CSF-1R) ligands. There are two known CSF-1R ligands: colony-stimulating factor-1 (CSF-1) and interleukin-34 (IL-34) (Fig. 2c) [51, 52]. CSF-1 induces a NO-producing, phagocytic, and antibacterial macrophage activation state [51], resembling mammalian M1 macrophages. IL-34 causes an arginase-1-expressing macrophage activation state (Fig. 2c). However, this activation state is hallmarked by production of cytokines and ROS and is considered as an antiviral macrophage activation [45, 51, 53]. We lack understanding how the antibacterial and antiviral macrophage functions are inhibited in amphibia. One possible candidate is a neuropeptide FF (NPFF) signaling, a putative pancreatic hormone in *Xenopus laevis* [42]. NPFF is a so-called RF-amide (RFa) belonging to the family of neuropeptides with a conserved N-terminal amidated arginine-phenylalanine motif. These peptides have diverse functions, including the control of M2 macrophage activation in mammals. Genetic ablation of the NPFF receptor 1

(NPFFR1) reduces the gene expression of macrophage activation genes in *Xenopus*, suggesting that there is an endocrine control of macrophage activation in amphibia [42].

6 M2 Macrophages in Reptiles, Birds, and in the Platypus

Reptiles provide insight into some unique regulatory mechanisms that control M1 or M2 traits of macrophages. For instance, the ultraviolet radiation of sunlight has been shown to suppress immune functions in mammals [54] as skin exposure to ultraviolet radiation determines vitamin D level, which indirectly affects macrophage functions. Reptiles apply "sunbath" or sun basking behavior to maintain their body temperature, so they may be useful models of understanding the immune effects of prolonged UV-B exposure. The green anole *Anolis carolinensis* is a northerly distributed reptile that utilizes sun basking behavior instead of hibernation during winter. Prolonged UV-B radiation, however, does not suppress macrophage immune surveillance in this species [55].

Intriguingly, reptile macrophages show marked sexual dimorphism. The splenic macrophages of female wall lizards show higher phagocytosis activity than their male counterparts [56], possibly because of the effects of estrogen; estrogen is known to have antiinflammatory activity in mammals, promoting an IL-10-dependent M2-like macrophage phenotype [57] and enhancing IL-4-induced M2 activation. Allergic asthma is a chronic Th2 inflammation associated with M2 macrophages in mammals. The prevalence of asthma also shows a marked dependence on sex hormones: it mostly affects boys in childhood and women in adulthood, suggesting the role of estrogens. Alveolar macrophages from female mice have a greater expression of IL-4Rα and estrogen receptor compared with macrophages from male mice following allergen challenge. Moreover, IL-4-stimulated macrophages from female mice exhibit more transcriptionally active histone modifications at M2 gene promoters than macrophages from male mice [58]. Estrogen seems to favor an M2 activation in mammals, and it is plausible that the evolutionary roots of this signal mechanism can be found in reptiles.

As detailed in the chapter "What Is an M2 Macrophage? Historical Overview of the Macrophage Polarization Model. The Th1/Th2 and M1/M2 Paradigm, the Arginine Fork," the testicle in mammals is an immunosuppressed organ, with M2 macrophages that tolerate the developing gametes. Similarly, reptile testicular macrophages support testosterone production from Leydig cells, probably by the secretion of a steroidogenic factor. In contrast, LPS-activated testicular macrophages inhibit testosterone production [59, 60]. The physiologically beneficial testicular macrophages are hence nonactivated or resemble mammalian M2 macrophages. The maintenance of this macrophage phenotype is probably ensured by histamine, released by testicular mast cells [59, 60].

The evagination of the roof of the diencephalon forms the pineal complex, which is well developed in reptiles. This brain region is often seen as a parietal eye in many reptiles, and it has a light- and temperature-controlled endocrine function. Immune responses are synchronized with the reproductive cycle in reptiles and are dependent on annual photoperiod [61]. The pineal complex (the pineal gland in mammals) releases melatonin in a light-dependent manner, and melatonin serves as a key hormone of the circadian clock. Melatonin is a known signal of M2 macrophage activation and inhibitor of M1 activation in mammals [62–64]. In reptiles, and in teleost fishes, melatonin seems to have an immune stimulant effect: it inhibits peroxidase activity but increases respiratory burst activity in teleost fish head kidney macrophages [65], and also increases nitrite release (indicative of NO synthesis) and superoxide production by splenic macrophages in reptiles [66]. Given the impact of melatonin in chronobiology, it would be intriguing to study further the immune effect of melatonin in reptiles, in which the pineal complex is much more developed than in mammals.

In birds, macrophages have preserved the amphibian-like CSF-R1 and IL-34 signaling [67], show a strong M1-like macrophage immune response toward viruses [68], but lack an M2-equivalent macrophage phenotype.

The duck-billed platypus (*Ornithorhynchus anatinus*) is a unique egg-laying mammal, an evolutionary link between birds and mammals. Mycotic skin lesions and skin parasite infections induce macrophage infiltration in the platypus [69, 70]; however, there is a lack of M1/M2 phenotyping of these cells. Tingible body macrophages, which engulf apoptotic debris in lymph nodes, have been shown in the platypus [71]. Whether these macrophages are equivalents of M2 macrophages of the mammalian spleen is still unknown. Also, M2 marker genes, such as C-type lectin [72] and *Cd163* [73], are known in the platypus genome. It is still unclear whether these genes are associated with macrophage activation. In the case of *Cd163*, it is plausible that its expression is a consequence of the unusually high level of hemoglobin in platypus that results from adaptation to the hypoxic conditions in their burrows, in which they spend extended periods [73].

7 What Can We Learn from Chordate Macrophages?

There are many gaps in our understanding of macrophage biology in nonmammalian chordates (Fig. 2d). However, some unique traits of these animals can help understand biological problems with universal impact. For instance, temperature- and light-sensitive macrophage activation of elasmobranch fishes and reptiles can help us understand better how climatic changes have shaped immunity in the course of evolution, and what kind of molecular fingerprint the glacial–interglacial cycles have left in the human innate immune system. Similarly, chronobiology of innate immunity can be better understood by these mechanisms. Because innate immunity directly controls metabolic performance in humans, this can also guide us to better understand the origin of the globally widespread metabolic diseases of today. The unique ability of amphibia to discriminate larval and adult antigens in the course of metamorphosis, and to discard the massive amount of apoptotic cells without rendering the body vulnerable to infections, is another intriguing model for better understanding of the roots of autoimmunity.

References

1. Delsuc F, Brinkmann H, Chourrout D, Philippe H. Tunicates and not cephalochordates are the closest living relatives of vertebrates. Nature. 2006;439:965–8.
2. Cima F, Ballarin L, Gasparini F, Burighel P. External amebocytes guard the pharynx entry in a tunicate (Ascidiacea). Dev Comp Immunol. 2006;30:463–72.
3. Ballarin L, Cima F, Sabbadin A. Histoenzymatic staining and characterization of the colonial ascidian *Botryllus schlösseri* hemocytes. Bollettino di Zoologia. 1993;60:19–24.
4. Cima F, Peronato A, Ballarin L. The haemocytes of the colonial aplousobranch ascidian *Diplosoma listerianum*: structural, cytochemical and functional analyses. Micron. 2017;102:51–64.
5. Franchi N, Ballarin L. Immunity in protochordates: the tunicate perspective. Front Immunol. 2017;8:674.
6. Franchini A, Conte A, Ottaviani E. Nitric oxide: an ancestral immunocyte effector molecule. Adv Neuroimmunol. 1995;5:463–78.
7. Ottaviani E, Franchini A, Franceschi C. Presence of several cytokine-like molecules in molluscan hemocytes. Biochem Biophys Res Commun. 1993;195:984–8.
8. Voskoboynik A, Rinkevich B, Weiss A, Moiseeva E, Reznick AZ. Macrophage involvement for successful degeneration of apoptotic organs in the colonial urochordate *Botryllus schlosseri*. J Exp Biol. 2004;207:2409–16.
9. Franchi N, Ballin F, Ballarin L. Protection from oxidative stress in immunocytes of the colonial ascidian *Botryllus schlosseri*: transcript characterization and expression studies. Biol Bull. 2017;232:45–57.
10. Blanchoud S, Zondag L, Lamare MD, Wilson MJ. Hematological analysis of the ascidian *Botrylloides leachii* (Savigny, 1816) during whole-body regeneration. Biol Bull. 2017;232:143–57.
11. Cotelle N, Moreau S, Bernier JL, Catteau JP, Hénichart JP. Antioxidant properties of natural hydroquinones from the marine colonial tunicate *Aplidium californicum*. Free Radic Biol Med. 1991;11:63–8.
12. Sato A, Shindo T, Kasanuki N, Hasegawa K. Antioxidant metabolites from the tunicate *Amaroucium multiplicatum*. J Nat Prod. 1989;52:975–81.
13. Knight J, Taylor GW, Wright P, Clare AS, Rowley AF. Eicosanoid biosynthesis in an advanced deuterostomate invertebrate, the sea squirt (*Ciona intestinalis*). Biochim Biophys Acta Mol Cell Biol Lipid. 1999;1436:467–78.
14. Du J, Xie X, Chen H, Yang W, Dong M, Su J, Wang Y, Yu C, Zhang S, Xu A. Macrophage migration inhibitory factor (MIF) in Chinese amphioxus as a molecular marker of immune evolution during the transition of invertebrate/vertebrate. Dev Comp Immunol. 2004;28:961–71.
15. Yaddanapudi K, Putty K, Rendon BE, Lamont GJ, Faughn JD, Satoskar A, Lasnik A, Eaton JW, Mitchell RA. Control of tumor-associated macrophage alternative activation by macrophage migration inhibitory factor. J Immunol. 2013;190:2984–93.
16. Rhodes C, Ratcliffe N, Rowley A. Presence of coelomocytes in the primitive chordate amphioxus (*Branchiostoma lanceolatum*). Science. 1982;217:263–5.
17. Xu A-L. Amphioxus immunity: tracing the origins of human immunity. Amsterdam: Elsevier; 2015.
18. Han Y, Huang G, Zhang Q, Yuan S, Liu J, Zheng T, Fan L, Chen S, Xu A. The primitive immune system of amphioxus provides insights into the ancestral structure of the vertebrate immune system. Dev Comp Immunol. 2010;34:791–6.

19. Röszer T. The invertebrate midintestinal gland ("hepatopancreas") is an evolutionary forerunner in the integration of immunity and metabolism. Cell Tissue Res. 2014;358:685–95.
20. He C, Han T, Liao X, Zhou Y, Wang X, Guan R, Tian T, Li Y, Bi C, Lu N, He Z, Hu B, Zhou Q, Hu Y, Lu Z, Chen JY. Phagocytic intracellular digestion in amphioxus (*Branchiostoma*). Proc Biol Sci. 2018;285:20180438.
21. Franz A, Wood W, Martin P. Fat body cells are motile and actively migrate to wounds to drive repair and prevent infection. Dev Cell. 2018;44:460–470.e3.
22. Happonen KE, Tran S, Morgelin M, Prince R, Calzavarini S, Angelillo-Scherrer A, Dahlback B. The Gas6–Axl protein interaction mediates endothelial uptake of platelet microparticles. J Biol Chem. 2016;291:10586–601.
23. McKinney EC, Haynes L, Droese AL. Macrophage-like effector of spontaneous cytotoxicity from the shark. Dev Comp Immunol. 1986;10:497–508.
24. Pettey CL, Churchill McKinney E. Mitogen induced cytotoxicity in the nurse shark. Dev Comp Immunol. 1981;5:53–64.
25. McKinney EC, Flajnik MF. IgM-mediated opsonization and cytotoxicity in the shark. J Leukoc Biol. 1997;61:141–6.
26. Marra NJ, Richards VP, Early A, Bogdanowicz SM, Pavinski Bitar PD, Stanhope MJ, Shivji MS. Comparative transcriptomics of elasmobranchs and teleosts highlight important processes in adaptive immunity and regional endothermy. BMC Genomics. 2017;18:87.
27. Hyder SL, Cayer ML, Pettey CL. Cell types in peripheral blood of the nurse shark: an approach to structure and function. Tissue Cell. 1983;15:437–55.
28. McKinney EC. Shark cytotoxic macrophages interact with target membrane amino groups. Cell Immunol. 1990;127:506–13.
29. Bernard AC, Fitzpatrick EA, Maley ME, Gellin GL, Tsuei BJ, Arden WA, Boulanger BR, Kearney PA, Ochoa JB. Beta adrenoceptor regulation of macrophage arginase activity. Surgery (St. Louis). 2000;127:412–8.
30. Boomershine CS, Lafuse WP, Zwilling BS. Beta2-adrenergic receptor stimulation inhibits nitric oxide generation by *Mycobacterium avium*-infected macrophages. J Neuroimmunol. 1999;101:68–75.
31. Delgado M, Leceta J, Sun W, Gomariz RP, Ganea D. VIP and PACAP induce shift to a Th2 response by upregulating B7.2 expression. Ann N Y Acad Sci. 2000;921:68–78.
32. Waqas SFH, Hoang A, Lin Y, Ampem G, et al. Neuropeptide FF increases M2 activation and self-renewal of adipose tissue macrophages. J Clin Invest. 2017;127:2842–54.
33. Agius C, Roberts RJ. Melano-macrophage centres and their role in fish pathology. J Fish Dis. 2003;26:499–509.
34. Franco-Belussi L, de Oliveira C. The spleen of *Physalaemus nattereri* (Amphibia: Anura): morphology, melanomacrophage pigment compounds and responses to α-melanocyte stimulating hormone. Italian J Zool. 2016;83:298–305.
35. Amlacher E. Taschenbuch der Fischkrankheiten. Jena: VEB Gustav Fisher Verlag; 1976.
36. Nemeth E, Tuttle MS, Powelson J, Vaughn MB, Donovan A, Ward DM, Ganz T, Kaplan J. Hepcidin regulates cellular iron efflux by binding to ferroportin and inducing its internalization. Science. 2004;306:2090–3.
37. Alam MZ, Devalaraja S, Haldar M. The heme connection: linking erythrocytes and macrophage biology. Front Immunol. 2017;8:33.
38. Hodgkinson JW, Grayfer L, Belosevic M. Biology of bony fish macrophages. Biology (Basel). 2015;4:881–906.
39. Godwin JW, Pinto AR, Rosenthal NA. Macrophages are required for adult salamander limb regeneration. Proc Natl Acad Sci USA. 2013;110:9415–20.
40. Blanco RE, Vega-Meléndez GS, De La Rosa-Reyes V, del Cueto C, Blagburn JM. Application of CNTF or FGF-2 increases the number of M2-like macrophages after optic nerve injury in adult *Rana pipiens*. PLoS One. 2019;14:e0209733.
41. Ishizuya-Oka A, Hasebe T, Shi Y-B. Apoptosis in amphibian organs during metamorphosis. Apoptosis. 2010;15:350–64.

42. Waqas SFH, Noble A, Hoang A, Ampem G, Popp M, Strauß S, Guille M, Röszer T. Adipose tissue macrophages develop from bone marrow-independent progenitors in *Xenopus laevis* and mouse. J Leukoc Biol. 2017;102:845–55.
43. Chen XD, Turpen JB. Intraembryonic origin of hepatic hematopoiesis in *Xenopus laevis*. J Immunol. 1995;154:2557–67.
44. Hadji-Azimi I, Coosemans V, Canicatti C. Atlas of adult *Xenopus laevis laevis* hematology. Dev Comp Immunol. 1987;11:807–74.
45. Grayfer L, Robert J. Amphibian macrophage development and antiviral defenses. Dev Comp Immunol. 2016;58:60–7.
46. Costa RM, Soto X, Chen Y, Zorn AM, Amaya E. Spib is required for primitive myeloid development in *Xenopus*. Blood. 2008;112:2287–96.
47. Costa R, Chen Y, Paredes R, Amaya E. Labeling primitive myeloid progenitor cells in *Xenopus*. Methods Mol Biol. 2012;916:141–55.
48. Scalia M, Di Pietro C, Poma M, Ragusa M, Sichel G, Corsaro C. The spleen pigment cells in some amphibia. Pigment Cell Res. 2004;17:119–27.
49. Corsaro C, Scalia M, Leotta N, Mondio F, Sichel G. Characterisation of Kupffer cells in some Amphibia. J Anat. 2000;196(Pt 2):249–61.
50. Sichel G, Scalia M, Mondio F, Corsaro C. The amphibian Kupffer cells build and demolish melanosomes: an ultrastructural point of view. Pigment Cell Res. 1997;10:271–87.
51. Grayfer L, Robert J. Distinct functional roles of amphibian (*Xenopus laevis*) colony-stimulating factor-1- and interleukin-34-derived macrophages. J Leukoc Biol. 2015;98:641–9.
52. Edholm E-S, Rhoo KH, Robert J. Evolutionary aspects of macrophages polarization. Results Probl Cell Differ. 2017;62:3–22.
53. Grayfer L, De Jesus Andino F, Robert J. Prominent amphibian (*Xenopus laevis*) tadpole type III interferon response to the frog virus 3 ranavirus. J Virol. 2015;89:5072–82.
54. Schwarz T. Ultraviolette Strahlung: Immunantwort. JDDG J Dtsch Dermatol Ges. 2005;3:S11–8.
55. Cope RB, Fabacher DL, Lieske C, Miller CA. Resistance of a lizard (the green anole, *Anolis carolinensis*; Polychridae) to ultraviolet radiation-induced immunosuppression. Photochem Photobiol. 2001;74:46–54.
56. Mondal S, Rai U. Sexual dimorphism in phagocytic activity of wall lizard's splenic macrophages and its control by sex steroids. Gen Comp Endocrinol. 1999;116:291–8.
57. Villa A, Rizzi N, Vegeto E, Ciana P, Maggi A. Estrogen accelerates the resolution of inflammation in macrophagic cells. Sci Rep. 2015;5:15224.
58. Keselman A, Fang X, White PB, Heller NM. Estrogen signaling contributes to sex differences in macrophage polarization during asthma. J Immunol. 2017;199:1573–83.
59. Khan UW, Rai U. Paracrine role of testicular macrophages in control of Leydig cell activities in the wall lizard, *Hemidactylus flaviviridis*. Gen Comp Endocrinol. 2008;156:44–50.
60. Khan UW, Rai U. Differential effects of histamine on Leydig cell and testicular macrophage activities in wall lizards: precise role of H1/H2 receptor subtypes. J Endocrinol. 2007;194:441–8.
61. Tripathi MK, Singh R, Pati AK. Daily and seasonal rhythms in immune responses of splenocytes in the freshwater snake, *Natrix piscator*. PLoS One. 2015;10:e0116588.
62. Markus RP, Fernandes PA, Kinker GS, da Silveira Cruz-Machado S, Marçola M. Immune-pineal axis – acute inflammatory responses coordinate melatonin synthesis by pinealocytes and phagocytes. Br J Pharmacol. 2018;175:3239–50.
63. Yi W-J, Kim TS. Melatonin protects mice against stress-induced inflammation through enhancement of M2 macrophage polarization. Int Immunopharmacol. 2017;48:146–58.
64. Liu Z, Gan L, Zhang T, Ren Q, Sun C. Melatonin alleviates adipose inflammation through elevating α-ketoglutarate and diverting adipose-derived exosomes to macrophages in mice. J Pineal Res. 2018;64:e12455.

65. Cuesta A, Rodríguez A, Calderón MV, Meseguer J, Esteban MÁ. Effect of the pineal hormone melatonin on teleost fish phagocyte innate immune responses after in vitro treatment. J Exp Zool A Ecol Genet Physiol. 2007;307A:509–15.
66. Tripathi M, Singh R. Melatonin modulates splenocyte immune response in the freshwater snake, *Natrix piscator*. Herpetol Conserv Biol. 2014;9:257–66.
67. Hume DA, Gutowska-Ding MW, Garcia-Morales C, Kebede A, Bamidele O, Trujillo AV, Gheyas AA, Smith J. Functional evolution of the colony-stimulating factor 1 receptor (CSF1R) and its ligands in birds. J Leukoc Biol. 2020;107(2):237–50.
68. Liu A, Li H, Qi X, Wang Q, Yang B, Wu T, Yan N, Li Y, Pan Q, Gao Y, Gao L, Liu C, Zhang Y, Cui H, Li K, Wang Y, Wang X. Macrophage migration inhibitory factor triggers inflammatory responses during very virulent infectious bursal disease virus infection. Front Microbiol. 2019;10:2225.
69. Whittington RJ, Spratt DM. Lesions associated with metazoan parasites of wild platypuses (*Ornithorhynchus anatinus*). J Wildl Dis. 1989;25:521–6.
70. Connolly JH, Canfield PJ, Obendorf DL. Gross, histological and immunohistochemical features of mucormycosis in the platypus. J Comp Pathol. 2000;123:36–46.
71. Osogoe B, Tsujii T, Liszczynsky HR, Naora H, Seno S. Intestinal mucosa and intra-abdominal lymphoid tissues of the platypus, *Ornithorhynchus anatinus*. Okajimas Folia Anat Jpn. 1991;67:457–66.
72. Wong ESW, Sanderson CE, Deakin JE, Whittington CM, Papenfuss AT, Belov K. Identification of natural killer cell receptor clusters in the platypus genome reveals an expansion of C-type lectin genes. Immunogenetics. 2009;61:565–79.
73. Warren WC, Hillier LW, Marshall Graves JA, Birney E, Ponting CP, Grützner F, Belov K, Miller W, Clarke L, Chinwalla AT, Yang S-P, Heger A, Locke DP, Miethke P, Waters PD, Veyrunes F, Fulton L, Fulton B, Graves T, Wallis J, Puente XS, López-Otín C, Ordóñez GR, Eichler EE, Chen L, Cheng Z, Deakin JE, Alsop A, Thompson K, Kirby P, Papenfuss AT, Wakefield MJ, Olender T, Lancet D, Huttley GA, Smit AFA, Pask A, Temple-Smith P, Batzer MA, Walker JA, Konkel MK, Harris RS, Whittington CM, Wong ESW, Gemmell NJ, Buschiazzo E, Vargas Jentzsch IM, Merkel A, Schmitz J, Zemann A, Churakov G, Ole Kriegs J, Brosius J, Murchison EP, Sachidanandam R, Smith C, Hannon GJ, Tsend-Ayush E, McMillan D, Attenborough R, Rens W, Ferguson-Smith M, Lefèvre CM, Sharp JA, Nicholas KR, Ray DA, Kube M, Reinhardt R, Pringle TH, Taylor J, Jones RC, Nixon B, Dacheux J-L, Niwa H, Sekita Y, Huang X, Stark A, Kheradpour P, Kellis M, Flicek P, Chen Y, Webber C, Hardison R, Nelson J, Hallsworth-Pepin K, Delehaunty K, Markovic C, Minx P, Feng Y, Kremitzki C, Mitreva M, Glasscock J, Wylie T, Wohldmann P, Thiru P, Nhan MN, Pohl CS, Smith SM, Hou S, Renfree MB, Mardis ER, Wilson RK. Genome analysis of the platypus reveals unique signatures of evolution. Nature. 2008;453:175–83.

Signal Mechanisms of M2 Macrophage Activation

Keywords M2 polarization · STAT6 · IL-4 · IL-10 · IL-6 · TGFβ · Th2 · Cytokines · Nuclear receptors · Apoptotic cells · Efferocytosis · Macrophage

1 Introduction

In the previous chapters, we have reviewed the definition and major functions of M2 macrophages of mammals, the evolutionary origin of the M2 macrophages, and possible M2 macrophage equivalents in invertebrates and nonmammalian vertebrates. This chapter is dedicated to the molecular mechanisms that drive M2 macrophage activation, or as often called, M2 macrophage polarization. There are prototypical M2 activating signals: Th2 cytokines, and the uptake of apoptotic cells. Th2 cytokines act through signal transducer and activator of transcription 6 (STAT6) by inducing its phosphorylation, nuclear translocation, and promoting transcription of STAT6-responsive genes. Apoptotic cells are metabolized within the macrophages, eventually activating lipid-sensing nuclear receptors. Excess M2 macrophage number and unwanted M2 macrophage activation cause disease; hence, mechanisms that inactivate STAT6 signaling are important to balance macrophage activation. Inactivation of STAT6 signaling can occur via STAT6 ubiquitination by the ubiquitin ligase ring finger protein 128 [RNF128, also known as gene related to anergy in lymphocytes (GRAIL)]. Ubiquitination of STAT6 leads to its proteasomal degradation, hence allowing an M2 activation state to be switched off. There are noncanonical inducers of M2 genes, such as certain pathogen products, inflammatory mediators, lipid metabolites, hormones, and neurotransmitters. Epigenetic modifications also affect M2 activation, and there is an autocrine cytokine loop that sustains M2 activation.

T. Röszer, *The M2 Macrophage*, Progress in Inflammation Research 86,
https://doi.org/10.1007/978-3-030-50480-9_4

2 M2 Activation: A Net Result of a Signal Meshwork

The Th2 cytokines interleukin-4 (IL-4) and interleukin-13 (IL-13) and apoptotic cells are prototypical M2-activating signals. IL-4 and IL-13 act through signal transducer and activator of transcription 6 (STAT6) signaling, whereas apoptotic cells act by activating lipid-sensing nuclear receptors (NRs). Pathogen-derived molecules, such as helminth products and intracellular parasites, also elicit M2 activation: this pathogen-induced M2 activation is detailed in chapter "Immune Functions of the M2 Macrophages: Host Defense, Self-Tolerance, and Autoimmunity". More recently, several further mechanisms have been identified that support M2 macrophage functions: these include signals of the nervous system and the endocrine organs, NR ligands of the diet, paracrine signals of the tissue environment, and epigenetic modifications (Fig. 4). The M2 macrophages also emit signal molecules that provide feedback to immune cells, the nervous system, and the tissue environment. The M2 macrophages can also sustain their activation state by an autocrine cytokine loop (Fig. 1).

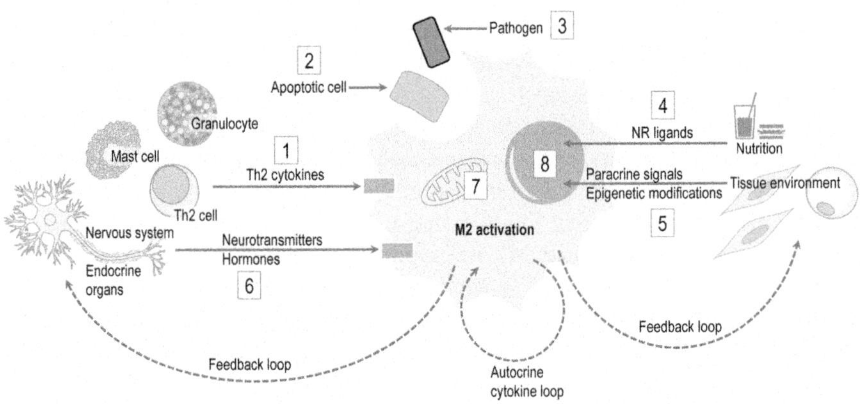

Fig. 1 Network of signals that determine M2 activation. Key mechanisms can elicit M2 activation and shape M2 macrophage responses: (1) Th2 cytokine signaling through STAT6 and cAMP, (2) signaling by apoptotic cell content, (3) pathogen products, (4) nutritional factors such as lipid metabolites and vitamins, which may be nuclear receptor (NR) ligands or act through other signal pathways, (5) tissue-specific cues and epigenetic factors, (6) neurotransmitters and hormones, (7) mitochondrial content (e.g., nucleic acids and reactive oxygen species), and (8) various transcription factors and tumor suppressors. M2 macrophages have feedback mechanisms to the neuroendocrine system and tissue environment and also have autocrine cytokine loops

3 Th2 Cytokines and STAT6 Signaling Are M2 Activators

The Th2 cytokines IL-4 and IL-13 induce M2 macrophage activation [1]. IL-4 is a secreted glycoprotein of 153 amino acids in the human, 140 amino acids in mouse, and 147 amino acids in rat. It was discovered in 1982 as a factor that induces activated mouse B cell proliferation and IgG1 secretion. Studies in the early 1990s have defined the molecular structure of IL-4: it is built up from four left-handed α-helices and short β-sheets, fixed with three disulfide bridges. IL-4 shares 20–25% sequence identity with IL-13, and its structure resembles granulocyte-macrophage colony-stimulating factor (GM-CSF), macrophage colony-stimulating factor (MCSF), and growth hormone (GH). Interestingly, MCSF potentiates M2 activation, and both GM-CSF and MCSF induce the expression of certain genes associated with M2 activation [2–4]. Also, GH has a part in antagonizing inflammatory (M1) activation, which effect is dependent on the macrophage GH receptor [5]. In the context of M2 polarization, the cellular sources of IL-4 are Th2 cells, mast cells, and basophil granulocytes. IL-4 binds to its receptor, IL-4 receptor alpha (IL-4Rα). It leads to the activation of Janus kinase (JAK), a tyrosine kinase that transduces cytokine-mediated signals. In response to IL-4, activated JAK causes the phosphorylation of tyrosine residues on STAT6. The phosphorylated, hence activated, STAT6 forms a homodimer, translocates to the nucleus, and induces gene transcription [6] (Fig. 2).

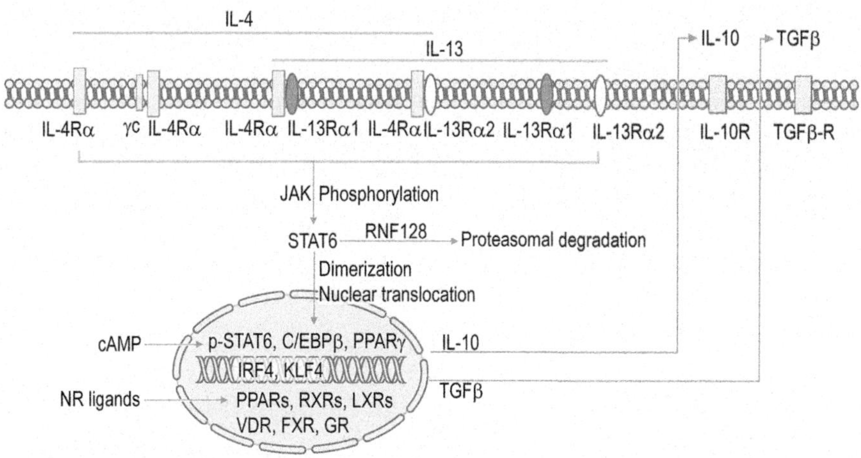

Fig. 2 STAT6 and NR signaling of M2 macrophage activation
The key signals of M2 activation are IL-4 and IL-13, which bind to their respective receptors. The downstream signaling involves JAK-STAT6, leading to STAT6 phosphorylation, dimerization, and nuclear translocation. RNF128 (GRAIL) ubiquitinates STAT6, which is eventually targeted for proteasomal degradation. RNF128 is hence a possible switch-off mechanism of M2 activation. A wide range of signals work in parallel with STAT6 signaling to determine macrophage responses, such as cAMP (activates C/EBPβ), and various nuclear receptor (NR) ligands, such as lipid metabolites, vitamins, and hormones. M2 activation is further shaped by autocrine IL-10 and transforming growth factor-β (TGFβ) signaling

STAT6 also functions as a transcriptional repressor and suppresses inflammatory macrophage responses [7]. It is also shown that phosphorylation of serine residues can cause STAT6 activation in response to IL-4 [8]. Human IL-13 has approximately 25% similarity to IL-4, and it can also activate STAT6 signaling. IL-13 can bind to its own receptors (IL-13Rα1 and IL-13Rα2), or to IL-4Rα. Moreover, IL-4Rα can heterodimerize with IL-13Rα1 or IL-13Rα2. The STAT6 pathway is a hub for M2 macrophage activation [9]. Activation of the STAT6 signaling pathway induces the transcription of typical M2 marker genes, such as *Arg1*, *Ym1 (Chil3)*, and *Cd206 (Mrc1)* [4]. Signaling through STAT6 is complemented with transcriptional regulators, such as CCAAT-enhancer-binding protein beta (C/EBPβ) and peroxisome proliferator-activated receptor gamma (PPARγ) [10–12]. Activation of C/EBPβ by cAMP increases the transcription of M2 genes [10, 12–15]. A cognate protein, C/EBPα, is required for both M1 and M2 responses [14]. Similarly, ligands of PPARγ increase the transcription of M2 activation genes [16, 17]. STAT6 signaling also increases the transcription of secondary transcription factors, such as interferon regulatory factor 4 (IRF4) and Krüppel-like factor 4 (KLF4), which elicit M2 activation [18–20]. These transcription factors support the metabolic changes associated with M2 activation and inhibit the NF-κB pathway of M1 activation [18, 21].

M2 activation is a double-edged sword, as detailed in chapter "What Is an M2 Macrophage? Historical Overview of the Macrophage Polarization Model. The Th1/Th2 and M1/M2 Paradigm, the Arginine Fork". Although it is favorable for many aspects of tissue development and renewal, it may make tissues vulnerable to malignant cell growth, can cause tissue fibrosis, or may induce various immune pathological conditions. It is intriguing that, despite its impact, little is known about the signals that minimize the unfavorable actions of excess M2 macrophage activation and inactivate STAT6 signaling. Recently, an E3 ubiquitin ligase, the ring finger protein 128 (RNF128, or GRAIL), has been identified as a STAT6 ubiquitinating enzyme that inactivates STAT6 signaling [22]. Ubiquitination of STAT6 or phosphorylated STAT6 (p-STAT6) leads to their proteasomal degradation [22, 23]. Treatment of macrophages with IL-4 reduces the protein and mRNA levels of RNF128 and sustains the activation of STAT6; similarly, signals that maintain IL-4/STAT6 signaling also inhibit *Rnf128* transcription [22].

4 Autocrine Loop of M2 Activation Through Interleukin-10 and TGFβ

Both M1 and M2 macrophages secrete interleukin-10 (IL-10), because its expression is increased by both M1 and M2 polarizing stimuli: lipopolysaccharide (LPS), tumor necrosis factor-α (TNFα) (as M1-activating signals), and α-melanocyte-stimulating hormone, prostaglandin E2, catecholamines, and many other signals that increase intracellular cAMP level (these are today considered as M2-activating signals) [24]. The link between cAMP and IL-10 expression is C/EBPβ,which is activated through the cAMP-activated transcription factor, cAMP response element-binding protein (CREB) [12]. IL-10 was discovered as a cytokine that inhibits Th1 cytokine production and suppresses inflammation [25].[1] Accordingly, IL-10 inhibits M1 cytokine expression, and IL-10 upregulates the expression of the M2 genes *Arg1*, *Il4ra*, *Retnla*, and *Ym1/Chil3* in an autocrine loop in macrophages [26]. This autocrine signaling augments the IL-4-induced M2 activation [27], albeit the M2 activating signals IL-4 and IL-10 reduce IL-10 expression [24]. Autocrine IL-10 signaling is hijacked by the intracellular pathogen *Mycobacterium tuberculosis*. A mycobacterial chaperon protein belonging to the HSP70 chaperon family, named DnaK, polarizes macrophages to an M2-like phenotype [28] in an IL-10-dependent mechanism [26]: this allows the survival of *Mycobacterium* cells within the infected macrophages. *Mycobacterium* cells induce IFNβ response and augment cytosolic DNA sensing in macrophages; however, they also inhibit these mechanisms and suppress immune response [29–33]. As a result, the transcriptional landscape of *Mycobacterium*-infected macrophages shows hallmarks of both M1 and M2 activation [34]. Intestinal macrophages are constantly exposed to bacterial products of the gut microbiota. Resident $CD206^+$ (hence, M2) macrophages of the colon lamina propria constitutively secrete IL-10 [35]. M1 activation is hence limited by the autocrine IL-10 signaling to avoid the development of an immune response against the gut flora. Accordingly, loss of IL-10 signaling results in chronic intestinal inflammation [36, 37]. Similarly, $CD206^+$ human decidual macrophages produce IL-10 with a possible role in the maternal immunological tolerance of the fetus [38] (chapter "Immune Functions of the M2 Macrophages: Host Defense, Self-Tolerance, and Autoimmunity").

TGFβ is secreted by M2 macrophages and, accordingly, by some tissue-resident macrophage types [39–41]. TGFβ induces IL-10 expression in macrophages and inhibits the expression of the inflammatory cytokines TNFα and IL-12 [42]. This effect of TGFβ results from upregulation of zinc finger protein SNAI-1 (SNAIL). TGFβ also activates the Akt/FoxO1 pathway in LPS-stimulated macrophages, which induces IL-10 and arginase-1 expression [43].

5 Apoptotic Cell Uptake Induces M2 Activation[2]

Typical M2 macrophage function is the safe disposal of apoptotic cells, which trait is further detailed in chapter "Immune Functions of the M2 Macrophages: Host Defense, Self-Tolerance, and Autoimmunity". Apoptotic cells often should be

[1] Principles of cytokine biology, history of cytokines, and detailed explanation of mechanisms how cytokines act in the immune system are summarized in the milestone work *The Cytokine Handbook*, edited by Angus W. Thomson, and published by Academic Press in 1998 and in consecutive updated editions.

[2] This section is an edited reprint of my previously published review on the same topic (44. Röszer, T. (2017) Transcriptional control of apoptotic cell clearance by macrophage nuclear receptors. *Apoptosis* **22**, 284–294.) With permission of Springer Nature, license number 4793591127332.

cleared from damaged or inflamed tissue where macrophages are exposed to signals that can evoke the M1 activation state. However, macrophages exposed to apoptotic cells adopt an M2-like activation state [45] (Graphic Abstract 1). As outlined earlier, several signal mechanisms lead to M2 activation, and among these, apoptotic cell clearance causes an M2c-like state with increased production of IL-10 and TGFβ, high Mer receptor tyrosine kinase (MerTK) expression, and expression of CD163 and CD206 [45]. M2 activation caused by apoptotic cells may help the resolution of inflammation in injured or diseased tissue. The underlying mechanisms allowing apoptotic cells to promote an M2 state are complex, and involve diverse signals, such as complement, IL-10, and apoptotic cell-derived molecules. For example, complement C1q and opsonins increase M2 activation, with the augmentation of apoptotic cell internalization [45, 46] (Fig. 3). Apparently, apoptotic cell ingestion is autoregulated by MerTK and IL-10. Increased MerTK expression increases Gas6 secretion, and it augments IL-10 production in macrophages [45].

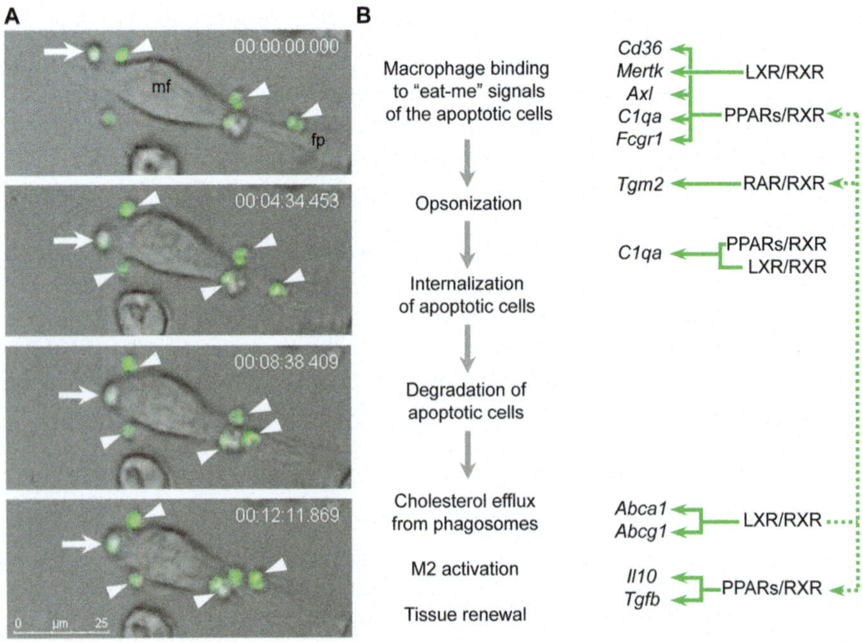

Fig. 3 Apoptotic cell clearance induces M2 activation and is controlled by nuclear receptors. (**a**) Image sequence of apoptotic cell engulfment by a mouse macrophage in vitro. Apoptotic cells are labeled with *green* fluorescence; *arrowheads* indicate apoptotic cells attached to the cell membrane of the macrophage; *arrow* shows an apoptotic cell during attachment and engulfment. (**b**) Stages of apoptotic cell uptake, and the role of various NRs in the control of these stages. (Images by author; the figure has been reprinted and adapted from [44])

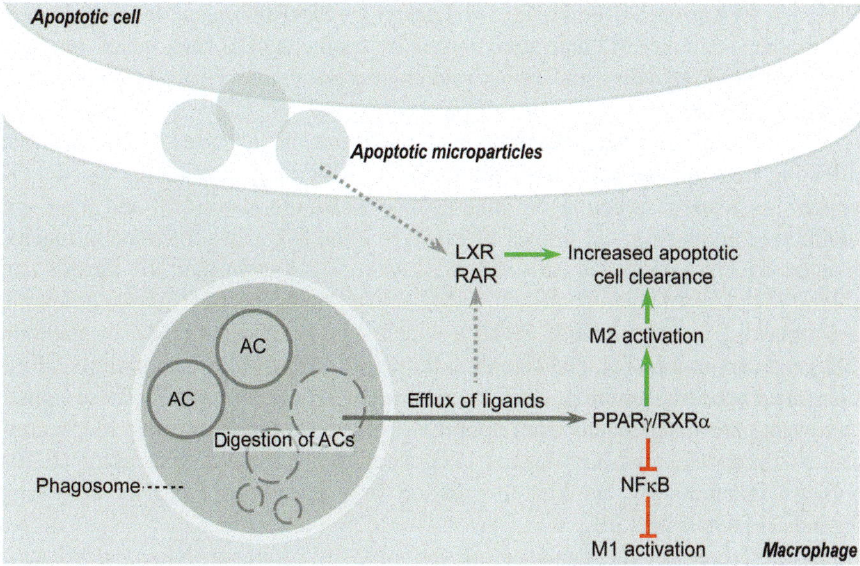

Fig. 4 The "apoptotic cell digestion model" to explain M2 activation in response to apoptotic cell uptake. Macrophages take up microvesicles released from apoptotic cells, or take up apoptotic cells and digest them in phagosomes. Lipid metabolites and various signal molecules may be released from the microvesicles or from the late phagosomes, which eventually activate NR signaling and inhibit M1 genes, while activating M2 genes and sustaining apoptotic cell clearance. (Reprinted from [44])

Microparticles released by apoptotic cells, and the apoptotic debris itself, also cause M2 activation [47] (Fig. 4). It is feasible that membrane lipids and lipid mediators packed into the microparticles and the apoptotic debris provide ligands for nuclear receptors (NR). NRs are transcription factors that regulate gene transcription in a ligand-dependent manner [48]. In response to their ligands, such as lipid metabolites, vitamins, or hormones, they suppress or activate transcription, allowing control of diverse biological functions.

Many NRs are central in the immune functions of macrophages [49].[3] Macrophages, as well as nonprofessional phagocytes, express several NRs, including PPARs (α, β/δ, and γ isotypes), LXR (α and β isotypes), RXR (α and β isotypes), RAR, vitamin D receptor (VDR), and glucocorticoid receptor (GR), which are all implicated in phagocytosis, a crucial mechanism for apoptotic cell clearance [50–52]. NR ligands influence the transcription of genes of lipid homeostasis, inflammatory cytokine production, resolution of inflammation, and synthesis of paracrine factors required for tissue healing [4, 48, 53]. RXRs are important signaling hubs in nuclear receptor-controlled transcription because they form heterodimers with other nuclear receptors, such as PPARs, LXR, RAR, and VDR (reviewed in [48]). Apoptotic cells fail to inhibit M1 activation and mitigate the acquisition of M2 activation in lack of macrophage

[3] A special issue of *Current Opinion in Pharmacology* is dedicated to the immune functions of NRs (Immunomodulation 2020, *Current Opinion in Pharmacology*, August 2020).

PPARβ/δ, PPARγor RXRα [51, 52, 54]. Lack of the PPARβ/δ-ligand-binding domain is sufficient to impair M2 activation caused by apoptotic cells [51], which suggests that PPAR/RXR signaling may be activated by the metabolites derived from internalized apoptotic cells.

Of note, increased clearance of apoptotic cells may be increased by NRs without affecting macrophage activation: for instance, LXR activation increases MerTK expression without affecting the macrophage activation state [55], and there are conflicting findings on the impact of PPARγ on the M2 activation state in macrophages ingesting apoptotic cells [52, 55]. Also, it is known that NR ligands can inhibit M1 activation and contribute to M2 activation independently from apoptotic cell uptake. Examples include PPARγ, which synergizes with STAT6 to maintain M2 gene expression [11], and signaling through PPARγ/RXRα also inhibits NFκB controlled gene transcription, a key mechanism in M1 activation [52]. These signaling events are independent from apoptotic cell uptake, and are most likely controlled by endogenous NR ligands [44]. Similarly, farnesoid X receptor (FXR) inhibits inflammatory cytokine expression in macrophages, and its activating ligands are bile acids [56].

In brief, three key mechanisms link apoptotic cell clearance, NRs, and M2 activation: (1) recognition and internalization of apoptotic cells, (2) opsonization of apoptotic cells, and (3) the late phagosome function of the macrophages, which can generate NR ligands from the apoptotic cell cargo (Fig. 3).

Recognition of apoptotic cells is the first event in the uptake of apoptotic cell remains [57]. The cell membrane of apoptotic cells displays "eat-me" signals for macrophages, such as nucleic acids, and most notably phosphatidylserine (PtdSer). In nonapoptotic cell membranes, PtdSer is present in the inner membrane layer and is missing from the outer cell surface [57]. The efficient interaction between PtdSer exposed by apoptotic cells and macrophages is ensured by various PtdSer recognition receptors and scavenger receptors [57], such as CD36, MerTK, and Axl. CD36 is a non-opsonic cell-surface receptor complex [58]. The *Cd36* promoter contains a response element for PPARγ/RXRα heterodimers [52], and PPARγ and RXR ligands enhance *Cd36* gene expression and consequently facilitate phagocytosis in vitro [59, 60]. Similarly, enhanced phagocytic capacity of PPARγ ligand-treated human alveolar macrophages, monocytes, and foam cells is coupled to elevated CD36 expression [59–62]. Lack of macrophage PPARγ or RXRα impairs recognition and internalization of apoptotic cells [52, 63], leading to accumulation of apoptotic cell debris in the tissues [52].

MerTK belongs to the Tyro3, Axl, and Mertk (TAM) family of receptor tyrosine kinases. MerTK binds to PtdSer exposed on the cell surface of the apoptotic cells; it tethers the apoptotic cells and allows their binding and eventual uptake by macrophages [57, 64]. It is expressed by a wide range of tissue-resident macrophages [65], thus allowing the homeostatic clearance of apoptotic cells in the tissues. Deficiency of MerTK impedes apoptotic cell uptake [52] and increases inflammation. MerTK binds the apoptotic cells by the growth arrest-specific gene 6 (GAS 6), protein-S, and other bridging molecules [57]. In response to PtdSer, rapid bridging or MerTK/

GAS6 occurs, leading to MerTK dimerization and phosphorylation in its tyrosine kinase domain [57, 64]. The mechanism by which MerTK is involved in phagocytosis is not fully understood. It is required for cytoskeletal rearrangement during phagocytosis and the proper attachment and migration of macrophages [66]. Interestingly, dendritic cells do not require MerTK for ingestion of apoptotic cells [67]. Transcription of *Mertk* is regulated by PPARβ/δ/RXR [51], PPARγ/RXRα [52], and LXR/RXR [50]. Apoptotic cells increase *Mertk* transcription, which requires functional PPAR/RXR and LXR/RXR signaling [50–52]. LXR, PPARγ, and RXR ligands increase apoptotic cell uptake by macrophages along with increased *Mertk* transcription [50–52]. Lack of PPARγ or RXRα results in deficient apoptotic cell clearance and disturbs the adherence and migration of macrophage progenitors [52, 68], resembling the consequences of MerTK deficiency. *Mertk* is also controlled indirectly by glucocorticoids [69]. It has been shown that glucocorticoid treatment upregulates LXR/RXR expression, eventually increasing apoptotic cell uptake [69]. Interestingly, MerTK/Gas6 has a feedback mechanism on LXR signaling [70]. Mice lacking MerTK signaling show reduced expression of LXRα and LXRβ, along with decreased expression of LXR target genes in peritoneal macrophages and in organs that are rich in resident macrophages (lung and spleen) [70]. Inhibition of MerTK/GAS6 bridging has a similar effect. In vitro exposure of RAW264.7 cells or primary peritoneal macrophages to GAS6 increases LXR abundance in an Akt-dependent manner [70]. MerTK/GAS6 increased LXR transcription helps the resolution of inflammation [70].

Axl tyrosine kinase is another TAM family member with a function in apoptotic cell uptake. Similarly to MerTK, it acts with protein-S and Gas6, and is involved in the recognition of PtdSer-exposing apoptotic cells [71]. Axl allows apoptotic cell uptake by macrophages and dendritic cells; it also allows apoptotic cell uptake by nonprofessional phagocytes, such as endothelial cells [72]. Also, Axl along with MertTK is expressed in macrophages of the amyloid plaques, where they are involved in amyloid removal [73]. In macrophages, *Axl* transcription is increased by agonists of PPARγ PPARβ/δ, and RXR [52, 63, 73]. Lack of macrophage PPARγ, PPARβ/δ, or RXRα [51, 52], or pharmacological blocking of PPARγ [73], impairs *Axl* transcription and inhibits apoptotic cell uptake.

Opsonization of the apoptotic cells aids their uptake. Several plasma proteins, such as IgG, IgM, and complement, can bind to apoptotic cells. Binding of IgG and IgM to apoptotic cells generates an "eat-me" signal for macrophages, in an analogue process of the opsonization of pathogens. Importantly, uptake of IgG- or IgM-coated particles is increased by PPARγ and RXR ligands, and in turn, lack of PPARγ or RXRα impairs their uptake by macrophages [52]. Moreover, genes encoding receptors for immunoglobulins are also upregulated by PPARγ/RXRα in macrophages [52]. In addition to immunoglobulins, the complement system is also involved in apoptotic cell clearance. The most studied complement allowing apoptotic cell recognition is C1q. Human deficiency in C1q leads to deficient apoptotic cell uptake, eventually causing autoimmunity against dying cells [74]. The apoptotic cell surface exposes molecules for C1q binding, such as nucleic acid fragments and calreticulin; however, direct complement binding requires cell membrane remodeling, which occurs only in the late phase of apoptosis [74, 75]. Complement binding of apoptotic cells thus may be a secondary consequence of the binding of other plasma proteins, most likely IgM [74]. Macrophage-specific deletion of PPARβ/δ, as well as PPARγ or RXRα, impairs the expression of complement factors such as C1q, which reduces the binding of apoptotic cells to macrophages [51, 52]. Transglutaminase-2 (TG2) is involved in clearance of apoptotic cells by macrophages [76–81]. TG2 is a protein cross-linking enzyme that acts as a coreceptor for integrin beta-3, and crosslinks syndecan-4, CD44 (phagocytic glycoprotein 1), and milk fat globulin EGF8 [76, 78]: all are involved in apoptotic cell binding by macrophages. TGM2 is involved in the formation of the phagocytic cup, which engulfs the bound apoptotic cells [76, 78]. Lack of TGM2 leads to defective clearance of apoptotic cells in mice, causing autoimmunity. Transcription of *Tgm2* is increased by PPARγ/RXRα, and also by RAR/RXRα [52]. Indirectly, LXR activation also increases *Tgm2* transcription via augmenting RARα signaling [50]. Interaction between apoptotic cells and macrophages also increases TGM2 expression [79], and it is plausible that RAR ligands generated by apoptotic cells may cause the upregulated *Tgm2* transcription [80, 81].

Late phagosome functions involve the degradation of the apoptotic cells, and the efflux of their components, mainly phospholipids and cholesterol from the phagosomes. Proper digestion of apoptotic cells induces intracellular signaling within the macrophage, and provides NR ligands, which are important for continued uptake of further apoptotic cells and also for the determination of macrophage activation state [51, 52, 57, 80, 82]. PPARγ, LXR, RXR, and RAR have impact on the late phagosome function and on macrophage activation state following ingestion of apoptotic cells. LXR not only controls the end of the chain in the apoptotic cell uptake, but also augments the transcription of other nuclear receptors, maintaining the continued binding and internalization of apoptotic cells [83].

6 Further M2 Activating Signals

The signal mechanisms that induce or maintain M2 macrophage traits are abundant. In addition to the IL-4/STAT6 signaling just detailed, the autocrine IL-10 loop, and the apoptotic cell clearance, many other signal mechanisms induce M2 macrophage activation. For instance, macrophage activation is induced by intracellular cAMP, which mechanism is evolutionarily conserved in chordates and vertebrates (see in chapter "Evolution of M2 Macrophage Functions in Chordates"). Neuropeptides such as vasoactive intestinal peptide (VIP), pituitary adenylate cyclase-activating peptide (PACAP), and neuropeptide FF (NPFF) induce M2 activation, and the underlying signal mechanisms involve cAMP as a second messenger [84, 85]. Increased intracellular cAMP activates M2 gene transcription through CREB-C/EBPβ signaling, which was detailed above. Alternatively, cAMP can act through protein kinase A (PKA), which sustains STAT3 stability and leads to M1-like activation [86]. NPFF, for instance, reduces intracellular cAMP level, diminishes STAT3 signaling, and increases M2 gene transcription [85].

Neurotransmitters, intercellular mediators, hormones, high-density lipoproteins [85, 87–89], and many other signals of the tissue environment can elicit M2 macrophage responses or inhibit M1 activation [4, 90–92]. Neuroendocrine signals act through neuropeptide receptors (e.g., G-protein-linked receptor 10, NPFF receptor, neuropeptide Y receptor) [93, 94], adenosine receptors [92], acetylcholine receptors [90, 91], capsaicin receptor (transient receptor potential cation channel subfamily V member 1, TrpV1) [95], and β-adrenergic receptor [96, 97]. Macrophages also affect neurotransmission: macrophages in adipose tissue can eliminate noradrenaline that is released from peripheral nerves [98]. M2 macrophages are constituents of the so-called diffuse or tissue neuroendocrine system, and release growth factors and neurotransmitters and hormones, such as substance P, neuropeptide Y, VIP, insulin-like growth factor, and adiponectin [84, 99–102].

Mitochondria also have an impact on macrophage activation [103]. In resident peritoneal macrophages, mitochondria are recruited to phagosomes and produce reactive oxygen species to ensure pathogen-eliminating function of the patrolling macrophages [104]. Intracellular pathogens, for instance, can induce macrophage apoptosis by destructing mitochondria [105]. M2 macrophages show mitochondrial fusion and mitochondrial biogenesis, and they have a mitochondria-dependent metabolism: a high degree of oxidative phosphorylation and fatty acid oxidation [106]. Mitophagy is a quality control mechanism that ensures the removal of damaged or nonfunctional mitochondria, which is necessary to avoid the release of mitochondrial content and subsequent inflammation. A tagging mechanism of the nondesired mitochondria is a complex formed by the mitochondrial PTEN-induced kinase-1 (PINK1) and Parkin [107]. When mitophagy is impaired, a release of mitochondrial components, such as nucleic acids into the cytoplasm, triggers interferon response and inflammation [108]. Mitophagy is thought to curb inflammation by removing potentially immunogenic mitochondrial content [109]. For instance, NPFF, which sustains the M2 activation state [85], increases Parkin transcription [110]. It is tempting to assume that mitophagy is inhibiting M1 and promoting M2 activation; however, it has also been shown that reducing mitophagy increases M2-like macrophage activation [111]. The precise role of mitophagy in M2 macrophages is to be explored.

Mitochondria, together with peroxisomes, the endoplasmic reticulum, and lipid droplets, are necessary for macrophage lipid metabolism. Macrophages synthesize both antibacterial and antiinflammatory lipids such as omega-3 fatty acids, lipoxins, eicosanoids, platelet-activating factor, and palmitoleic acid [112–118]. These lipids can also serve as intercellular lipid mediators or can have autocrine effects on macrophages [113, 119, 120]. M1 and M2 macrophages have characteristic lipid mediators [121, 122], and, for example, omega 3 fatty acids and lipoxin A4 stimulate apoptotic cell uptake by neutrophil granulocytes and reduce NF-κB-mediated inflammatory cytokine expression in macrophages [113, 123]. Lipid signaling hence can impede M1 and induce M2 activation. Moreover, M2 activation can lead to increased production of antiinflammatory lipid derivatives [124], as well as the elimination of precursors of inflammatory lipid mediators [85]. For instance, macrophages can synthesize platelet-activating factor (PAF), a potent inflammatory mediator from various lipid precursors, including alkylglycerol-type ether lipids [110, 125]. However, M2 activation increases the expression of an alkylglycerol catabolizing enzyme, alkylglycerol monooxygenase (AGMO, or TMEM195) [85]. Eventually, AGMO converts the PAF precursors into fatty acids, thus diminishing inflammatory PAF synthesis [117, 126].

The tumor suppressor p53 protein also controls macrophage activation. Macrophage p53 suppresses M1 activation, and accordingly, p53 deficiency increases Th1 cytokine secretion [127], and the inhibition or loss of p53 inhibits M2 activation [128]. The lack of p53 also leads to the generation of immunogenic mitochondrial dsRNA, which induces type I interferon response [128]. The p38 mitogen-activated protein kinase pathway also induces M2 activation of tumor-associated macrophages [129]. V-Maf musculoaponeurotic fibrosarcoma oncogene homologue B (MafB), a basic leucine zipper transcription factor that is necessary for the regulation of hematopoiesis, also has a role in macrophage responses. MafB is expressed during macrophage development and maturation [130] and promotes M2 activation [131]. M1-activated macrophages initiate a specific cell death-signaling mechanism, called pyroptosis, which leads to inflammasome activation and release of inflammatory cytokines [132, 133]. STAT6 and MafB inhibit pyroptosis and macrophage apoptosis, respectively [7, 134]. However, this effect may lead to a sustained macrophage pool that is unfavorable, as in atherosclerotic plaque [134].

7 The Effect of Epigenetic Factors on M2 Activation

Additional mechanisms, such as epigenetic modifications, shape M2 activation: for example, microRNA (miRNA) species, which can control the fate of mRNA species encoding effectors of the M2 activation. The miRs are short noncoding RNA species that can hybridize with mRNA species and eventually decrease their stability or block their translation into proteins. The function of miRNAs is a posttranscriptional control of gene expression. For instance, M2 macrophages express miR-181a, an miRNA that suppresses KLF6 and C/EBPα protein expression and is thought to promote M2 activation [13]. Krüppel-like factor 6 (KLF6) expression is induced by M1 stimuli (LPS and IFNγ) and suppressed by IL-4 and IL-13: it is necessary for proper M1 polarization and cooperates with NF-κB, a key regulator of inflammatory cytokine expression [135]. C/EBPα, as a macrophage transcription factor, is necessary for both M1 and M2 activation [12, 14]. In this scenario, inhibition of M1 activation occurs, with probably a secondary effect on the expression of M2 genes. Several miRNAs have been described as activators of M1 polarization [136] or inhibitors of M2 activation [137]. *Arg1* has several binding sequences for miRNAs, including miR-340-5p, as shown in the TargetScan database [138]. A recent study showed that downregulation of miR-340-5p promotes M2 activation of tumor-associated macrophages [139]. Accordingly, upregulation of miR-340-5p inhibits tumor growth, probably by inhibiting M2 activation [139]. A rapidly increasing number of miRs either induce or inhibit M2 activation

[140]. For instance, miR-146a increases M2 gene transcription (*Arg1*, *Ccl17*, *Ccl22*, *Cd206*) [141], and miR-34a and miR-720 suppress M2 activation [137, 142]. Macrophages secrete specific miRs in response to M1 or M2 polarizing stimuli [143].

Tissue-resident macrophages have their own unique developmental program, which is shaped by tissue-specific signals, transcription factors, posttranscriptional histone modifications and DNA methylation, and the unique lipidomic profile of the macrophages [122, 144, 145]. These factors together determine tissue-resident macrophage identity and responsiveness to polarizing stimuli. As a result, there are considerable differences between the responsiveness of the distinct tissue-resident macrophages to pathogens and polarizing stimuli. Also, when a tissue-resident macrophage pool is separated from its tissue environment and cultured in vitro, various traits, such as expression level of enzymes, and polarization state, are gradually changed. The in vitro macrophage culture hence can lack some features of the bona fide tissue-resident macrophages.

8 Inflammatory Mediators and the Expression of M2 Genes

In the course of M1 activation, most M2 genes are transcriptionally suppressed [146]. The M1 activation is an acute response to pathogen challenge. However, an acute phase of inflammation involves mechanisms that limit excessive inflammatory cytokine burst and eventual tissue damage [147]. Chronic inflammation has been shown to increase M2-like activation of tumor-associated macrophages [129]. Inflammation-associated signals such as IL-6 and IL-8 promote M2 activation in gastric cancer [148]. Several inflammatory settings increase M2 macrophage content of the affected tissues [149–152], making plausible the scenario that pro-resolving and repair M2 macrophages are elicited by inflammatory signaling [129]. Hence, M1-activating stimuli also cause a certain degree of upregulation of M2 genes in macrophages; moreover, M1 activation has the potential to evoke epigenetic changes that affect M2 gene expression [144].

A complex network of feedback mechanisms interconnect the pathways of M2 and M1 macrophage activation. For instance, M1 macrophages express receptors for pro-resolving signals [153]. TLR signaling activates *Arg1* promoter in macrophages, via a MyD88-C/EBPβ-dependent, STAT6-independent signal pathway [154]. TGFβ and IL-6 have inflammatory and chemotactic effects; however, they also stimulate the expression of M2 genes. IL-6 increases IL-10 secretion by Th cells [155], induces CD206 expression and *Il4ra* transcription [156, 157], and augments IL-4 response of macrophages [158]. Of note, today we know that in mouse, $CD206^+$ macrophages may be either M1 or M2 macrophages [159–161], and CD206 expression alone is not a definite diagnosis of M2 activation state in the mouse. Nevertheless, CD206 is a widely used M2 marker, especially in human macrophages. Because IL-6 increases *Il4ra* expression, it is tempting to assume that it increases IL-4 sensitivity, and again, serves as a feedback signal mechanism between the M1 and M2 activation machinery. IL-10, which is an M2-polarizing cytokine, can also increase IL-6 expression in macrophages [162]; IL-6 increases C/EBPβ expression [15], and IL-6 also promotes M2 activation in tumors through the JAK2/STAT3 signaling pathway [148]. It is plausible that IL-6 mimics the effect of IL-10 by activating STAT3, which eventually synergizes with STAT6 if IL-4 is provided to the macrophages [158]. PPARγ, which is an M2-promoting NR, can have an inverse function in the tumor environment, and can curb M2 activation and T lymphocyte suppression by macrophages [163]. Moreover, STAT1/IRF3 signaling is an M1-polarizing mechanism; however, it can also elicit the M2 state in tumor-associated macrophages [164].

GM-CSF increases macrophage responsiveness to M1 stimuli [2], and GM-CSF activates IRF4 and increases chemokine (C-C motif) ligand 17 (CCL17) expression in macrophages. CCL17 is associated with M2 macrophages [165, 166]; however, it also mediates the proinflammatory and analgesic effect of GM-CSF in arthritis [167]. Also, there are CCL17-independent effects of IRF4 signaling in macrophages [168], and when IRF4 is activated in a Th2 immune context, it increases M2 gene expression [20, 169]. It is plausible that STAT6 signaling increases IRF4 expression and that IRF4 acts as a secondary transcription factor in the STAT6 pathway of M2 activation. It has also been shown that Jumonji domain-containing 3, a histone demethylase, is necessary for IRF4 expression in macrophages [20, 169]. IRF4 signaling is also involved in metabolic reprogramming during M2 activation [21].

Pathogens can skew macrophage activation toward M2 to avoid immune response; hence TLR signaling, IL-6, and many pathogen-derived products can induce M2 gene expression [156]. M1 and M2 gene transcription may be induced simultaneously, as exemplified by *Mycobacterium*-infected macrophages [34]. In an experimental setup one should however consider that even though transcription of M2 macrophage genes may increase in response to inflammatory mediators, these mediators are most likely inducing a change that is magnitudes higher in the transcription of M1-associated genes at the same time. Exceptions exist, such as helminth products and obligate intracellular parasites, which have evolved strategies to avoid M1 activation and favor the M2 state of macrophages (see in chapter "Immune Functions of the M2 Macrophages: Host Defense, Self-Tolerance, and Autoimmunity").

9 M2 Activation: A Lifelong Commitment?

It is an intriguing question whether M2 activation is a terminal differentiation state of a macrophage and might be a permanent condition in the life cycle of a macrophage. Macrophages are wandering cells; they take up foreign particles and apoptotic material, accumulate lipids, and produce a wide range of autocrine mediators, all of which have potential to change macrophage activation. Macrophages hence acquire their activation in response to the signals they encounter, which makes plausible that M2 macrophages can change their activation state. As we detail in chapter

"Mechanisms Which Control the Size of M2 Macrophage-Dominated Tissue Macrophage Niches", M2 macrophages may develop from naïve macrophages and monocytes, and it is possible for an M2 macrophage to switch off its activation state (i.e., deactivate STAT6 signaling), or adopt an M1 activation state (e.g., through TLR signaling). Do M1 macrophages polarize toward the M2 state? In this regard one should consider that M1 macrophages generate cytotoxic molecules from which they themselves are largely unprotected. Indeed, apoptosis of pathogen-engulfing macrophages serves as a signal for the activation of other macrophages. In this scenario one may think that an M1 macrophage has severe cellular damage while fighting pathogens [4, 170], and one questions whether it may function further properly as an M2 macrophage. Moreover, M1 macrophage can undergo programmed necrosis as part of their activation program [133]. Instead of an M1–M2 repolarization of an individual macrophage, it is more plausible that in tissue environment M1 macrophages are replaced by M2 macrophages, and as a result, the macrophage pool, rather than the single macrophages, convert into an M2-dominated population. This possibility is discussed in chapter "Mechanisms Which Control the Size of M2 Macrophage-Dominated Tissue Macrophage Niches".

References

1. Ginhoux F, Schultze JL, Murray PJ, Ochando J, Biswas SK. New insights into the multidimensional concept of macrophage ontogeny, activation and function. Nat Immunol. 2015;17:34.
2. Hamilton TA, Zhao C, Pavicic PG Jr, Datta S. Myeloid colony-stimulating factors as regulators of macrophage polarization. Front Immunol. 2014;5:554.
3. Svensson J, Jenmalm MC, Matussek A, Geffers R, Berg G, Ernerudh J. Macrophages at the fetal–maternal interface express markers of alternative activation and are induced by M-CSF and IL-10. J Immunol. 2011;187:3671–82.
4. Röszer T. Understanding the mysterious M2 macrophage through activation markers and effector mechanisms. Mediat Inflamm. 2015;2015:16.
5. Lu C, Kumar PA, Sun J, Aggarwal A, Fan Y, Sperling MA, Lumeng CN, Menon RK. Targeted deletion of growth hormone (GH) receptor in macrophage reveals novel osteopontin-mediated effects of GH on glucose homeostasis and insulin sensitivity in diet-induced obesity. J Biol Chem. 2013;288:15725–35.
6. Wick KR, Berton MT. IL-4 induces serine phosphorylation of the STAT6 transactivation domain in B lymphocytes. Mol Immunol. 2000;37:641–52.
7. Czimmerer Z, Daniel B, Horvath A, Rückerl D, Nagy G, Kiss M, Peloquin M, Budai MM, Cuaranta-Monroy I, Simandi Z, Steiner L, Nagy B Jr, Poliska S, Banko C, Bacso Z, Schulman IG, Sauer S, Deleuze J-F, Allen JE, Benko S, Nagy L. The transcription factor STAT6 mediates direct repression of inflammatory enhancers and limits activation of alternatively polarized macrophages. Immunity. 2018;48:75–90.e76.
8. Wang Y, Malabarba MG, Nagy ZS, Kirken RA. Interleukin 4 regulates phosphorylation of serine 756 in the transactivation domain of Stat6: roles for multiple phosphorylation sites and Stat6 function. J Biol Chem. 2004;279:25196–203.
9. Röszer T. Understanding the mysterious M2 macrophage through activation markers and effector mechanisms. Mediat Inflamm. 2015;2015:816460.

10. Gray MJ, Poljakovic M, Kepka-Lenhart D, Morris SM. Induction of arginase I transcription by IL-4 requires a composite DNA response element for STAT6 and C/EBPβ. Gene (Amst). 2005;353:98–106.
11. Szanto A, Balint BL, Nagy ZS, Barta E, Dezso B, Pap A, Szeles L, Poliska S, Oros M, Evans RM, Barak Y, Schwabe J, Nagy L. STAT6 transcription factor is a facilitator of the nuclear receptor PPARγ-regulated gene expression in macrophages and dendritic cells. Immunity. 2010;33:699–712.
12. Ruffell D, Mourkioti F, Gambardella A, Kirstetter P, Lopez RG, Rosenthal N, Nerlov C. A CREB-C/EBPbeta cascade induces M2 macrophage-specific gene expression and promotes muscle injury repair. Proc Natl Acad Sci USA. 2009;106:17475–80.
13. Bi J, Zeng X, Zhao L, Wei Q, Yu L, Wang X, Yu Z, Cao Y, Shan F, Wei M. miR-181a induces macrophage polarized to M2 phenotype and promotes M2 macrophage-mediated tumor cell metastasis by targeting KLF6 and C/EBPα. Mol Ther Nucleic Acids. 2016;5:e368.
14. Lee B, Qiao L, Lu M, Yoo HS, Cheung W, Mak R, Schaack J, Feng G-S, Chi N-W, Olefsky JM, Shao J. C/EBPα regulates macrophage activation and systemic metabolism. Am J Physiol Endocrinol Metab. 2014;306:E1144–54.
15. Huber R, Pietsch D, Panterodt T, Brand K. Regulation of C/EBPβ and resulting functions in cells of the monocytic lineage. Cell Signal. 2012;24:1287–96.
16. Lefevre L, Gales A, Olagnier D, Bernad J, Perez L, Burcelin R, Valentin A, Auwerx J, Pipy B, Coste A. PPARgamma ligands switched high fat diet-induced macrophage M2b polarization toward M2a thereby improving intestinal *Candida* elimination. PLoS One. 2010;5:e12828.
17. Bouhlel MA, Derudas B, Rigamonti E, Dièvart R, Brozek J, Haulon S, Zawadzki C, Jude B, Torpier G, Marx N, Staels B, Chinetti-Gbaguidi G. PPARγ activation primes human monocytes into alternative M2 macrophages with anti-inflammatory properties. Cell Metab. 2007;6:137–43.
18. Liao X, Sharma N, Kapadia F, Zhou G, Lu Y, Hong H, Paruchuri K, Mahabeleshwar GH, Dalmas E, Venteclef N, Flask CA, Kim J, Doreian BW, Lu KQ, Kaestner KH, Hamik A, Clément K, Jain MK. Krüppel-like factor 4 regulates macrophage polarization. J Clin Invest. 2011;121:2736–49.
19. Kapoor N, Niu J, Saad Y, Kumar S, Sirakova T, Becerra E, Li X, Kolattukudy PE. Transcription factors STAT6 and KLF4 implement macrophage polarization via the dual catalytic powers of MCPIP. J Immunol (Baltim) (1950). 2015;194:6011–23.
20. El Chartouni C, Schwarzfischer L, Rehli M. Interleukin-4 induced interferon regulatory factor (Irf) 4 participates in the regulation of alternative macrophage priming. Immunobiology. 2010;215:821–5.
21. Huang SC-C, Smith AM, Everts B, Colonna M, Pearce EL, Schilling JD, Pearce EJ. Metabolic reprogramming mediated by the mTORC2-IRF4 signaling axis is essential for macrophage alternative activation. Immunity. 2016;45:817–30.
22. Waqas SFH, Hoang AC, Lin YT, Ampem G, Azegrouz H, Balogh L, Thuroczy J, Chen JC, Gerling IC, Nam S, Lim JS, Martinez-Ibanez J, Real JT, Paschke S, Quillet R, Ayachi S, Simonin F, Schneider EM, Brinkman JA, Lamming DW, Seroogy CM, Röszer T. Neuropeptide FF increases M2 activation and self-renewal of adipose tissue macrophages. J Clin Invest. 2017;127:2842–54.
23. Hanson EM, Dickensheets H, Qu C-K, Donnelly RP, Keegan AD. Regulation of the dephosphorylation of Stat6: participation of Tyr-713 in the interleukin-4 receptor α, the tyrosine phosphatase shp-1, and the proteasome. J Biol Chem. 2003;278:3903–11.
24. Malefyt RDW, Moore WK. Interleukin-10. In: Thomson A, editor. The cytokine handbook. San Diego: Academic Press; 1998. p. 333–64.
25. de Waal Malefyt R, Hans Y, Roncarolo M-G, Spits H, de Vries JE. Interleukin-10. Curr Opin Immunol. 1992;4:314–20.
26. Schreiber T, Ehlers S, Heitmann L, Rausch A, Mages J, Murray PJ, Lang R, Hölscher C. Autocrine IL-10 induces hallmarks of alternative activation in macrophages and sup-

presses antituberculosis effector mechanisms without compromising T cell immunity. J Immunol (Baltim) (1950). 2009;183:1301–12.

27. Makita N, Hizukuri Y, Yamashiro K, Murakawa M, Hayashi Y. IL-10 enhances the phenotype of M2 macrophages induced by IL-4 and confers the ability to increase eosinophil migration. Int Immunol. 2014;27:131–41.
28. Lopes RL, Borges TJ, Zanin RF, Bonorino C. IL-10 is required for polarization of macrophages to M2-like phenotype by mycobacterial DnaK (heat shock protein 70). Cytokine. 2016;85:123–9.
29. Kahnert A, Seiler P, Stein M, Bandermann S, Hahnke K, Mollenkopf H, Kaufmann SHE. Alternative activation deprives macrophages of a coordinated defense program to *Mycobacterium tuberculosis*. Eur J Immunol. 2006;36:631–47.
30. Banks DA, Ahlbrand SE, Keith Hughitt V, Shah S, Vogel SN, El-Sayed NM, Briken V. *Mycobacterium tuberculosis* inhibits autocrine type I interferon signaling to increase intracellular survival. bioRxiv. 2018; https://doi.org/10.1101/425116.
31. Chunfa L, Xin S, Qiang L, Sreevatsan S, Yang L, Zhao D, Zhou X. The central role of IFI204 in IFN-β release and autophagy activation during *Mycobacterium bovis* infection. Front Cell Infect Microbiol. 2017;7:169.
32. Manzanillo PS, Shiloh MU, Portnoy DA, Cox JS. *Mycobacterium tuberculosis* activates the DNA-dependent cytosolic surveillance pathway within macrophages. Cell Host Microbe. 2012;11:469–80.
33. Andreu N, Phelan J, de Sessions PF, Cliff JM, Clark TG, Hibberd ML. Primary macrophages and J774 cells respond differently to infection with *Mycobacterium tuberculosis*. Sci Rep. 2017;7:42225.
34. Roy S, Schmeier S, Kaczkowski B, Arner E, Alam T, Ozturk M, Tamgue O, Parihar SP, Kawaji H, Itoh M, Lassmann T, Carninci P, Hayashizaki Y, Forrest ARR, Guler R, Bajic VB, Brombacher F, Suzuki H. Transcriptional landscape of *Mycobacterium tuberculosis* infection in macrophages. Sci Rep. 2018;8:6758.
35. Rivollier A, He J, Kole A, Valatas V, Kelsall BL. Inflammation switches the differentiation program of Ly6Chi monocytes from antiinflammatory macrophages to inflammatory dendritic cells in the colon. J Exp Med. 2012;209:139–55.
36. Simon JM, Davis JP, Lee SE, Schaner MR, Gipson GR, Weiser M, Sartor RB, Herfarth HH, Rahbar R, Sadiq TS, Koruda MJ, McGovern DP, Lieb JD, Mohlke KL, Furey TS, Sheikh SZ. Alterations to chromatin in intestinal macrophages link IL-10 deficiency to inappropriate inflammatory responses. Eur J Immunol. 2016;46:1912–25.
37. Ueda Y, Kayama H, Jeon SG, Kusu T, Isaka Y, Rakugi H, Yamamoto M, Takeda K. Commensal microbiota induce LPS hyporesponsiveness in colonic macrophages via the production of IL-10. Int Immunol. 2010;22:953–62.
38. Svensson-Arvelund J, Mehta RB, Lindau R, Mirrasekhian E, Rodriguez-Martinez H, Berg G, Lash GE, Jenmalm MC, Ernerudh J. The human fetal placenta promotes tolerance against the semiallogeneic fetus by inducing regulatory T cells and homeostatic M2 macrophages. J Immunol. 2015;194:1534–44.
39. Liu Z, Kuang W, Zhou Q, Zhang Y. TGF-β1 secreted by M2 phenotype macrophages enhances the stemness and migration of glioma cells via the SMAD2/3 signalling pathway. Int J Mol Med. 2018;42:3395–403.
40. Spittau B, Wullkopf L, Zhou X, Rilka J, Pfeifer D, Krieglstein K. Endogenous transforming growth factor-beta promotes quiescence of primary microglia in vitro. Glia. 2013;61:287–300.
41. Dupasquier M, Stoitzner P, Wan H, Cerqueira D, van Oudenaren A, Voerman JS, Denda-Nagai K, Irimura T, Raes G, Romani N, Leenen PJ. The dermal microenvironment induces the expression of the alternative activation marker CD301/mMGL in mononuclear phagocytes, independent of IL-4/IL-13 signaling. J Leukoc Biol. 2006;80:838–49.
42. Zhang F, Wang H, Wang X, Jiang G, Liu H, Zhang G, Wang H, Fang R, Bu X, Cai S, Du J. TGF-β induces M2-like macrophage polarization via SNAIL-mediated suppression of a pro-inflammatory phenotype. Oncotarget. 2016;7:52294–306.

43. Liu F, Qiu H, Xue M, Zhang S, Zhang X, Xu J, Chen J, Yang Y, Xie J. MSC-secreted TGF-β regulates lipopolysaccharide-stimulated macrophage M2-like polarization via the Akt/FoxO1 pathway. Stem Cell Res Ther. 2019;10:345.
44. Röszer T. Transcriptional control of apoptotic cell clearance by macrophage nuclear receptors. Apoptosis. 2017;22:284–94.
45. Zizzo G, Hilliard BA, Monestier M, Cohen PL. Efficient clearance of early apoptotic cells by human macrophages requires M2c polarization and MerTK induction. J Immunol. 2012;189:3508–20.
46. Bohlson SS, O'Conner SD, Hulsebus HJ, Ho M-M, Fraser DA. Complement, C1q, and C1q-related molecules regulate macrophage polarization. Front Immunol. 2014;5:402.
47. Vasina EM, Cauwenberghs S, Feijge MA, Heemskerk JW, Weber C, Koenen RR. Microparticles from apoptotic platelets promote resident macrophage differentiation. Cell Death Dis. 2011;2:e211.
48. Röszer T, Menendez-Gutierrez MP, Cedenilla M, Ricote M. Retinoid X receptors in macrophage biology. Trends Endocrinol Metab. 2013;24(9):460–8.
49. Rigamonti E, Chinetti-Gbaguidi G, Staels B. Regulation of macrophage functions by PPAR-alpha, PPAR-gamma, and LXRs in mice and men. Arterioscler Thromb Vasc Biol. 2008;28:1050–9.
50. Gonzalez NA, Bensinger SJ, Hong C, Beceiro S, Bradley MN, Zelcer N, Deniz J, Ramirez C, Diaz M, Gallardo G, de Galarreta CR, Salazar J, Lopez F, Edwards P, Parks J, Andujar M, Tontonoz P, Castrillo A. Apoptotic cells promote their own clearance and immune tolerance through activation of the nuclear receptor LXR. Immunity. 2009;31:245–58.
51. Mukundan L, Odegaard JI, Morel CR, Heredia JE, Mwangi JW, Ricardo-Gonzalez RR, Goh YP, Eagle AR, Dunn SE, Awakuni JU, Nguyen KD, Steinman L, Michie SA, Chawla A. PPAR-delta senses and orchestrates clearance of apoptotic cells to promote tolerance. Nat Med. 2009;15:1266–72.
52. Röszer T, Menendez-Gutierrez MP, Lefterova MI, Alameda D, Nunez V, Lazar MA, Fischer T, Ricote M. Autoimmune kidney disease and impaired engulfment of apoptotic cells in mice with macrophage peroxisome proliferator-activated receptor gamma or retinoid X receptor alpha deficiency. J Immunol. 2011;186:621–31.
53. Menendez-Gutierrez MP, Röszer T, Ricote M. Biology and therapeutic applications of peroxisome proliferator- activated receptors. Curr Top Med Chem. 2012;12:548–84.
54. Yoon YS, Kim SY, Kim MJ, Lim JH, Cho MS, Kang JL. PPARgamma activation following apoptotic cell instillation promotes resolution of lung inflammation and fibrosis via regulation of efferocytosis and proresolving cytokines. Mucosal Immunol. 2015;8:1031–46.
55. Zizzo G, Cohen PL. The PPAR-gamma antagonist GW9662 elicits differentiation of M2c-like cells and upregulation of the MerTK/Gas6 axis: a key role for PPAR-gamma in human macrophage polarization. J Inflamm (Lond). 2015;12:36.
56. Fiorucci S, Biagioli M, Zampella A, Distrutti E. Bile acids activated receptors regulate innate immunity. Front Immunol. 2018;9:1853.
57. Penberthy KK, Ravichandran KS. Apoptotic cell recognition receptors and scavenger receptors. Immunol Rev. 2016;269:44–59.
58. Silverstein RL, Febbraio M. CD36, a scavenger receptor involved in immunity, metabolism, angiogenesis, and behavior. Sci Signal. 2009;2:re3.
59. Patel SN, Serghides L, Smith TG, Febbraio M, Silverstein RL, Kurtz TW, Pravenec M, Kain KC. CD36 mediates the phagocytosis of *Plasmodium falciparum*-infected erythrocytes by rodent macrophages. J Infect Dis. 2004;189:204–13.
60. Smith TG, Serghides L, Patel SN, Febbraio M, Silverstein RL, Kain KC. CD36-mediated nonopsonic phagocytosis of erythrocytes infected with stage I and IIA gametocytes of *Plasmodium falciparum*. Infect Immun. 2003;71:393–400.
61. Aronoff DM, Serezani CH, Carstens JK, Marshall T, Gangireddy SR, Peters-Golden M, Reddy RC. Stimulatory effects of peroxisome proliferator-activated receptor-gamma on Fcgamma receptor-mediated phagocytosis by alveolar macrophages. PPAR Res. 2007;2007:52546.

62. Berry A, Balard P, Coste A, Olagnier D, Lagane C, Authier H, Benoit-Vical F, Lepert JC, Seguela JP, Magnaval JF, Chambon P, Metzger D, Desvergne B, Wahli W, Auwerx J, Pipy B. IL-13 induces expression of CD36 in human monocytes through PPARgamma activation. Eur J Immunol. 2007;37:1642–52.
63. Majai G, Sarang Z, Csomos K, Zahuczky G, Fesus L. PPARgamma-dependent regulation of human macrophages in phagocytosis of apoptotic cells. Eur J Immunol. 2007;37:1343–54.
64. Dransfield I, Zagorska A, Lew ED, Michail K, Lemke G. Mer receptor tyrosine kinase mediates both tethering and phagocytosis of apoptotic cells. Cell Death Dis. 2015;6:e1646.
65. Epelman S, Lavine KJ, Beaudin AE, Sojka DK, Carrero JA, Calderon B, Brija T, Gautier EL, Ivanov S, Satpathy AT, Schilling JD, Schwendener R, Sergin I, Razani B, Forsberg EC, Yokoyama WM, Unanue ER, Colonna M, Randolph GJ, Mann DL. Embryonic and adult-derived resident cardiac macrophages are maintained through distinct mechanisms at steady state and during inflammation. Immunity. 2014;40:91–104.
66. Tang Y, Wu S, Liu Q, Xie J, Zhang J, Han D, Lu Q, Lu Q. Mertk deficiency affects macrophage directional migration via disruption of cytoskeletal organization. PLoS One. 2015;10:e0117787.
67. Behrens EM, Gadue P, Gong SY, Garrett S, Stein PL, Cohen PL. The mer receptor tyrosine kinase: expression and function suggest a role in innate immunity. Eur J Immunol. 2003;33:2160–7.
68. Menendez-Gutierrez MP, Röszer T, Fuentes L, Nunez V, Escolano A, Redondo JM, De Clerck N, Metzger D, Valledor AF, Ricote M. Retinoid X receptors orchestrate osteoclast differentiation and postnatal bone remodeling. J Clin Invest. 2015;125(2):809–23.
69. Garabuczi E, Sarang Z, Szondy Z. Glucocorticoids enhance prolonged clearance of apoptotic cells by upregulating liver X receptor, peroxisome proliferator-activated receptor-delta and UCP2. Biochim Biophys Acta. 2015;1853:573–82.
70. Choi JY, Seo JY, Yoon YS, Lee YJ, Kim HS, Kang JL. Mer signaling increases the abundance of the transcription factor LXR to promote the resolution of acute sterile inflammation. Sci Signal. 2015;8:ra21.
71. Scott RS, McMahon EJ, Pop SM, Reap EA, Caricchio R, Cohen PL, Earp HS, Matsushima GK. Phagocytosis and clearance of apoptotic cells is mediated by MER. Nature. 2001;411:207–11.
72. Happonen KE, Tran S, Morgelin M, Prince R, Calzavarini S, Angelillo-Scherrer A, Dahlback B. The Gas6-Axl protein interaction mediates endothelial uptake of platelet microparticles. J Biol Chem. 2016;291:10586–601.
73. Savage JC, Jay T, Goduni E, Quigley C, Mariani MM, Malm T, Ransohoff RM, Lamb BT, Landreth GE. Nuclear receptors license phagocytosis by trem2+ myeloid cells in mouse models of Alzheimer's disease. J Neurosci. 2015;35:6532–43.
74. Trouw LA, Blom AM, Gasque P. Role of complement and complement regulators in the removal of apoptotic cells. Mol Immunol. 2008;45:1199–207.
75. Vandivier RW, Ogden CA, Fadok VA, Hoffmann PR, Brown KK, Botto M, Walport MJ, Fisher JH, Henson PM, Greene KE. Role of surfactant proteins A, D, and C1q in the clearance of apoptotic cells in vivo and in vitro: calreticulin and CD91 as a common collectin receptor complex. J Immunol. 2002;169:3978–86.
76. Toth B, Garabuczi E, Sarang Z, Vereb G, Vamosi G, Aeschlimann D, Blasko B, Becsi B, Erdodi F, Lacy-Hulbert A, Zhang A, Falasca L, Birge RB, Balajthy Z, Melino G, Fesus L, Szondy Z. Transglutaminase 2 is needed for the formation of an efficient phagocyte portal in macrophages engulfing apoptotic cells. J Immunol. 2009;182:2084–92.
77. Kawabe K, Takano K, Moriyama M, Nakamura Y. Lipopolysaccharide-stimulated transglutaminase 2 expression enhances endocytosis activity in the mouse microglial cell line BV-2. Neuroimmunomodulation. 2015;22:243–9.
78. Nadella V, Wang Z, Johnson TS, Griffin M, Devitt A. Transglutaminase 2 interacts with syndecan-4 and CD44 at the surface of human macrophages to promote removal of apoptotic cells. Biochim Biophys Acta. 2015;1853:201–12.

79. Sandor K, Pallai A, Duro E, Legendre P, Couillin I, Saghy T, Szondy Z. Adenosine produced from adenine nucleotides through an interaction between apoptotic cells and engulfing macrophages contributes to the appearance of transglutaminase 2 in dying thymocytes. Amino Acids. 2016;49:671–81.
80. Garabuczi E, Kiss B, Felszeghy S, Tsay GJ, Fesus L, Szondy Z. Retinoids produced by macrophages engulfing apoptotic cells contribute to the appearance of transglutaminase 2 in apoptotic thymocytes. Amino Acids. 2013;44:235–44.
81. Sarang Z, Joos G, Garabuczi E, Ruhl R, Gregory CD, Szondy Z. Macrophages engulfing apoptotic cells produce nonclassical retinoids to enhance their phagocytic capacity. J Immunol. 2014;192:5730–8.
82. Sarang Z, Garabuczi E, Joos G, Kiss B, Toth K, Ruhl R, Szondy Z. Macrophages engulfing apoptotic thymocytes produce retinoids to promote selection, differentiation, removal and replacement of double positive thymocytes. Immunobiology. 2013;218:1354–60.
83. Rebe C, Raveneau M, Chevriaux A, Lakomy D, Sberna AL, Costa A, Bessede G, Athias A, Steinmetz E, Lobaccaro JM, Alves G, Menicacci A, Vachenc S, Solary E, Gambert P, Masson D. Induction of transglutaminase 2 by a liver X receptor/retinoic acid receptor alpha pathway increases the clearance of apoptotic cells by human macrophages. Circ Res. 2009;105:393–401.
84. Delgado M, Leceta J, Sun W, Gomariz RP, Ganea D. VIP and PACAP induce shift to a Th2 response by upregulating B7.2 expression. Ann N Y Acad Sci. 2000;921:68–78.
85. Waqas SFH, Hoang A, Lin Y, Ampem G, Azegrouz H, Balogh L, Thuróczy J, Chen JC, Gerling IC, Nam S, Lim JS, Martinez-Ibañez J, Real JT, Paschke S, Quillet R, Ayachi S, Simonin F, Schneider EM, Brinkman JA, Lamming DW, Seroogy CM, Röszer T. Neuropeptide FF increases M2 activation and self-renewal of adipose tissue macrophages. J Clin Invest. 2017;127:2842–54.
86. Tanabe K, Kozawa O, Iida H. cAMP/PKA enhances interleukin-1β-induced interleukin-6 synthesis through STAT3 in glial cells. Cell Signal. 2016;28:19–24.
87. Wei LH, Jacobs AT, Morris SM, Ignarro LJ. IL-4 and IL-13 upregulate arginase I expression by cAMP and JAK/STAT6 pathways in vascular smooth muscle cells. Am J Phys Cell Phys. 2000;279:C248–56.
88. Sanson M, Distel E, Fisher EA. HDL induces the expression of the M2 macrophage markers arginase 1 and Fizz-1 in a STAT6-dependent process. PLoS One. 2013;8:e74676.
89. Star RA, Rajora N, Huang J, Stock RC, Catania A, Lipton JM. Evidence of autocrine modulation of macrophage nitric oxide synthase by alpha-melanocyte-stimulating hormone. Proc Natl Acad Sci USA. 1995;92:8016–20.
90. Tsuchida Y, Hatao F, Fujisawa M, Murata T, Kaminishi M, Seto Y, Hori M, Ozaki H. Neuronal stimulation with 5-hydroxytryptamine 4 receptor induces anti-inflammatory actions via alpha7nACh receptors on muscularis macrophages associated with postoperative ileus. Gut. 2011;60:638–47.
91. Nemethova A, Michel K, Gomez-Pinilla PJ, Boeckxstaens GE, Schemann M. Nicotine attenuates activation of tissue resident macrophages in the mouse stomach through the beta2 nicotinic acetylcholine receptor. PLoS One. 2013;8:e79264.
92. Csoka B, Selmeczy Z, Koscso B, Nemeth ZH, Pacher P, Murray PJ, Kepka-Lenhart D, Morris SM Jr, Gause WC, Leibovich SJ, Hasko G. Adenosine promotes alternative macrophage activation via A2A and A2B receptors. FASEB J. 2012;26:376–86.
93. Lee JK, Chung J, Kannarkat GT, Tansey MG. Critical role of regulator G-protein signaling 10 (RGS10) in modulating macrophage M1/M2 activation. PLoS One. 2013;8:e81785.
94. Ganea D, Delgado M. Inhibitory neuropeptide receptors on macrophages. Microbes Infect. 2001;3:141–7.
95. Bok E, Chung YC, Kim K-S, Baik HH, Shin W-H, Jin BK. Modulation of M1/M2 polarization by capsaicin contributes to the survival of dopaminergic neurons in the lipopolysaccharide-lesioned substantia nigra in vivo. Exp Mol Med. 2018;50:76.

96. Grailer JJ, Haggadone MD, Sarma JV, Zetoune FS, Ward PA. Induction of M2 regulatory macrophages through the β2-adrenergic receptor with protection during endotoxemia and acute lung injury. J Innate Immun. 2014;6:607–18.
97. Boomershine CS, Lafuse WP, Zwilling BS. Beta2-adrenergic receptor stimulation inhibits nitric oxide generation by *Mycobacterium avium* infected macrophages. J Neuroimmunol. 1999;101:68–75.
98. Pirzgalska RM, Seixas E, Seidman JS, Link VM, Sánchez NM, Mahú I, Mendes R, Gres V, Kubasova N, Morris I, Arús BA, Larabee CM, Vasques M, Tortosa F, Sousa AL, Anandan S, Tranfield E, Hahn MK, Iannacone M, Spann NJ, Glass CK, Domingos AI. Sympathetic neuron–associated macrophages contribute to obesity by importing and metabolizing norepinephrine. Nat Med. 2017;23:1309.
99. Blum A, Setiawan T, Hang L, Stoyanoff K, Weinstock JV. Interleukin-12 (IL-12) and IL-23 induction of substance p synthesis in murine T cells and macrophages is subject to IL-10 and transforming growth factor beta regulation. Infect Immun. 2008;76:3651–6.
100. Odegaard JI, Ricardo-Gonzalez RR, Red Eagle A, Vats D, Morel CR, Goforth MH, Subramanian V, Mukundan L, Ferrante AW, Chawla A. Alternative M2 activation of Kupffer cells by PPARdelta ameliorates obesity-induced insulin resistance. Cell Metab. 2008;7:496–507.
101. Sakaguchi S, Shono J, Suzuki T, Sawano S, Anderson JE, Do MK, Ohtsubo H, Mizunoya W, Sato Y, Nakamura M, Furuse M, Yamada K, Ikeuchi Y, Tatsumi R. Implication of anti-inflammatory macrophages in regenerative moto-neuritogenesis: promotion of myoblast migration and neural chemorepellent semaphorin 3A expression in injured muscle. Int J Biochem Cell Biol. 2014;54:272–85.
102. Singer K, Morris DL, Oatmen KE, Wang T, DelProposto J, Mergian T, Cho KW, Lumeng CN. Neuropeptide Y is produced by adipose tissue macrophages and regulates obesity-induced inflammation. PLoS One. 2013;8:e57929.
103. Angajala A, Lim S, Phillips JB, Kim J-H, Yates C, You Z, Tan M. Diverse roles of mitochondria in immune responses: novel insights into Immuno-metabolism. Front Immunol. 2018;9:1605.
104. Davies LC, Rice CM, Palmieri EM, Taylor PR, Kuhns DB, McVicar DW. Peritoneal tissue-resident macrophages are metabolically poised to engage microbes using tissue-niche fuels. Nat Commun. 2017;8:2074.
105. Ramond E, Jamet A, Coureuil M, Charbit A. Pivotal role of mitochondria in macrophage response to bacterial pathogens. Front Immunol. 2019;10:2461.
106. Thapa B, Lee K. Metabolic influence on macrophage polarization and pathogenesis. BMB Rep. 2019;52:360–72.
107. Sato S, Furuya N. Induction of PINK1/parkin-mediated mitophagy. In: Hattori N, Saiki S, editors. Mitophagy: methods and protocols. New York: Springer; 2018. p. 9–17.
108. Harris J, Deen N, Zamani S, Hasnat MA. Mitophagy and the release of inflammatory cytokines. Mitochondrion. 2018;41:2–8.
109. Minton K. Anti-inflammatory effect of mitophagy. Nat Rev Immunol. 2016;16:206.
110. Yu H, Dilbaz S, Coßmann J, Hoang AC, Diedrich V, Herwig A, Harauma A, Hoshi Y, Moriguchi T, Landgraf K, Körner A, Lucas C, Brodesser S, Balogh L, Thuróczy J, Karemore G, Kuefner MS, Park EA, Rapp C, Travers JB, Röszer T. Breast milk alkylglycerols sustain beige adipocytes through adipose tissue macrophages. J Clin Invest. 2019;129:2485–99.
111. Bhatia D, Chung K-P, Nakahira K, Patino E, Rice MC, Torres LK, Muthukumar T, Choi AM, Akchurin OM, Choi ME. Mitophagy-dependent macrophage reprogramming protects against kidney fibrosis. JCI Insight. 2019;4:e132826.
112. Kim SJ. Formation of lipoxins by alveolar macrophages. Biochem Biophys Res Commun. 1988;150:870–6.
113. Li P, Spann NJ, Kaikkonen MU, Lu M, Da Young OH, Fox JN, Bandyopadhyay G, Talukdar S, Xu J, Lagakos WS, Patsouris D, Armando A, Quehenberger O, Dennis EA, Watkins SM,

Auwerx J, Glass CK, Olefsky JM. NCoR repression of LXRs restricts macrophage biosynthesis of insulin-sensitizing omega 3 fatty acids. Cell. 2013;155:200–14.
114. Kim SJ. Elevated formation of lipoxins in viral antibody-positive rat alveolar macrophages. Am J Respir Cell Mol Biol. 1990;3:113–8.
115. Moreno C, Prieto P, Macias A, Pimentel-Santillana M, de la Cruz A, Traves PG, Bosca L, Valenzuela C. Modulation of voltage-dependent and inward rectifier potassium channels by 15-epi-lipoxin-A4 in activated murine macrophages: implications in innate immunity. J Immunol. 2013;191:6136–46.
116. Borgeson E, McGillicuddy FC, Harford KA, Corrigan N, Higgins DF, Maderna P, Roche HM, Godson C. Lipoxin A4 attenuates adipose inflammation. FASEB J. 2012;26:4287–94.
117. Tokuoka SM, Kita Y, Shindou H, Shimizu T. Alkylglycerol monooxygenase as a potential modulator for PAF synthesis in macrophages. Biochem Biophys Res Commun. 2013;436:306–12.
118. Reyes-Quiroz ME, Alba G, Santa-Maria C, Saenz J, Geniz I, Jimenez J, Ramirez R, Martin-Nieto J, Pintado E, Sobrino F. Platelet-activating factor downregulates the expression of liver X receptor-alpha and its target genes in human neutrophils. FEBS J. 2014;281:970–82.
119. Godson C, Mitchell S, Harvey K, Petasis NA, Hogg N, Brady HR. Cutting edge: lipoxins rapidly stimulate nonphlogistic phagocytosis of apoptotic neutrophils by monocyte-derived macrophages. J Immunol. 2000;164:1663–7.
120. Filgueiras LR, Koga MM, Quaresma PG, Ishizuka EK, Montes MB, Prada PO, Saad MJ, Jancar S, Rios FJ. PAFR in adipose tissue macrophages is associated with anti-inflammatory phenotype and metabolic homoeostasis. Clin Sci. 2016;130:601–12.
121. Dalli J, Serhan CN. Specific lipid mediator signatures of human phagocytes: microparticles stimulate macrophage efferocytosis and pro-resolving mediators. Blood. 2012;120:e60–72.
122. Norris PC, Dennis EA. A lipidomic perspective on inflammatory macrophage eicosanoid signaling. Adv Biol Regul. 2014;54:99–110.
123. Huang YH, Wang HM, Cai ZY, Xu FY, Zhou XY. Lipoxin A4 inhibits NF-kappaB activation and cell cycle progression in RAW264.7 cells. Inflammation. 2014;37:1084–90.
124. Nassar GM, Morrow JD, Roberts LJ 2nd, Lakkis FG, Badr KF. Induction of 15-lipoxygenase by interleukin-13 in human blood monocytes. J Biol Chem. 1994;269:27631–4.
125. Watschinger K, Werner ER. Orphan enzymes in ether lipid metabolism. Biochimie. 2013;(Paris) 95:59–65.
126. Watschinger K, Keller MA, McNeill E, Alam MT, Lai S, Sailer S, Rauch V, Patel J, Hermetter A, Golderer G, Geley S, Werner-Felmayer G, Plumb RS, Astarita G, Ralser M, Channon KM, Werner ER. Tetrahydrobiopterin and alkylglycerol monooxygenase substantially alter the murine macrophage lipidome. Proc Natl Acad Sci USA. 2015;112:2431–6.
127. Komarova EA, Krivokrysenko V, Wang K, Neznanov N, Chernov MV, Komarov PG, Brennan M-L, Golovkina TV, Rokhlin OW, Kuprash DV, Nedospasov SA, Hazen SL, Feinstein E, Gudkov AV. p53 is a suppressor of inflammatory response in mice. FASEB J. 2005;19:1030–2.
128. Mukhopadhyay S, Antalis TM, Nguyen KP, Hoofnagle MH, Sarkar R. Myeloid p53 regulates macrophage polarization and venous thrombus resolution by inflammatory vascular remodeling in mice. Blood. 2017;129:3245–55.
129. Suarez-Lopez L, Sriram G, Kong YW, Morandell S, Merrick KA, Hernandez Y, Haigis KM, Yaffe MB. MK2 contributes to tumor progression by promoting M2 macrophage polarization and tumor angiogenesis. Proc Natl Acad Sci USA. 2018;115:E4236–44.
130. Moriguchi T, Hamada M, Morito N, Terunuma T, Hasegawa K, Zhang C, Yokomizo T, Esaki R, Kuroda E, Yoh K, Kudo T, Nagata M, Greaves DR, Engel JD, Yamamoto M, Takahashi S. MafB is essential for renal development and F4/80 expression in macrophages. Mol Cell Biol. 2006;26:5715–27.
131. Kim H. The transcription factor MafB promotes anti-inflammatory M2 polarization and cholesterol efflux in macrophages. Sci Rep. 2017;7:7591.
132. Bergsbaken T, Fink SL, Cookson BT. Pyroptosis: host cell death and inflammation. Nat Rev Microbiol. 2009;7:99–109.

133. Robinson N, Ganesan R, Hegedűs C, Kovács K, Kufer TA, Virág L. Programmed necrotic cell death of macrophages: focus on pyroptosis, necroptosis, and parthanatos. Redox Biol. 2019;26:101239.
134. Hamada M, Nakamura M, Tran MT, Moriguchi T, Hong C, Ohsumi T, Dinh TT, Kusakabe M, Hattori M, Katsumata T, Arai S, Nakashima K, Kudo T, Kuroda E, Wu CH, Kao PH, Sakai M, Shimano H, Miyazaki T, Tontonoz P, Takahashi S. MafB promotes atherosclerosis by inhibiting foam-cell apoptosis. Nat Commun. 2014;5:3147.
135. Date D, Das R, Narla G, Simon DI, Jain MK, Mahabeleshwar GH. Kruppel-like transcription factor 6 regulates inflammatory macrophage polarization. J Biol Chem. 2014;289:10318–29.
136. Hu F, Tong J, Deng B, Zheng J, Lu C. MiR-495 regulates macrophage M1/M2 polarization and insulin resistance in high-fat diet-fed mice via targeting FTO. Pflugers Arch Eur J Physiol. 2019;471:1529–37.
137. Pan Y, Hui X, Hoo RLC, Ye D, Chan CYC, Feng T, Wang Y, Lam KSL, Xu A. Adipocyte-secreted exosomal microRNA-34a inhibits M2 macrophage polarization to promote obesity-induced adipose inflammation. J Clin Invest. 2019;129:834–49.
138. Agarwal V, Bell GW, Nam J-W, Bartel DP. Predicting effective microRNA target sites in mammalian mRNAs. eLife. 2015;4:e05005.
139. Liu Y, Li X, Zhang Y, Wang H, Rong X, Peng J, He L, Peng Y. An miR-340-5p-macrophage feedback loop modulates the progression and tumor microenvironment of glioblastoma multiforme. Oncogene. 2019;38:7399–415.
140. Curtale G, Rubino M, Locati M. MicroRNAs as molecular switches in macrophage activation. Front Immunol. 2019;10:799.
141. Li D, Duan M, Feng Y, Geng L, Li X, Zhang W. MiR-146a modulates macrophage polarization in systemic juvenile idiopathic arthritis by targeting INHBA. Mol Immunol. 2016;77:205–12.
142. Zhong Y, Yi C. MicroRNA-720 suppresses M2 macrophage polarization by targeting GATA3. Biosci Rep. 2016;36:e00363.
143. Cobos Jiménez V, Willemsen AM, Bradley EJ, Baas F, van Kampen AHC, Kootstra NA. Next-generation sequencing of microRNAs in primary human polarized macrophages. Genom Data. 2014;2:181–3.
144. Chen S, Yang J, Wei Y, Wei X. Epigenetic regulation of macrophages: from homeostasis maintenance to host defense. Cell Mol Immunol. 2019;17:36–49.
145. Huynh K, Pernes G, Mellett NA, Meikle PJ, Murphy AJ, Lancaster GI. Lipidomic profiling of murine macrophages treated with fatty acids of varying chain length and saturation status. Meta. 2018;8:29.
146. Lawrence T, Natoli G. Transcriptional regulation of macrophage polarization: enabling diversity with identity. Nat Rev Immunol. 2011;11:750–61.
147. Ivashkiv LB. Inflammatory signaling in macrophages: transitions from acute to tolerant and alternative activation states. Eur J Immunol. 2011;41:2477–81.
148. Li W, Zhang X, Wu F, Zhou Y, Bao Z, Li H, Zheng P, Zhao S. Gastric cancer-derived mesenchymal stromal cells trigger M2 macrophage polarization that promotes metastasis and EMT in gastric cancer. Cell Death Dis. 2019;10:918.
149. Nair MG, Guild KJ, Artis D. Novel effector molecules in type 2 inflammation: lessons drawn from helminth infection and allergy. J Immunol. 2006;177:1393–9.
150. Colton CA, Mott RT, Sharpe H, Xu Q, Van Nostrand WE, Vitek MP. Expression profiles for macrophage alternative activation genes in AD and in mouse models of AD. J Neuroinflammation. 2006;3:27.
151. Zhang W, Baban B, Rojas M, Tofigh S, Virmani SK, Patel C, Behzadian MA, Romero MJ, Caldwell RW, Caldwell RB. Arginase activity mediates retinal inflammation in endotoxin-induced uveitis. Am J Pathol. 2009;175:891–902.
152. Liu J, Copland DA, Horie S, Wu WK, Chen M, Xu Y, Paul Morgan B, Mack M, Xu H, Nicholson LB, Dick AD. Myeloid cells expressing VEGF and arginase-1 following uptake

of damaged retinal pigment epithelium suggests potential mechanism that drives the onset of choroidal angiogenesis in mice. PLoS One. 2013;8:e72935.
153. Herová M, Schmid M, Gemperle C, Hersberger M. ChemR23, the receptor for Chemerin and Resolvin E1, is expressed and functional on M1 but not on M2 macrophages. J Immunol. 2015;194:2330–7.
154. El Kasmi KC, Qualls JE, Pesce JT, Smith AM, Thompson RW, Henao-Tamayo M, Basaraba RJ, König T, Schleicher U, Koo M-S, Kaplan G, Fitzgerald KA, Tuomanen EI, Orme IM, Kanneganti T-D, Bogdan C, Wynn TA, Murray PJ. Toll-like receptor-induced arginase 1 in macrophages thwarts effective immunity against intracellular pathogens. Nat Immunol. 2008;9:1399–406.
155. Andres-Hernando A, Okamura K, Bhargava R, Kiekhaefer CM, Soranno D, Kirkbride-Romeo LA, Gil H-W, Altmann C, Faubel S. Circulating IL-6 upregulates IL-10 production in splenic $CD4^+$ T cells and limits acute kidney injury; induced lung inflammation. Kidney Int. 2017;91:1057–69.
156. Chen L, Wang S, Wang Y, Zhang W, Ma K, Hu C, Zhu H, Liang S, Liu M, Xu N. IL-6 influences the polarization of macrophages and the formation and growth of colorectal tumor. Oncotarget. 2018;9:17443–54.
157. Braune J, Weyer U, Hobusch C, Mauer J, Brüning JC, Bechmann I, Gericke M. IL-6 regulates M2 polarization and local proliferation of adipose tissue macrophages in obesity. J Immunol. 2017;198:2927–34.
158. Gupta S, Jain A, Syed SN, Snodgrass RG, Pflüger-Müller B, Leisegang MS, Weigert A, Brandes RP, Ebersberger I, Brüne B, Namgaladze D. IL-6 augments IL-4-induced polarization of primary human macrophages through synergy of STAT3, STAT6 and BATF transcription factors. Onco Targets Ther. 2018;7:e1494110.
159. Patsouris D, Li PP, Thapar D, Chapman J, Olefsky JM, Neels JG. Ablation of CD11c-positive cells normalizes insulin sensitivity in obese insulin resistant animals. Cell Metab. 2008;8:301–9.
160. Fujisaka S, Usui I, Bukhari A, Ikutani M, Oya T, Kanatani Y, Tsuneyama K, Nagai Y, Takatsu K, Urakaze M, Kobayashi M, Tobe K. Regulatory mechanisms for adipose tissue M1 and M2 macrophages in diet-induced obese mice. Diabetes. 2009;58:2574–82.
161. Zhuang G, Meng C, Guo X, Cheruku PS, Shi L, Xu H, Li H, Wang G, Evans AR, Safe S, Wu C, Zhou B. A novel regulator of macrophage activation: miR-223 in obesity-associated adipose tissue inflammation. Circulation. 2012;125:2892–903.
162. Chomarat P, Rybak ME, Benchereau J. Interleukin-4. In: Thomson AW, editor. The cytokine handbook. San Diego: Academic Press; 1998. p. 133–74.
163. Van Ginderachter JA, Meerschaut S, Liu Y, Brys L, De Groeve K, Hassanzadeh Ghassabeh G, Raes G, De Baetselier P. Peroxisome proliferator-activated receptor γ (PPARγ) ligands reverse CTL suppression by alternatively activated (M2) macrophages in cancer. Blood. 2006;108:525–35.
164. Biswas SK, Gangi L, Paul S, Schioppa T, Saccani A, Sironi M, Bottazzi B, Doni A, Vincenzo B, Pasqualini F, Vago L, Nebuloni M, Mantovani A, Sica A. A distinct and unique transcriptional program expressed by tumor-associated macrophages (defective NF-κB and enhanced IRF-3/STAT1 activation). Blood. 2006;107:2112–22.
165. Katakura T, Miyazaki M, Kobayashi M, Herndon DN, Suzuki F. CCL17 and IL-10 as effectors that enable alternatively activated macrophages to inhibit the generation of classically activated macrophages. J Immunol. 2004;172:1407–13.
166. Wilkinson H, Roberts E, Stafford A, Banyard K, Matteucci P, Mace K, Hardman M. Tissue iron promotes wound repair via M2 macrophage polarisation and the chemokines CCL17 and CCL22. Am J Pathol. 2019;189:2196–208.
167. Achuthan A, Cook AD, Lee M-C, Saleh R, Khiew H-W, Chang MWN, Louis C, Fleetwood AJ, Lacey DC, Christensen AD, Frye AT, Lam PY, Kusano H, Nomura K, Steiner N, Förster I, Nutt SL, Olshansky M, Turner SJ, Hamilton JA. Granulocyte macrophage colony-stimulating factor induces CCL17 production via IRF4 to mediate inflammation. J Clin Invest. 2016;126:3453–66.

168. Lee M-C, Lacey DC, Fleetwood AJ, Achuthan A, Hamilton JA, Cook AD. GM-CSF- and IRF4-dependent signaling can regulate myeloid cell numbers and the macrophage phenotype during inflammation. J Immunol. 2019;202:3033–40.
169. Satoh T, Takeuchi O, Vandenbon A, Yasuda K, Tanaka Y, Kumagai Y, Miyake T, Matsushita K, Okazaki T, Saitoh T, Honma K, Matsuyama T, Yui K, Tsujimura T, Standley DM, Nakanishi K, Nakai K, Akira S. The Jmjd3-Irf4 axis regulates M2 macrophage polarization and host responses against helminth infection. Nat Immunol. 2010;11:936–44.
170. Röszer T. Phagosomal and lysosomal no synthesis. In: The biology of subcellular nitric oxide. Dordrecht: Springer; 2012. p. 145–55.

Mechanisms Which Control the Size of M2 Macrophage-Dominated Tissue Macrophage Niches

Keywords Myelopoiesis · Hematopoiesis · Macrophage proliferation · Yolk sac · Fetal liver monocyte · Bone marrow · Macrophage development · Macrophage differentiation · Self-renewal · Tissue-resident macrophage

1 Introduction

The number of M2 macrophages in a tissue is a key to determine how the macrophage niche impacts tissue function. Enrichment of M2 macrophages can be temporal, such as in a healing wound; or can be sustained, such as in tissues that are rich in M2 macrophages under physiological conditions. Examples include the nervous system, the liver, and adult-type adipose tissue, where macrophages show hallmarks of M2 activation. Dominance of M2 macrophages, however, can have unfavorable consequences, such as in the tumor microenvironment, during immune evasion of certain pathogens, tissue fibrosis, asthma, allergy, and other diseases. The intriguing question is how the number of M2 macrophages is controlled in tissues. Three complementary mechanisms maintain the number of M2 macrophages: development from hematopoietic progenitors, in the course of either the embryonic or the adult-type hematopoiesis, replenishment by self-renewal, that is, by local proliferation within the tissues, and elimination of excess M2 macrophages (Fig. 1). This chapter aims to overview these mechanisms.

T. Röszer, *The M2 Macrophage*, Progress in Inflammation Research 86,
https://doi.org/10.1007/978-3-030-50480-9_5

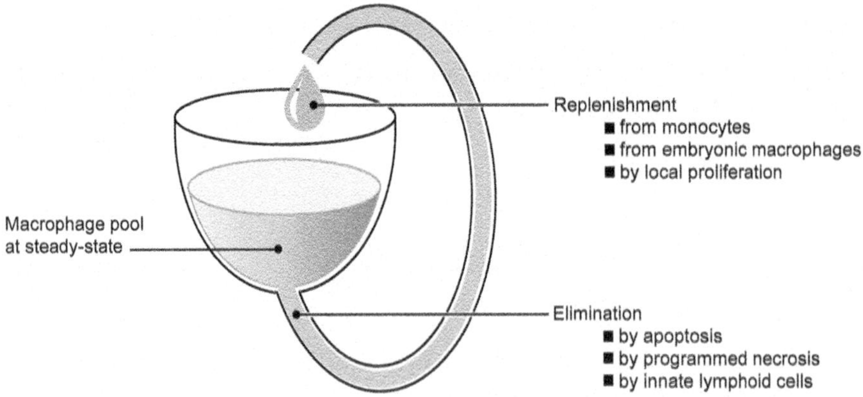

Fig. 1 Schematic representation of key mechanisms maintaining a steady-state tissue macrophage pool. Conceptually, there are three complementary mechanisms to maintain macrophage number in tissues: macrophages develop from hematopoietic progenitors, in the course of either embryonic or adult-type hematopoiesis; they are replenished by local proliferation within the tissues; and the excess macrophages are eliminated by apoptosis, programmed necrosis by innate lymphoid cells

2 Replenishment of M2 Macrophages from Blood Monocytes, from Embryonic Hematopoietic Compartments, and by Local Proliferation

2.1 Monocyte Origin of M2 Macrophages

Macrophages, similarly to other leukocytes, develop from hematopoietic progenitors [1]. For decades, circulating monocytes were considered to be the sole progenitors of tissue-resident macrophages, which settle in the tissues and eventually differentiate into resident macrophages [2] (Figs. 1 and 2). Leukocytes, including monocytes, are prone to migrate into tissues, especially when attracted by tissue damage or inflammation [1]. This process requires the attachment of circulating monocytes to the surface of endothelial cells, and eventually the extravasation of monocytes through intercellular gaps between adjacent endothelial cells and migration through the basement membrane of the endothelial layer. Extravasation and enrichment of monocytes in tissues is a major source of macrophages, and indeed, the first historical accounts of tissue lesion, inflammation, and infection describe macrophage evasion from vessels within the affected tissues [3]. In this setting, adult hematopoiesis (bone marrow hematopoiesis) is the source of monocytes that develop into macrophages.

Macrophage activation is eventually determined by tissue-specific cues and by mediators of other immune cells (Fig. 2; also see chapter "Signal mechanisms of M2 macrophage activation"). Because the extravasation of monocytes is enhanced in inflamed tissues, the resulting macrophages are likely to be activated toward an M1 phenotype in response to the inflammatory tissue microenvironment. Moreover,

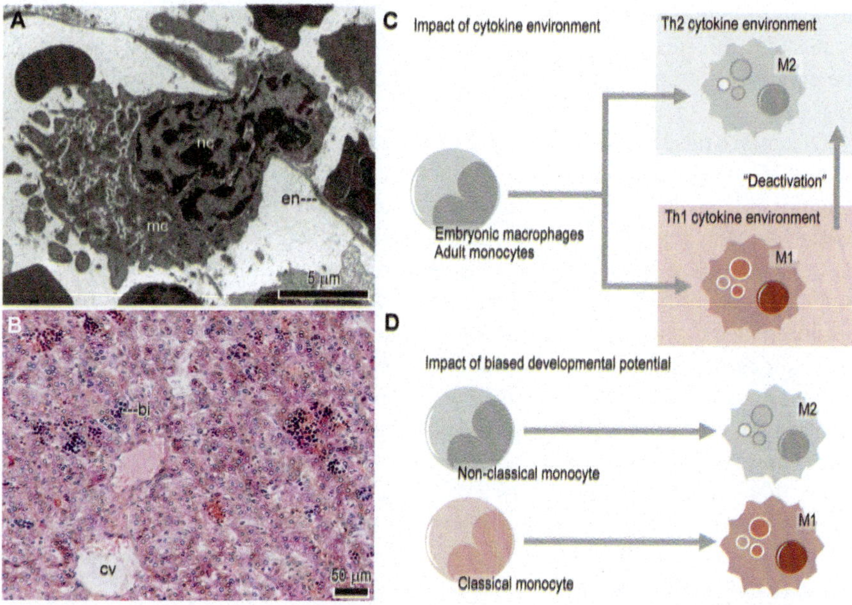

Fig. 2 Ontogeny of M2 macrophages. (**a**) Transmission electron microscopy image of an extravasating monocyte. The monocyte is evading the capillary lumen by migrating through an intercellular space between two adjacent endothelial cells. (Image by author.) Extravasation of monocyte requires the existence of such discontinuities between the endothelial cells, and the discontinuity can increase in response to inflammation, eventually facilitating monocyte extravasation. Capillaries that have continuous cell–cell connections (e.g., brain capillaries), do not allow extravasation of monocytes. *mc* monocyte cell body, *en* endothelial cell, *nc* nucleus. (**b**) Hematoxylin and eosin-stained section from liver of the human fetus. *cv* central vein of the hepatic lobule, *bi* blood island with hematopoietic niche. (Image by author.) (**c**) Embryonic macrophages and adult monocytes adopt M1 or M2 traits in response to appropriate stimuli. (**d**) Blood monocytes can be nonclassical or classical. Nonclassical monocytes are prone to develop into M2 macrophages

circulating inflammatory monocytes, called classical monocytes, traffic to the site of inflammation and are key for the early inflammatory response. These monocytes, following their extravasation, differentiate into M1 macrophages [4]. In human blood these monocytes express high levels of CD14 and lack the expression of CD16 ($CD14^{high}$, $CD16^{-}$ monocytes). Murine equivalents of these cells are called inflammatory or classical monocytes: these monocytes are $CD11b^{+}$, $CD115^{+}$, $CCR2^{+}$, and $CD62L^{+}$, and their hallmark is a strong Ly6C expression [4–6]. Ly6c is a differentiation marker glycoprotein, lymphocyte antigen 6 complex, locus C1, a marker used for discrimination of monocyte subsets and monocyte-derived macrophages [7]. Monocytes express Ly6C before leaving the bone marrow niche, and the majority of the bone marrow-derived macrophages are $Ly6C^{high}$ [8]. Tissues that are exposed to pathogenic stimuli, such as the intestinal mucosa, have high macrophage turnover and are infiltrated by monocyte-derived $Ly6C^{high}$ macrophages [9], and human inflammatory diseases, such as arthritis, atherosclerosis, and hemophagocytic syndrome, are associated with increased inflammatory monocyte counts [5]. This model of macrophage origin is easy to comprehend and can explain the enrichment of pathogen-fighting macrophages at the sites of tissue lesions and infections.

One might raise the question how M2 macrophages can appear in a regenerating wound where antiinflammatory and profibrotic macrophage functions are prevalent. One possible scenario is that extravasated blood monocytes settle in the damaged tissue, develop into macrophages, and later adopt M2 activation in response to tissue-specific cues such as apoptotic cells and apoptotic microvesicles (as detailed in the chapter entitled "Signal mechanisms of M2 macrophage activation"). Locally produced mediators may also guide the extravasating monocytes toward an M2 state [10]. For instance, in vitro, it is possible to elicit M2 activation of human blood-derived monocytes by providing them with appropriate stimuli [11]. A direct monocyte origin of M2 macrophages is supported by the presence of nonclassical monocytes in the circulation, which show antiinflammatory behavior [4]. In human blood these monocytes are CD16 expressing ($CD14^+$, $CD16^+$). In mouse, these $CD11b^+$, $CD115^+$ monocytes are called resident or patrolling monocytes, express fractalkine receptor CX_3CR1, and have low Ly6C expression [5]. These nonclassical monocytes are hence $Ly6C^-$, $CD43^+$, $CCR2^-$, $CD62L^-$, CX_3CR1^{high}, patrol on the endothelial surface of the vessels, and may settle in damaged tissues [8, 12]. The expression of Ly6C is lost in a differentiation program of monocytes, controlled by C/EBPβ, Krüppel-like factor 2, and nuclear receptor subfamily 4 group A member 1, also known as Nur77 [8, 13]. C/EBPβ is involved in M2 macrophage activation (detailed in the chapter "Signal mechanisms of M2 macrophage activation"), and the lack of Nur77 leads to an inflammatory macrophage phenotype [14]. It has been shown that nonclassical monocytes are biased toward an M2-like macrophage development [15]. The nonclassical monocytes settle in damaged tissue and differentiate into M2 macrophages [15]. Still, future research should answer whether there are further lineage determinants of M2 macrophages, that is, do tissue-specific M2 progenitors exist? These examples illustrate that M2 macrophages can develop from circulating monocytes; hence, adult bone marrow hematopoiesis is a source of M2 macrophages (Figs. 2 and 3).

2.2 Development of M2 Macrophages from Embryonic Hematopoiesis

More recently it has been shown that the embryonic hematopoiesis can also generate macrophages, and this developmental program lacks a circulating monocyte intermediate. The early progenitors of this macrophage lineage develop from endothelia-derived erythromyeloid progenitors of the yolk sac [16]. The responsible hematopoietic compartment resides in the blood islands of the yolk sac, and later in the hematopoietic tissue of the fetal liver [17]. Macrophages, as wandering cells,

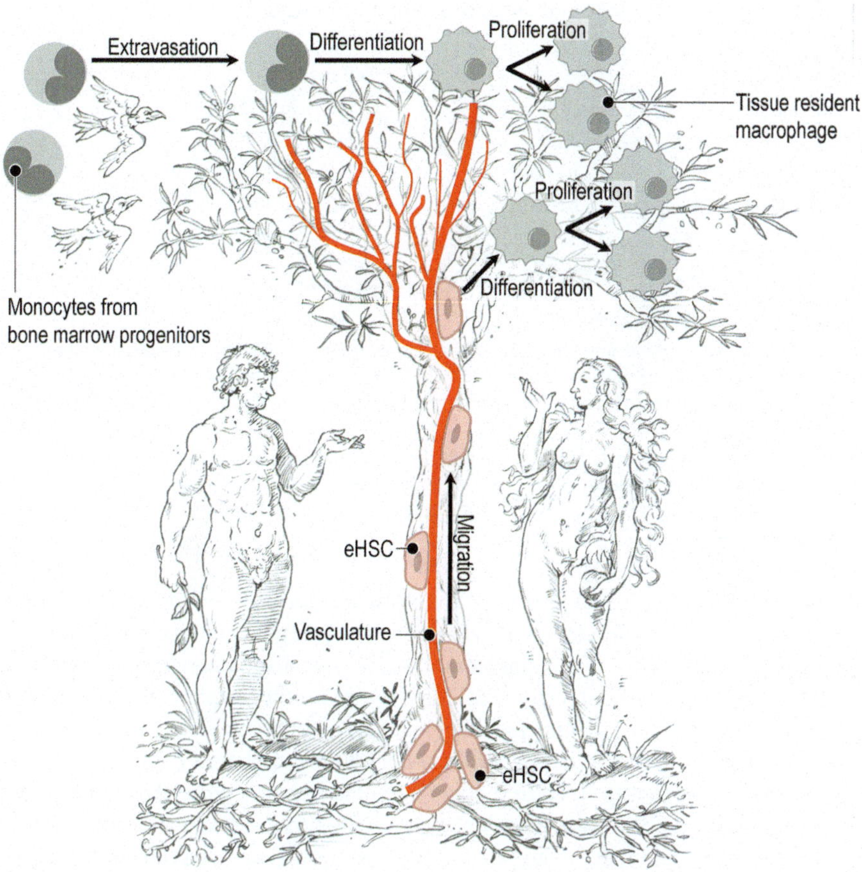

Fig. 3 Origin of tissue-resident macrophages. Tissue-resident macrophages may be descendants of embryonic hematopoietic stem cells (eHSC) of the yolk sac and the fetal liver. These progenitors give rise to macrophages that settle in the developing embryonic tissues by traveling with the growing vasculature. The embryonic macrophages proliferate and maintain tissue-resident macrophage niches. The adult-type bone marrow hematopoiesis also contributes to the tissue macrophage pool: bone marrow-derived monocytes travel with the bloodstream and settle in the tissues after extravasation. Monocyte-derived macrophages can proliferate and increase the macrophage niche size. The adult central nervous system has capillaries with continuous endothelia that impede monocyte extravasation; thus, the resident macrophage pool is exclusively eHSC derived. In contrast, organs that develop after the cessation of embryonic hematopoiesis (e.g., the visceral adipose tissue in the mouse) lack eHSC-derived macrophages. (Artwork by Péter Dráviczky)

colonize embryonic tissues by migrating and spreading within the tissues [18]. In those vertebrates that lack bone marrow hematopoiesis, the head kidney (in bony fishes), yolk sac-equivalent hematopoietic cells, or hematopoietic tissues of the liver and the spleen, remain the sources of tissue-resident macrophages [19]. In amphibia, early macrophage progenitors develop from the anterior ventral blood island, which is the equivalent of yolk sac hematopoietic cells [20].

Table 1 Identification of embryonic macrophages using specific markers

Yolk sac stem cells with macrophage lineage commitment	Cell markers
Maternal yolk sac macrophages[a] E7.5–E9.5	$CD45^+$ Kit^- $CD31^-$ $Mac1^+$ $F4/80^+$
Stage 1 yolk sac macrophages E8	$CD45^-$ Kit^+ $CD31^+$
Stage 2 yolk sac macrophages E8–E10	$CD45^+$ Kit^+ $CD31^+$ $Mac1^+$
Stage 3 yolk sac macrophages E8–E10	$CD45^+$ Kit^{low} $CD31^+$ $Mac1^+$ $MCSFR^+$ $CX3CR1^+$
Stage 4 yolk sac macrophages from E10.5[b]	$CD45^+$ Kit^{low} $CX3CR1^{bright}$ $MCSFR^+$ $F4/80^{bright}$

Source: Adapted from Ref. [25]

CD45 leukocyte common antigen, *Kit* c-kit, *CD31* platelet-endothelial cell adhesion molecule [PECAM-1], *Mac1* macrophage-1 antigen, *CX_3CR1* CX3C chemokine receptor-1, *MCSFR* macrophage colony-stimulating factor receptor, *F4/80* murine macrophage marker, so-called F4/80 antigen

[a]This population disappears at E9.5, and is therefore not analyzed because of its lack of effect on postnatal macrophage progeny

[b]This population persists in adult tissues and retains proliferative ability, and is thus considered a monopotent stem cell pool

In the mouse, the development of the yolk sac-derived macrophage lineage is Myb independent, whereas monocyte-derived macrophages develop in a Myb-dependent manner [21]. The human embryo yolk sac also generates macrophages, similarly to its mouse counterpart [22]. Induced pluripotent stem cells develop into macrophages in an MYB-independent manner in human, which resembles the developmental program of mouse yolk sac-derived macrophages [23]. These MYB-independent progenitors can differentiate into macrophages when they are exposed to tissue-specific cues [24]. In recent years, cell-surface markers of the various waves of embryonic macrophages have been well characterized, and some of the early lineage markers are summarized in Table 1.

Macrophages that develop from the embryonic hematopoietic compartment may be long-lived cells, such as the microglia in the brain [21]. In a newborn, most tissues are inhabited by macrophages that are descendants of the embryonic hematopoietic cells of the yolk sac or the fetal liver [17, 21, 25–27]. With age, however, the yolk sac- or fetal liver-derived macrophages are gradually replaced by bone marrow-derived monocytes [25, 28–30], which differentiate into macrophages as described earlier. This replacement of embryonic macrophages with monocyte-derived macrophages is rapid in organs constantly stimulated with pathogens, such as in the gut mucosa and in the skin. In these organs, the majority of resident macrophages are monocyte derived [28, 31]. In other organs, such as the liver, the heart, the serous cavities, the testicles, or adipose tissue, a varying degree of embryonic macrophages remain long persistent, unless tissue damage or inflammation replaces them with extravasated monocytes [9, 29, 30, 32–35]. We also know examples in which both embryonic macrophages and embryonic monocytes contribute to the establishment of the tissue macrophage niche. In the lung, alveolar macrophages develop from $F4/80^{high}$ $CD11b^{low}$ primitive macrophages and $Ly6C^{high}$ $CD11b^{high}$ fetal monocytes.

Those macrophages that inhabit embryonic tissues are necessary for normal development of tissues and organs through their role in extracellular matrix remodeling and vasculogenesis [9] and have immunometabolic functions specific to perinatal life [36]. The placenta contains maternal bone marrow-derived decidual macrophages and fetal hematopoiesis-derived placental macrophages, or Hofbauer cells [37]. In mouse, the subcutaneous white adipose tissue depot expands immediately after birth because it has a key metabolic function in infancy, and it is inhabited by macrophages that are descendants of late yolk sac progenitors [25]. The visceral white adipose tissue depot, however, which is missing at birth in mice, lacks yolk sac-derived macrophages and is seeded by bone marrow-derived macrophages [38].

3 Local Proliferation of Macrophages

The macrophages of mammals have long been considered as terminally differentiated immune cells that develop from monocytes and are unable to enter into the cell cycle [32, 39, 40]. In certain cases, however, both yolk sac-derived and bone marrow-derived macrophages retain the ability of proliferation and can replenish the macrophage pool without the need of monocyte supply [9] (Fig. 3). Because these macrophages serve as a stem cell reserve of tissue macrophage replenishment, we may consider this process as a self-renewal mechanism of the macrophage pool. The locally generated macrophages show some unique features that may distinguish them from monocyte-derived macrophages. For example, functional differences exist between yolk sac-derived and monocyte-derived macrophages in the peritoneum, such as in their levels of expression of CD11b, F4/80, Gr-1, AA4.1, CD40, CD80, CD86, CD11c, TLR4, MHC-II, phagocytosis activity, and NO production [41]. Furthermore, the IL-4 responsiveness of yolk sac- and monocyte-derived macrophages is distinct, and only monocyte-derived macrophages upregulate *Raldh2* and PD-L2 (programmed cell death ligand 2) in response to IL-4 [10]. Monocyte-derived macrophages also have high levels of aldehyde dehydrogenase activity and produce retinoic acid, whereas tissue-derived macrophages express high levels of mitochondrial uncoupling protein 1 [10]. In the lung, monocyte-derived macrophages promote allergic lung inflammation, whereas macrophages that are generated by self-renewal, and are probably descendants of long-lived embryonic macrophages, protect against inflammation [30, 42]. Still, there is much we should learn about the development and origin of M2 macrophages.

Recently I have reviewed the mechanisms that control macrophage proliferation, and herein I refer to that article [9]. Mitogenic signals for macrophages include M-CSF and GM-CSF, IL-1α, and erythropoietin, which are most likely provided by neighboring cells [43–50]. The adventitial layer of the large arteries contains resident macrophages, which develop from CX_3CR1^+ yolk sac macrophage precursors [51]. These adventitial macrophages are maintained by local proliferation. Because the arteries contain mesenchymal cells that produce the CX_3CR1 ligand CX_3CL1 (fractalkine or neurotactin), a paracrine signaling through CX_3CR1 may be responsible for maintaining the macrophage pool [52]. The tissue-specific cues, which are necessary for the maintenance of macrophage self-renewal, may be harmful and lead to excess macrophage proliferation in disease settings. For instance, the CD11b^{low} CD206^{+} MHC-II^{low} self-renewing macrophages are present in the mammary gland, and they receive a mitogenic M-CSF signal from gland cell epithelia [53]. However, excess M-CSF, as well as CCL2 produced by cancer cells and myeloid cells, can increase macrophage proliferation in breast cancer. This can promote tumor growth and metastasis by various mechanism [54].

What is important to note here is that IL-4/IL-4Rα signaling promotes macrophage proliferation [55], and the signals that augment this signal pathway secondarily contribute to macrophage proliferation [9]. A Th2 cytokine-dominated microenvironment favors M2 macrophage proliferation. For instance, *Leishmania* egg-induced granuloma is a Th2-dominated milieu, where M2 macrophage number is increased by local proliferation (see the chapter "Immune functions of the M2 macrophages: host defense, self-tolerance and autoimmunity"). Similarly, parasitism, such as *Giardia duodenalis* infection, causes local macrophage proliferation and increases a mixed M1/M2 (arginase-1^{+}, NOS2^{+}) macrophage population [56]. In certain cases, such as *Mycobacterim* infection, although the macrophages are skewed toward an M2 phenotype, the cell-cycle progression is inhibited by the intracellular pathogen to ensure its survival (see the chapter "Signal mechanisms of M2 macrophage activation"). Also, IL-4 may inhibit the mitogenic stimuli of MCSF [57].

The underlying mechanism of M2 macrophage proliferation is however not directly regulated by STAT6 activation. We have previously deciphered a complex signal machinery in macrophages that sustains the STAT6 level by the action of a metabolic regulator hormone, the neuropeptide FF (NPFF). Albeit NPFF sustains STAT6 signaling, and eventually M2 activation of macrophages, it increases macrophage proliferation by inhibition of STAT3 and interferon-stimulated gene expression, reducing the level of some cell-cycle inhibitor gene products (the Ifi200 family or p200 family members, such as Ifi204) [58]. Hence, M2 activation and cell-cycle entry are switched on by two distinct signal cascades. In the context of M2 macrophages, survival of *Mycobacterium* within macrophages is mentioned often, because this intracellular pathogen has evolved several mechanisms to avoid M1 activation, interferon secretion, and phagosome maturation, and skew macrophages towards an M2 state (see the chapters "What is an M2 macrophage? Historical overview of the macrophage polarization model. The Th1/Th2 and M1/M2 paradigm, the arginine fork" and "Signal mechanisms of M2 macrophage activation"). *Mycobacterium* also arrests macrophage cell cycle at the G_0/S transition [59] and increases the expression of Ifi204 [60], a known cell-cycle inhibitor [61]. Ifi204, also known as IFI16 in human, is involved in intracellular DNA sensing, and eventually can activate IFNβ secretion [60]. Although an increased Ifi204 level can increase interferon response, it also prolongs macrophage life cycle and thus supports the survival of *Mycobacterium*. Moreover, *Mycobacterium* cells inhibit IFNβ target genes [62].

Last, it is important to note that M2 macrophages have higher Myc expression than M1 macrophages, and they are prone to enter the cell cycle. In the context of tumor development, this ability of M2 macrophages is unfavorable as it increases the number of tumor-tolerating M2-like macrophages [63].

4 Elimination of M2 Macrophages

The size of the macrophage pool is also controlled by the elimination of unwanted macrophages (Fig. 1). In M1 macrophages, the life cycle of the activated macrophage ends with programmed cell death. For instance, pathogen-engulfing macrophages undergo apoptosis. Also, M1 macrophages can undergo programmed necrotic cell death, such as pyroptosis, necroptosis, and parthanatos [64]. These necrotic cell death mechanisms allow pathogen elimination, for instance, by secreting IL-1β. Apoptotic macrophages are eventually taken up by other macrophages, which transmit pathogen-associated and inflammation-modulating molecules from one macrophage to the other [65].

In the case of M2 macrophages, we also know a unique mechanism that limits the excessive increase of M2 macrophage numbers in tissues. In adult adipose tissues, there is a balanced self-renewal and removal of macrophages. Replenishment is supported by local self-renewal and at varying degres by monocyte extravasation [25, 58], whereas the elimination of excess macrophages is accomplished by type 1 lymphoid cells of the adipose tissue [66]. Type 1 lymphoid cells belong to the family of innate lymphoid cells, such as natural killer cells. They have cytotoxic activity and also have a cytokine profile resembling Th1 cells. Hence, it is a plausible scenario that, while eliciting a cytotoxic (i.e., antiviral) and Th1 (i.e., M1-activating) immune response, they limit the number of M2 macrophages.

References

1. Abbas A, Lichtman AH, Pillai S. Chapter 3: Leukocyte circulation and migration into tissues. In: Cellular and molecular immunology. Philadelphia: Elsevier; 2017.
2. Carr I. The macrophage: a review of ultrastructure and function. New York: Academic; 1973.
3. Masson P. Diagnósticos de laboratorio II. In: Tumores. Diagnósticos Histológicos, vol. XXVII. Barcelona: Editorial Pubul; 1927.
4. Narasimhan PB, Marcovecchio P, Hamers AAJ, Hedrick CC. Nonclassical monocytes in health and disease. Annu Rev Immunol. 2019;37:439–56.
5. Yang J, Zhang L, Yu C, Yang X-F, Wang H. Monocyte and macrophage differentiation: circulation inflammatory monocyte as biomarker for inflammatory diseases. Biomark Res. 2014;2:1–1.
6. Ziegler-Heitbrock L, Ancuta P, Crowe S, Dalod M, Grau V, Hart DN, Leenen PJM, Liu Y-J, MacPherson G, Randolph GJ, Scherberich J, Schmitz J, Shortman K, Sozzani S, Strobl H, Zembala M, Austyn JM, Lutz MB. Nomenclature of monocytes and dendritic cells in blood. Blood. 2010;116:e74–80.
7. Passlick B, Flieger D, Ziegler-Heitbrock H. Identification and characterization of a novel monocyte subpopulation in human peripheral blood. Blood. 1989;74:2527–34.

8. Polletti S, Natoli G. Understanding spontaneous conversion: the case of the Ly6C⁻ monocyte. Immunity. 2017;46:764–6.
9. Röszer T. Understanding the biology of self-renewing macrophages. Cell. 2018;7:103.
10. Gundra UM, Girgis NM, Ruckerl D, Jenkins S, Ward LN, Kurtz ZD, Wiens KE, Tang MS, Basu-Roy U, Mansukhani A, Allen JE, Loke PN. Alternatively activated macrophages derived from monocytes and tissue macrophages are phenotypically and functionally distinct. Blood. 2014;123:e110–22.
11. Singla DK, Wang J, Singla R. Primary human monocytes differentiate into M2 macrophages and involve Notch-1 pathway. Can J Physiol Pharmacol. 2017;95:288–94.
12. Weisheit C, Zhang Y, Faron A, Köpke O, Weisheit G, Steinsträsser A, Frede S, Meyer R, Boehm O, Hoeft A, Kurts C, Baumgarten G. Ly6Clow and not Ly6Chigh macrophages accumulate first in the heart in a model of murine pressure-overload. PLoS One. 2014;9:e112710.
13. Mildner A, Schönheit J, Giladi A, David E, Lara-Astiaso D, Lorenzo-Vivas E, Paul F, Chappell-Maor L, Priller J, Leutz A, Amit I, Jung S. Genomic characterization of murine monocytes reveals C/EBPβ transcription factor dependence of Ly6C⁻ cells. Immunity. 2017;46:849–862.e847.
14. Hanna RN, Shaked I, Hubbeling HG, Punt JA, Wu R, Herrley E, Zaugg C, Pei H, Geissmann F, Ley K, Hedrick CC. NR4A1 (Nur77) deletion polarizes macrophages toward an inflammatory phenotype and increases atherosclerosis. Circ Res. 2012;110:416–27.
15. Olingy CE, San Emeterio CL, Ogle ME, Krieger JR, Bruce AC, Pfau DD, Jordan BT, Peirce SM, Botchwey EA. Non-classical monocytes are biased progenitors of wound healing macrophages during soft tissue injury. Sci Rep. 2017;7:447.
16. Kasaai B, Caolo V, Peacock HM, Lehoux S, Gomez-Perdiguero E, Luttun A, Jones EAV. Erythro-myeloid progenitors can differentiate from endothelial cells and modulate embryonic vascular remodeling. Sci Rep. 2017;7:43817.
17. Sheng J, Ruedl C, Karjalainen K. Most tissue-resident macrophages except microglia are derived from fetal hematopoietic stem cells. Immunity. 2015;43:382–93.
18. Horn K. Biologie der Gewebe. Freie Zellen. In: Rolf S, editor. Allgemeine Zoologie, Band 2, Spezielle Zoologie. Stuttgart/New York: Gustav Fisher; 1980. p. 168–9.
19. Costa R, Chen Y, Paredes R, Amaya E. Labeling primitive myeloid progenitor cells in *Xenopus*. Methods Mol Biol. 2012;916:141–55.
20. Smith SJ, Kotecha S, Towers N, Latinkic BV, Mohun TJ. XPOX2-peroxidase expression and the XLURP-1 promoter reveal the site of embryonic myeloid cell development in Xenopus. Mech Dev. 2002;117:173–86.
21. Perdiguero EG, Klapproth K, Schulz C, Busch K, Azzoni E, Crozet L, Garner H, Trouillet C, de Bruijn MF, Geissmann F, Rodewald HR. Tissue-resident macrophages originate from yolk-sac-derived erythro-myeloid progenitors. Nature. 2014;518(7540):547–51.
22. Enzan H. Electron microscopic studies of macrophages in early human yolk sacs. Acta Pathol Jpn. 1986;36:49–64.
23. Buchrieser J, James W, Moore MD. Human induced pluripotent stem cell-derived macrophages share ontogeny with MYB-independent tissue-resident macrophages. Stem Cell Rep. 2017;8:334–45.
24. Takata K, Kozaki T, Lee CZW, Thion MS, Otsuka M, Lim S, Utami KH, Fidan K, Park DS, Malleret B, Chakarov S, See P, Low D, Low G, Garcia-Miralles M, Zeng R, Zhang J, Goh CC, Gul A, Hubert S, Lee B, Chen J, Low I, Shadan NB, Lum J, Wei TS, Mok E, Kawanishi S, Kitamura Y, Larbi A, Poidinger M, Renia L, Ng LG, Wolf Y, Jung S, Önder T, Newell E, Huber T, Ashihara E, Garel S, Pouladi MA, Ginhoux F. Induced-pluripotent-stem-cell-derived primitive macrophages provide a platform for modeling tissue-resident macrophage differentiation and function. Immunity. 2017;47:183–198.e6.
25. Waqas SFH, Noble A, Hoang A, Ampem G, Popp M, Strauß S, Guille M, Röszer T. Adipose tissue macrophages develop from bone marrow-independent progenitors in *Xenopus laevis* and mouse. J Leukoc Biol. 2017;102:845–55.

26. Stremmel C, Schuchert R, Wagner F, Thaler R, Weinberger T, Pick R, Mass E, Ishikawa-Ankerhold HC, Margraf A, Hutter S, Vagnozzi R, Klapproth S, Frampton J, Yona S, Scheiermann C, Molkentin JD, Jeschke U, Moser M, Sperandio M, Massberg S, Geissmann F, Schulz C. Yolk sac macrophage progenitors traffic to the embryo during defined stages of development. Nat Commun. 2018;9:75.
27. DeFalco T, Bhattacharya I, Williams AV, Sams DM, Capel B. Yolk-sac-derived macrophages regulate fetal testis vascularization and morphogenesis. Proc Natl Acad Sci USA. 2014;111:E2384–93.
28. Bain CC, Bravo-Blas A, Scott CL, Gomez Perdiguero E, Geissmann F, Henri S, Malissen B, Osborne LC, Artis D, Mowat AM. Constant replenishment from circulating monocytes maintains the macrophage pool in the intestine of adult mice. Nat Immunol. 2014;15:929–37.
29. Bain CC, Hawley CA, Garner H, Scott CL, Schridde A, Steers NJ, Mack M, Joshi A, Guilliams M, Mowat AM, Geissmann F, Jenkins SJ. Long-lived self-renewing bone marrow-derived macrophages displace embryo-derived cells to inhabit adult serous cavities. Nat Commun. 2016;7:ncomms11852.
30. Röszer T. Understanding the biology of self-renewing macrophages. Cell. 2018;7(8):103.
31. Wang Y, Szretter KJ, Vermi W, Gilfillan S, Rossini C, Cella M, Barrow AD, Diamond MS, Colonna M. IL-34 is a tissue-restricted ligand of CSF1R required for the development of Langerhans cells and microglia. Nat Immunol. 2012;13:753–60.
32. Ginhoux F, Schultze JL, Murray PJ, Ochando J, Biswas SK. New insights into the multidimensional concept of macrophage ontogeny, activation and function. Nat Immunol. 2015;17:34.
33. Campbell SM, Knipper JA, Ruckerl D, Finlay CM, Logan N, Minutti CM, Mack M, Jenkins SJ, Taylor MD, Allen JE. Myeloid cell recruitment versus local proliferation differentiates susceptibility from resistance to filarial infection. eife. 2018;7:e30947.
34. Molawi K, Wolf Y, Kandalla PK, Favret J, Hagemeyer N, Frenzel K, Pinto AR, Klapproth K, Henri S, Malissen B, Rodewald HR, Rosenthal NA, Bajenoff M, Prinz M, Jung S, Sieweke MH. Progressive replacement of embryo-derived cardiac macrophages with age. J Exp Med. 2014;211:2151–8.
35. Epelman S, Lavine KJ, Beaudin AE, Sojka DK, Carrero JA, Calderon B, Brija T, Gautier EL, Ivanov S, Satpathy AT, Schilling JD, Schwendener R, Sergin I, Razani B, Forsberg EC, Yokoyama WM, Unanue ER, Colonna M, Randolph GJ, Mann DL. Embryonic and adult-derived resident cardiac macrophages are maintained through distinct mechanisms at steady state and during inflammation. Immunity. 2014;40:91–104.
36. Yu H, Dilbaz S, Coßmann J, Hoang AC, Diedrich V, Herwig A, Harauma A, Hoshi Y, Moriguchi T, Landgraf K, Körner A, Lucas C, Brodesser S, Balogh L, Thuróczy J, Karemore G, Kuefner MS, Park EA, Rapp C, Travers JB, Röszer T. Breast milk alkylglycerols sustain beige adipocytes through adipose tissue macrophages. J Clin Invest. 2019;129:2485–99.
37. Reyes L, Wolfe B, Golos T. Hofbauer cells: placental macrophages of fetal origin. In: Kloc M, editor. Macrophages: origin, functions and biointervention. Cham: Springer International; 2017. p. 45–60.
38. Chen Q, Ruedl C. Obesity retunes turnover kinetics of tissue-resident macrophages in fat. J Leukoc Biol. 2020:n/a. https://doi.org/10.1002/JLB.1MA1219.
39. Hashimoto D, Chow A, Noizat C, Teo P, Beasley MB, Leboeuf M, Becker CD, See P, Price J, Lucas D, Greter M, Mortha A, Boyer SW, Forsberg EC, Tanaka M, van Rooijen N, García-Sastre A, Stanley ER, Ginhoux F, Frenette PS, Merad M. Tissue resident macrophages self-maintain locally throughout adult life with minimal contribution from circulating monocytes. Immunity. 2013;38:792–804. https://doi.org/10.1016/j.immuni.2013.1004.1004.
40. Chang ZL. Recent development of the mononuclear phagocyte system: in memory of Metchnikoff and Ehrlich on the 100th anniversary of the 1908 Nobel Prize in Physiology or Medicine. Biol Cell. 2009;101:709–21.
41. Ghosn EEB, Cassado AA, Govoni GR, Fukuhara T, Yang Y, Monack DM, Bortoluci KR, Almeida SR, Herzenberg LA, Herzenberg LA. Two physically, functionally, and developmentally distinct peritoneal macrophage subsets. Proc Natl Acad Sci USA. 2010;107:2568–73.

42. Zasłona Z, Przybranowski S, Wilke C, van Rooijen N, Teitz-Tennenbaum S, Osterholzer JJ, Wilkinson JE, Moore BB, Peters-Golden M. Resident alveolar macrophages suppress, whereas recruited monocytes promote, allergic lung inflammation in murine models of asthma. J Immunol. 2014;193:4245–53.
43. Lehnert BE, Valdez YE, Lehnert NM, Park MS, Englen MD. Stimulation of rat and murine alveolar macrophage proliferation by lung fibroblasts. Am J Respir Cell Mol Biol. 1994;11:375–85.
44. Hawgood S, Akiyama J, Brown C, Allen L, Li G, Poulain FR. GM-CSF mediates alveolar macrophage proliferation and type II cell hypertrophy in SP-D gene-targeted mice. Am J Physiol Lung Cell Mol. 2001;280:L1148–56.
45. Worgall S, Singh R, Leopold PL, Kaner RJ, Hackett NR, Topf N, Moore MAS, Crystal RG. Selective expansion of alveolar macrophages in vivo by adenovirus-mediated transfer of the murine granulocyte-macrophage colony-stimulating factor cDNA. Blood. 1999;93:655–66.
46. François H, Sandra LR, Giulia G, Francine U, Raynal D, Yousof Y, Nadtha P, Mihaly PP, Virginie R, Monique D, Etienne M, Nicolas D, Isabelle C, Bernhard R, Jean-Christophe R, Dominique L. IL-1α induces CD11b low alveolar macrophage proliferation and maturation during granuloma formation. J Pathol. 2015;235:698–709.
47. Makoto N, Go H, Kiyoshi T. Development, differentiation, and maturation of Kupffer cells. Microsc Res Tech. 1997;39:350–64.
48. Pridans C, Sauter KA, Irvine KM, Davis GM, Lefevre L, Raper A, Rojo R, Nirmal AJ, Beard P, Cheeseman M, Hume DA. Macrophage colony-stimulating factor increases hepatic macrophage content, liver growth, and lipid accumulation in neonatal rats. Am J Physiol Gastrointest Liver Physiol. 2018;314:G388–98.
49. Gilboa D, Haim-Ohana Y, Deshet-Unger N, Ben-Califa N, Hiram-Bab S, Reuveni D, Zigmond E, Gassmann M, Gabet Y, Varol C, Neumann D. Erythropoietin enhances Kupffer cell number and activity in the challenged liver. Sci Rep. 2017;7:10379.
50. Williams CB, Yeh ES, Soloff AC. Tumor-associated macrophages: unwitting accomplices in breast cancer malignancy. NPJ Breast Cancer. 2016;2:15025.
51. Ensan S, Li A, Besla R, Degousee N, Cosme J, Roufaiel M, Shikatani EA, El-Maklizi M, Williams JW, Robins L, Li C, Lewis B, Yun TJ, Lee JS, Wieghofer P, Khattar R, Farrokhi K, Byrne J, Ouzounian M, Zavitz CC, Levy GA, Bauer CM, Libby P, Husain M, Swirski FK, Cheong C, Prinz M, Hilgendorf I, Randolph GJ, Epelman S, Gramolini AO, Cybulsky MI, Rubin BB, Robbins CS. Self-renewing resident arterial macrophages arise from embryonic CX3CR1(+) precursors and circulating monocytes immediately after birth. Nat Immunol. 2016;17:159–68.
52. Swirski FK, Robbins CS, Nahrendorf M. Development and function of arterial and cardiac macrophages. Trends Immunol. 2016;37:32–40.
53. Linde N, Casanova-Acebes M, Sosa MS, Mortha A, Rahman A, Farias E, Harper K, Tardio E, Reyes Torres I, Jones J, Condeelis J, Merad M, Aguirre-Ghiso JA. Macrophages orchestrate breast cancer early dissemination and metastasis. Nat Commun. 2018;9:21.
54. Fleming JM, Miller TC, Kidacki M, Ginsburg E, Stuelten CH, Stewart DA, Troester MA, Vonderhaar BK. Paracrine interactions between primary human macrophages and human fibroblasts enhance murine mammary gland humanization in vivo. Breast Cancer Res (BCR). 2012;14:R97.
55. Jenkins SJ, Ruckerl D, Cook PC, Jones LH, Finkelman FD, van Rooijen N, MacDonald AS, Allen JE. Local macrophage proliferation, rather than recruitment from the blood, is a signature of TH2 inflammation. Science. 2011;332:1284–8.
56. Fink MY, Maloney J, Keselman A, Li E, Menegas S, Staniorski C, Singer SM. Proliferation of resident macrophages is dispensable for protection during *Giardia duodenalis* infections. ImmunoHorizons. 2019;3:412–21.
57. Arpa L, Valledor AF, Lloberas J, Celada A. IL-4 blocks M-CSF-dependent macrophage proliferation by inducing p21Waf1 in a STAT6-dependent way. Eur J Immunol. 2009;39:514–26.

58. SFH W, Hoang A, Lin Y, Ampem G, et al. Neuropeptide FF increases M2 activation and self-renewal of adipose tissue macrophages. J Clin Invest. 2017;127:2842–54.
59. Cumming BM, Rahman MA, Lamprecht DA, Rohde KH, Saini V, Adamson JH, Russell DG, Steyn AJC. *Mycobacterium tuberculosis* arrests host cycle at the G(1)/S transition to establish long term infection. PLoS Pathog. 2017;13:e1006389.
60. Chunfa L, Xin S, Qiang L, Sreevatsan S, Yang L, Zhao D, Zhou X. The central role of IFI204 in IFN-β release and autophagy activation during *Mycobacterium bovis* infection. Front Cell Infect Microbiol. 2017;7:619.
61. Bourette RP, Mouchiroud G. The biological role of interferon-inducible P204 protein in the development of the mononuclear phagocyte system. Front Biosci. 2008;13:879–86.
62. Banks DA, Ahlbrand SE, Keith Hughitt V, Shah S, Vogel SN, El-Sayed NM, Briken V. *Mycobacterium tuberculosis* inhibits autocrine type I interferon signaling to increase intracellular survival. bioRxiv. 2018:425116.
63. Lolo FN, Rius C, Casas-Tinto S. Elimination of classically-activated macrophages in tumor-conditioned medium by alternatively-activated macrophages. Biology Open. 2017;6:1897–903.
64. Robinson N, Ganesan R, Hegedűs C, Kovács K, Kufer TA, Virág L. Programmed necrotic cell death of macrophages: focus on pyroptosis, necroptosis, and parthanatos. Redox Biol. 2019;26:101239.
65. Röszer T. Transcriptional control of apoptotic cell clearance by macrophage nuclear receptors. Apoptosis. 2017;22:284–94.
66. Boulenouar S, Michelet X, Duquette D, Alvarez D, Hogan AE, Dold C, O'Connor D, Stutte S, Tavakkoli A, Winters D, Exley MA, O'Shea D, Brenner MB, von Andrian U, Lynch L. Adipose type one innate lymphoid cells regulate macrophage homeostasis through targeted cytotoxicity. Immunity. 2017;46:273–86.

Part II
Specific M2 Macrophage Functions

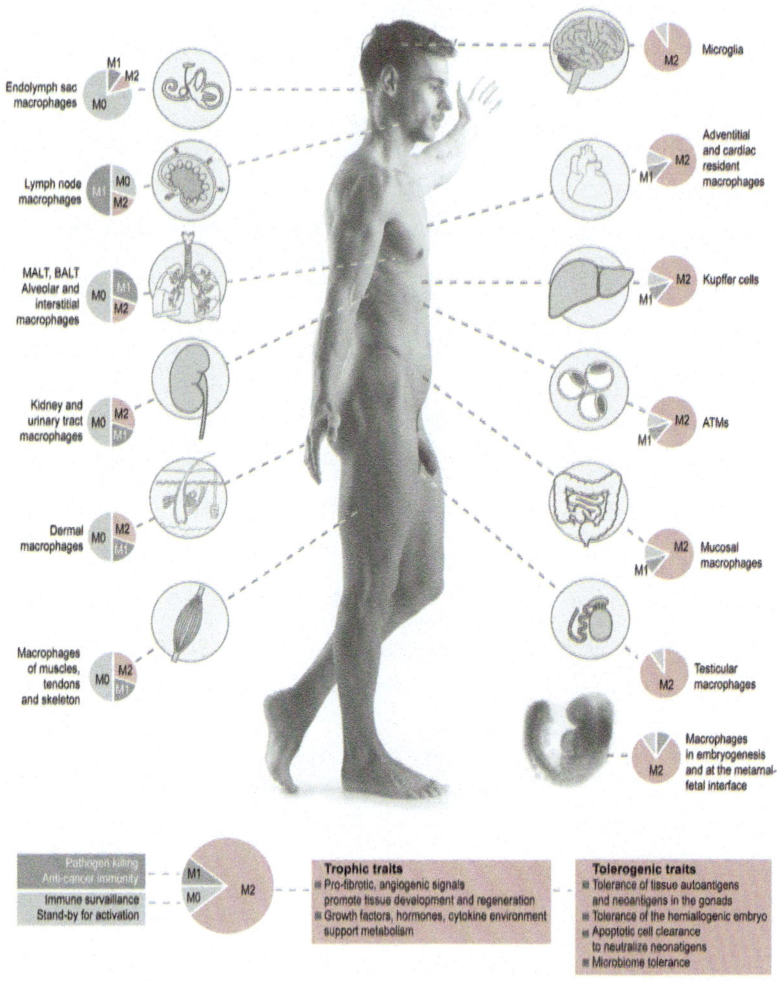

Graphic Abstract of Part II

M2 Macrophages Occupy Functional Niches in the Human Organ Systems

Each tissue contains naïve or nonactivated (M0) macrophages, and varying degrees of pathogen-fighting M1 macrophages and homeostatic, tolerogenic, pro-resolving M2 macrophages. During morphogenesis, and under physiological conditions, some organs are rich in M2 macrophages: the immune-privileged organs and organs in which there is a demand of neoantigen tolerance, microbial tolerance, and apoptotic cell clearance. In contrast, immune vigilance requires M1 macrophages.

Immune Functions of the M2 Macrophages: Host Defense, Self-tolerance, and Autoimmunity

Keywords M2 macrophage functions · Immunity · Host defense · Th2 · Cytokines · Allergy · Helminths · *Mycobacterium* · *Leishmania* · *Cryptococcus* · Parasitism · Self-tolerance · Wound healing · Macrophage

1 Introduction

This chapter summarizes the immune functions of M2 macrophages. A bona fide M2 macrophage activation is associated with Th2 immune response against helminth infections. Some intracellular pathogens can hijack macrophage activation, induce Th2/M2 immune response, and survive within M2 macrophages. This macrophage activation also shows some elements of an incomplete M1 activation, skewed toward an M2 state by the intracellular pathogen. The term tolerogenic macrophage is sometimes used to tag M2 macrophages. This term reflects one important trait of M2 macrophages, the maintenance of immunological self-tolerance, that is, to avoid the development of immunity against self-antigens, such as nucleic acids, nuclear fragments, cell debris, apoptotic bodies, and apoptotic cells. This function has been linked to the phagocytosing ability of macrophages, which clears potential self-antigens from the tissues. This function avoids the activation of other immune cells, and eventually acts against the development of inflammation and autoimmunity. Tolerogenic immune behavior of M2 macrophages is also present in the tumor environment, where M2 traits impede the recognition and elimination of tumor cells. This chapter discusses these immune functions of M2 macrophages.

T. Röszer, *The M2 Macrophage*, Progress in Inflammation Research 86,
https://doi.org/10.1007/978-3-030-50480-9_6

2 Th2 Immune Milieu of Helminth Infections Generates M2 Macrophages

Parasitic helminths, such as intestinal and filarial nematode, cestode, and trematode worms, elicit a Th2 immune response with prominent IL-4 and IL-13 release. These worms belong to two phyla: Platyhelminthes, or flatworms [Cestoda (tapeworms) and Trematoda (flukes)], and Nemathelminthes (roundworms) [1]. Helminth-derived molecules can suppress the host immune response to escape host defense, which may lead to chronic helminth infections. For instance, glycocalyx proteins, also called secretory/excretory (ES) products, of the tapeworm *Hymenolepis diminuta* inhibit T cell activation and M1 macrophage polarization [2–5]. To date several helminth ES products have been identified that induce Th2 immune response [6]. The trematode worm *Schistosoma* sp. lays eggs with polysaccharide antigens that polarize the immune response toward Th2 by increasing IL-10 production from B cells [7]. IL-10, in turn, promotes Th2 development (Fig. 1). Skewing immune response toward Th2 is a survival strategy of the helminth to subvert immune surveillance. However, elimination of enteric filarial nematodes is dependent on Th2 response, and the lack or

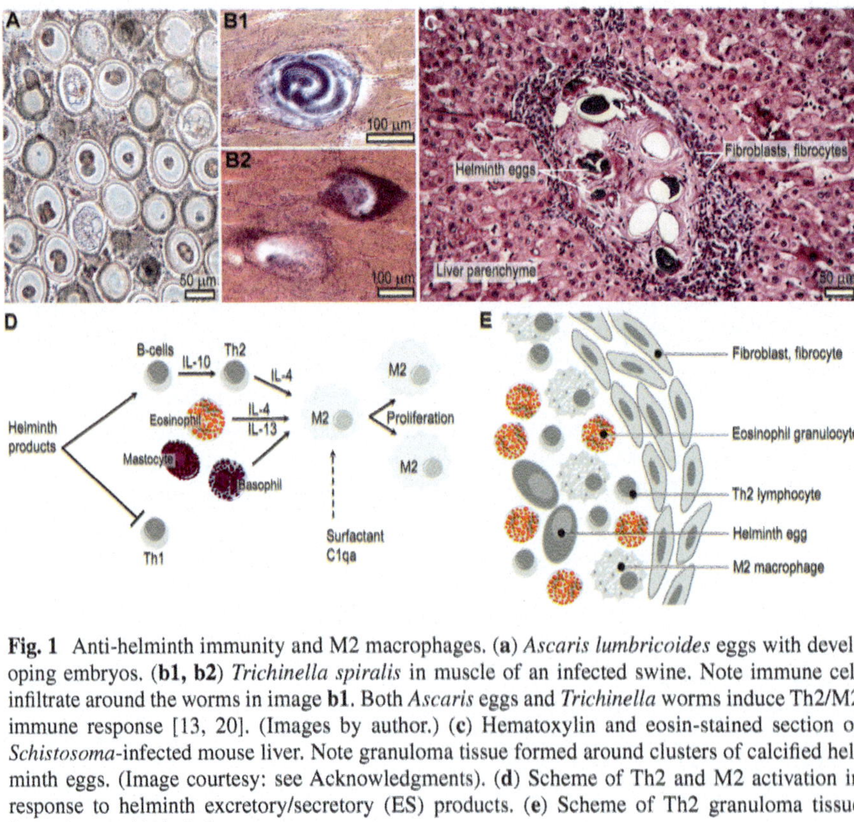

Fig. 1 Anti-helminth immunity and M2 macrophages. (**a**) *Ascaris lumbricoides* eggs with developing embryos. (**b1, b2**) *Trichinella spiralis* in muscle of an infected swine. Note immune cell infiltrate around the worms in image **b1**. Both *Ascaris* eggs and *Trichinella* worms induce Th2/M2 immune response [13, 20]. (Images by author.) (**c**) Hematoxylin and eosin-stained section of *Schistosoma*-infected mouse liver. Note granuloma tissue formed around clusters of calcified helminth eggs. (Image courtesy: see Acknowledgments). (**d**) Scheme of Th2 and M2 activation in response to helminth excretory/secretory (ES) products. (**e**) Scheme of Th2 granuloma tissue formed around helminth eggs

blocking of IL-4 signaling leads to persistent helminth infection in mice whereas IL-4 leads to parasite elimination [7, 8]. Indeed, the distinct Th1 or Th2 responsiveness of laboratory mouse strains determines whether they are resistant or susceptible to certain helminth infections. This dichotomy of immune response led to the discovery of M1/M2 macrophage polarization and the so-called arginine fork; as detailed in chapter "What Is an M2 Macrophage? Historical Overview of the Macrophage Polarization Model. The Th1/Th2 and M1/M2 Paradigm, the Arginine Fork".

Not surprisingly, the helminth-induced Th2 immune response leads to the activation of macrophages toward the M2 state. Filarial nematode infections recruit macrophages to the site of infection that express arginase-1, Ym1, and FIZZ1/RELM-α [9–11]. The function of these nematode-elicited, M2 macrophages may protect the host from an excessive Th1 response, aid parasite expulsion, and develop Th2 memory cells but also may aid immune evasion of the parasites (reviewed in [12, 13]). Expulsion of several intestinal nematode parasites is diminished by clodronate liposome depletion of macrophages and by arginase inhibition [14]. M2 activation is locally enhanced in the lung by surfactant protein A (SP-A) and in the peritoneal cavity by complement C1q. Complement is a group of serum proteins that trigger a cascade of reactions in response to pathogen infection, and leads to the generation of inflammatory mediators. C1q is the subunit of C1, the first component of the so-called classical complement pathway [15]. The Th2 immune milieu also increases local macrophage proliferation and leads to monocyte-independent increase of macrophage number at the site of infection [16, 17] (Fig. 1). Enhanced IL-4-dependent proliferation and activation of alveolar macrophages accelerates parasite clearance and reduces pulmonary injury following infection with a lung-migrating helminth. In the peritoneal cavity and liver, C1q enhancement of M2 macrophage activation is required for liver repair following bacterial infection. Both SP-A and C1q generate their effects on macrophages via the unconventional myosin18A, which acts as a cell-surface receptor [17, 18].

Certain nematode infections, such as infection with species of *Schistosoma*, lead to granuloma formation, especially around worm eggs (Fig. 1). The periovum (i.e., egg-surrounding) granuloma tissue is built up from Th2 cells, eosinophil granulocytes and M2 macrophages, and bordered by a fibrocyte layer. The granuloma is hence a Th2 microenvironment in which M2 macrophages aid recruitment of fibroblasts and ultimately promote fibrosis and granuloma formation around the worm eggs (Fig. 1). The granuloma encapsulates the worm eggs, protecting the surrounding tissue from collateral damage [19].

3 Intracellular Pathogens Are Overseen by M2 Macrophages

Some intracellular pathogens that are relevant in human disease, such as protozoan parasites (*Leishmania donovani*, *Leishmania tropica*, *Leishmania brasiliensis*, *Toxoplasma gondii*), and bacteria (*Mycobacterium tuberculosis*), show tropism toward macrophages [1]. Some fungal pathogens, such as *Aspergillus fumigatus* or *Cryptococcus neoformans*, also have intracellular growth stages and can survive in macrophages [21]. These intracellular pathogens can survive within macrophages because they have evolved mechanisms that inhibit maturation of the phagosome and proper fusion of lysosomes with the phagosome. Hence, the pathogenic cells are protected from lysis within the phagosome. As an example, the major proliferative stage of *Leishmania* species resides in the mature phagolysosomes of macrophages, evading pathogen killing [22]. Indeed, the phagolysosomes, filled with developing *Leishmania* individuals, are also called Leishman-Donovan bodies [1, 23]. When taken up by the macrophages, *Mycobacterium tuberculosis* inhibits phagosome–lysosome fusion by secreting phosphatase SapM and serine/threonine kinase PknG. Moreover, *Mycobacterium* cells recruit the host protein coronin 1/TACO, which further prevents lysosome–phagosome fusion [24].

Intracellular pathogens may survive within the macrophage phagosome because they can avoid M1 activation [25], can neutralize the cytotoxic effects of the NO burst of an M1 macrophage [24], and sustain their iron access by host transferrin receptor [26]. Infection with *Mycobacterium* cells sustains M2 activation, and the pathogen survives and grows within M2 macrophages [26]. Also, some intracellular pathogens, such as *Plasmodium falciparum*, induce M2-like activation of monocytes [27]. The fungal pathogen *Aspergillus fumigatus* also promotes M2 activation [28]. Arginase-1 expression is induced by intracellular pathogens in an IL-4/STAT6-independent mechanism, using Toll-like receptor signaling [29]. Inhibiting arginase activity can reduce the immune evasion of *Toxoplasma gondii* and *Mycobacterium* in mouse [29]. *Mycobacteria* induce profound changes in the transcriptional landscape of macrophages, and increase the transcription of both M1 and M2 activation genes, as well as several transcription factors that serve as hubs for macrophage activation [30]. *Mycobacterium* cells also augment the previously acquired activation state of macrophages. Because *Mycobacterium* favors residing within M2 macrophages, further augmentation of M2 effector genes supports its persistence and survival within M2 macrophages [30].

Another intracellular pathogen, *Listeria monocytogenes*, shifts macrophage metabolism toward the biosynthesis of 25-hydroxy-cholesterol [31]. This cholesterol metabolite is a possible endogenous ligand of the nuclear receptor LXRs. LXRs are known to prolong macrophage survival [32] and suppress genes of M1 activation [33]. The combination of these two factors, prolonged survival and tolerogenic immune behavior, allows parasite survival and dissemination.

Intracellular parasites that hijack macrophage activation also benefit from the urea, L-ornithine, and polyamine synthesis of M2 macrophages. Because M2 macrophages have high expression level of arginase-1, their urea cycle is shifted toward the production of polyamines. This metabolic setting favors the growth and survival of the intracellular parasite *Leishmania* spp. [34].

However, whether intracellular pathogens are tolerated by macrophages depends on a combination of a set of factors, such as signals generated by the parasites, susceptibility of the host (e.g., immunocompromised state), virulence of the pathogen, genetic background in murine models of parasite infections, and the type of macrophage challenged by the infection [35]. For instance, *Toxoplasma gondii* can infect any kind of nucleated cell; macrophages and related mononuclear phagocytes are its preferred target cells in vivo [1]. The type I and III strains of *Toxoplasma gondii* skew immune response toward M2 macrophage activation, whereas type II triggers M1 macrophage activation [36], showing that the same pathogen may trigger strain-specific, distinct macrophage activation. There are also differences between distinct tissue macrophage types as to whether they allow the growth of intracellular pathogens or effectively eliminate them. Microglia are not efficient in fungal cell clearance, but lung alveolar macrophages are prone to kill fungal pathogens [37]. Of note, often granulocytes, rather than macrophages, are the effector cells that eliminate intracellular pathogens in many tissues.

Tolerance of intracellular pathogens is certainly an unfavorable aspect of M2 macrophage biology. However, because M2 macrophages express arginase-1, they metabolize arginine effectively, and thus limit the L-arginine supply of *Aspergillus* cells [28]. As L-arginine is essential for fungal cell growth, this metabolic setting may limit pathogen survival. Also, M2 macrophages can aid tissue healing after infection by intracellular parasites, as seen after *Listeria monocytogenes* infection in the liver wherein basophil granulocytes release IL-4, which in turn induces M2 polarization of recruited monocytes in the liver. This stage eventually promotes tissue repair after infection [38].

Moreover, thinking in an evolutionary scope, one should consider that eukaryote cell organelles are remnants of ancient intracellular parasites (Graphic Abstract 1). Paradoxically, the complexity of the eukaryote cell has evolved thanks to its ability to tolerate intracellular parasitism and convert it into a symbiosis. At least, the endosymbiont theory, which is our current model for eukaryote cell evolution, considers that cell organelles (i.e., mitochondria, plastids, cell nucleus, etc.) are descendants of ancient prokaryote cells (Fig. 2). This symbiosis would not be possible if the intracellular parasites were always recognized and eliminated by the host cell.

Could the first eukaryotic cells have been similar to M2 macrophage-like cells? To answer this possibility, one may consider diatoms, unicellular organisms that have evolved through the symbiosis of ancient bacteria and red algae, which also have adopted metazoan genes [39]. These species have a urea cycle similar to that of metazoans, and they also have a metazoan-type arginase (Fig. 2). The urea cycle can generate various metabolites to help recovery from prolonged nitrogen starvation of the diatom cell. Moreover, arginase generates substrates of polyamine synthesis, fueling the generation of highly modified long-chain polyamines used for building the diatom-specific siliceous cell wall [40]. Arginase activity is hence a key adaptive advantage of the symbiosis, and the urea cycle becomes a metabolic hub for carbon and nitrogen metabolism in diatoms [41].

4 Immunological Self-tolerance and M2 Macrophages

4.1 The Tolerogenic M2 Macrophage

For a multicellular organism, it is critical to discriminate self and foreign cells. Evolutionarily, the rejection of foreign ("non-self") cells served the elimination of cells that would compete for resources with the "self" cells. Recognition and

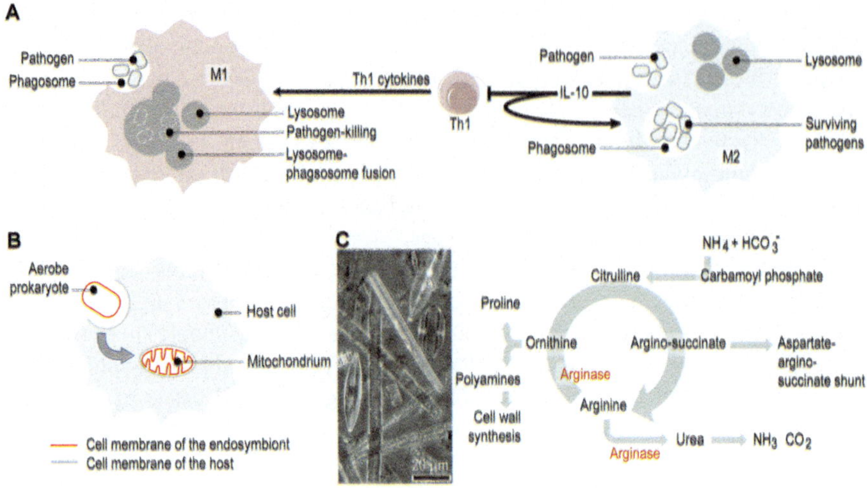

Fig. 2 Intracellular parasitism and endosymbiosis. (**a**) M1 macrophages engulf pathogens into phagosomes, which eventually fuse with lysosomes and degrade the pathogen cells. This process is triggered by pathogen-derived molecules and augmented by Th1 cytokines. However, immune evasion mechanisms of certain intracellular pathogens force IL-10 production by the infected macrophages, which leads to M2 activation and inhibits M1 activation. The intracellular pathogens can also inhibit lysosome fusion with the phagosomes. The pathogens are then taken up by phagocytosis but they survive and grow inside the phagosome. (**b**) Endosymbiotic origin of the eukaryote cell. Intracellular parasitism may lead to metabolic benefits to the host cell and establish symbiosis and metabolic co-dependence of the host and the parasite. (**c**) Diatom cells have evolved as a result of symbiosis among archaea, plant, and animal cells. The M2 trait arginase enzyme is necessary for both nitrogen metabolism and building the cell wall

tolerance of the self-antigens develops in a complex process. Macrophages also share the maintenance of self-tolerance: by secreting IL-10, they send tolerogenic signals to antigen-presenting cells [42]. IL-10 was first identified in 1989 as a molecule of Th2 cells that inhibits production of Th1 cytokines and was named a cytokine synthesis inhibitory factor [43]. Soon after its discovery, it was also recognized that IL-10 inhibits inflammatory cytokine production, that is, M1 activation, of macrophages [44]. As detailed in chapter "Signal Mechanisms of M2 Macrophage Activation", IL-10 is a signal for M2 activation, and IL-10 renders an antiinflammatory milieu in tissues. This action is exemplified by macrophages of the intestinal mucosa, which are challenged by signals of the gut microbiota; however, they acquire an antiinflammatory function in an IL-10R-dependent manner, and they also secrete IL-10 [45]. The local IL-10 signaling eventually develops tolerogenic macrophages that do not respond to the presence of microbiota in the gut. Disruption of this tolerogenic signaling leads to an immune response triggered by the normal gut flora, causing inflammatory bowel disease [46].

Tolerogenic M2 macrophages are present in organs that contain neoantigens. Neoantigens are "foreign" antigens, but they are generated by "self" cells (Graphic Abstract 1). Meiotic and postmeiotic germ cells, for instance, express a large variety of neoantigens. Accordingly, the testicles, following puberty, generate neoantigens,

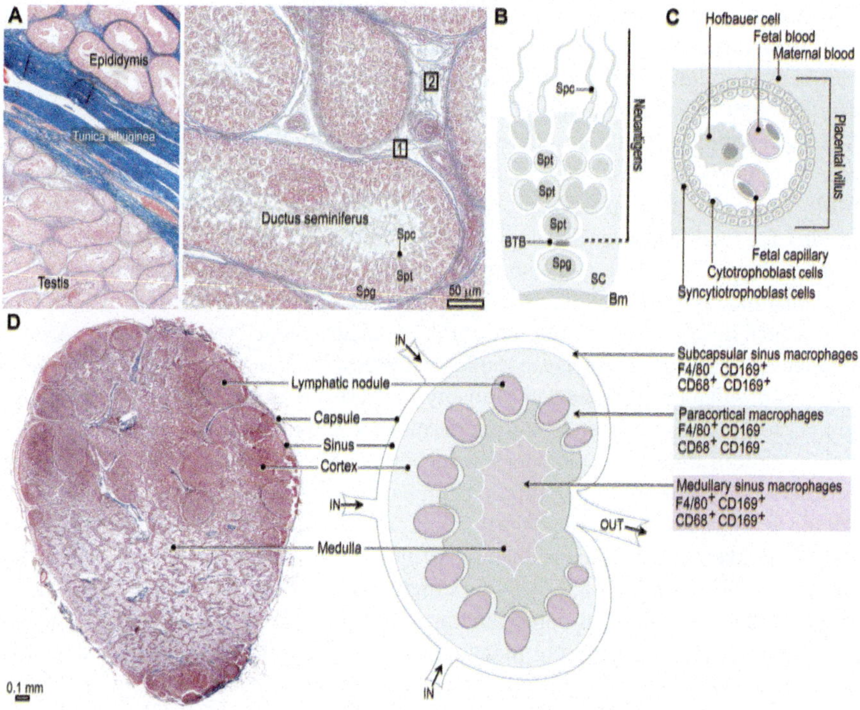

Fig. 3 A demand for M2 macrophages: tolerance of neoantigens. (**a**) Testis and epididymis show spermatocyte-generating seminiferous ducts. Collagenous sheaths covering the testicle and basal membrane of the seminiferous ducts are stained *blue*. *Spg* spermatogonia, *Spt* spermatids, *Spc* spermatocytes, *region 1* periductal space, *region 2* interstitial space. (**b**) The testis generates germ spermatocytes from the onset of puberty. The spermatocytes develop in the inner lining of the seminiferous ducts, and the developing cells are anchored by the Sertoli cells (*SC*). The spermatids are generated by meiosis and express neoantigens. The blood–testis barrier (BTB) is established by tight junctions between Sertoli cells and protects neoantigen exposure to immune cells. Moreover, testicular macrophages are tolerogenic M2 macrophages, which maintain an immune-privileged environment to support spermatogenesis. (**c**) Scheme of the maternal–fetal interface in the placental villus. Hofbauer cells are M2 macrophages that contribute to the tolerance of the fetal antigens. The fetal monocytes also adopt a tolerogenic phenotype. (**d**) Human lymph node histology *(left)* and schematic structure *(right)*. *IN* inward lymph flow, *OUT* outward lymph flow. Macrophages of the lymph node should recognize and eliminate neoantigens to establish anticancer immunity. However, medullary sinus macrophages safely dispose apoptotic cells, hence impeding the possible autoimmunity against neoantigens of dying cells. (Images by author)

because they produce male gametes that have neoantigens and are hence foreign for the body (Fig. 3). The testicle is thus an immune-privileged organ to avoid the development of an immune response against the developing spermatocytes. Spermatogenesis takes place in the epithelia of the seminiferous ducts of the testicles, and there is a so-called blood–testis barrier, which separates the neoantigen-containing maturing spermatocytes from the blood circulation, hiding the neoantigens from immune cells. The blood–testis barrier is formed by tight junctions, desmosomes, and gap junctions between cells of the epithelial lining of the seminiferous ducts, dividing the epithelia into a basal and an adluminal compartment. The latter contains the maturing spermatocytes and creates an immune-privileged microenvironment [47]. In addition to the blood–testis barrier, further factors maintain an immunosuppressed environment within the stroma of the testicle: high levels of androgens, corticosteroids, and enrichment of M2 macrophages [48]. Testicular macrophages are the largest leukocyte population in the testis, needed for fetal testicular development and maintenance of the immune-privileged state of the adult testicle and also for testosterone production [49]. Immunological self-tolerance is mainly maintained by the interstitial testicular macrophage population, which develops from embryonic yolk sac macrophage progenitors [50]. A peritubular testicular macrophage population develops during prepuberty from bone marrow-derived progenitors and probably supports the differentiation of spermatogonia [50]. Once established, both the peritubular and interstitial macrophage populations exhibit a long lifespan and low turnover in the steady state. However, as the proliferative capacity of interstitial macrophages declines, monocytes begin to contribute to the testicular macrophage population [51]. Testicular macrophages express the M2 marker CD163 and produce IL-10 and TGFβ, which contribute to testicular immunosuppression [52–55]. Although there is tolerogenic M2 macrophage dominance under physiological conditions in the testicle, upon infection the testicular macrophages and immigrating monocyte-derived macrophages readily adopt M1 activation and neutralize pathogens [47].

During pregnancy, the fetus represents a unique form of neoantigen for the maternal immune system: the fetus is semi-allogeneic, which means that has antigens identical with maternal antigens. However, it also has foreign, that is, paternal, antigens. In humans, as in most animals, sexual reproduction generates offspring that inherits half maternal and half paternal alleles. In mammals, the fetus develops in the uterus, where there is an interface between the circulation of the fetus and that of the mother, allowing the mutual recognition of each other's antigens. The maternal–fetal interface is however a tolerogenic, immunosuppressed environment, populated by M2 macrophages, and so the maternal immune system does not reject the fetal antigens.

The decidua of the placenta is rich in macrophages (Fig. 3), and following implantation of the embryo, M2 dominance hallmarks the decidual macrophage pool. Macrophages that are present at the maternal–fetal interface during the entire pregnancy are called Hofbauer cells, gathering within the placental villi [56]. They develop from fetal hematopoiesis [57] and impede the immune rejection of the fetus [58–62]. These cells secrete IL-10 and TGFβ, and express CD206, CD163, and CD209; they are implicated in placental vasculogenesis and morphogenesis, in HIV-1 and Zika virus transmission between mother and child [63, 64], and maintain maternal immunological tolerance of the fetus [58–60, 63].

Glucocorticoids and IL-10 induce Hofbauer cell activation toward the M2 state [65]; inflammatory signals and gestational diabetes do not curb M2 activation of Hofbauer cells [66, 67]. In turn, CD206-expressing human decidual macrophages produce IL-10 and CCL18 with a possible role in the maternal immunological tolerance of the fetus [62]. The DNA methylation profile of the Hofbauer cells also shows the hallmarks of M2 state: many immune response-related genes and M1-associated genes are hypermethylated, whereas genes of M2 macrophage activation are hypomethylated in Hofbauer cells [68]. Maternal antigens are also foreign for the fetus; however, genes associated with immune response are hypermethylated in fetal monocytes [68].

These examples show that M2 macrophages tolerate neoantigens and the semi-allogeneic fetus. Tumor cells also expose neoantigens, which can trigger immune response. Neoantigen-recognizing macrophages are enriched in the lymph nodes, which filter and ensure the immune vigilance of the lymph [69, 70]. The lymph nodes contain multiple macrophage populations that are functionally distinct and anatomically separated from each other. Among these, the subcapsular sinus contains macrophages, which are key for recognition and rejection of foreign antigens and neoantigens. The subcapsular sinus surrounds the B cell-rich lymphatic follicles, and this region receives the inward lymph flow (from afferent lymph draining vessels) [70]. Pathogen-derived antigens as a well as neoantigens from cancer cells are captured by these macrophages. The subcapsular sinus macrophages have tail-like cellular processes that project toward the lymphatic follicles. This morphology suggests that they can transfer the captured antigens by their membrane protrusion to the lymphatic follicles [69]. The subcapsular sinus macrophages are CD11b^{high}, CD169^{high}, CD11c^{low}, F4/80^{-} in mouse, and CD68^{+}, Siglec-1^{+} in human [69, 71]. Siglec-1 is a sialoadhesin molecule also called CD169. Reduced number or M2 state of the subcapsular sinus macrophages is a poor prognostic marker for malignant tumors because they impair neoantigen recognition by the lymph nodes. Macrophages associated with tumors often display M2 traits, and eventually they become tolerogenic toward the cancer cells. Because CD169 is expressed by the subcapsular sinus macrophages, ligands binding to it can be used for targeted drug delivery into the subcapsular sinus macrophages. Delivery of single-stranded synthetic DNA, for example, CpG oligodeoxynucleotides, to the subcapsular sinus macrophages elicits an antitumor immune response via activating TLR9 signaling and reduces lymph node metastasis [72]. However, a prolonged activation of lymph node macrophages can lead to immune cell exhaustion, destruction of lymphoid follicles, and exacerbation of autoimmune diseases [73].

The lymph nodes also contain tolerogenic macrophage pools. The medullary sinus macrophages phagocytose apoptotic granulocytes and B cells, and subsets of medullary sinus macrophages express the M2 marker mannose receptor. Of note, lymph node macrophages may also take up soluble mannose receptor, without expressing mRNA-encoding mannose receptor [69]. As detailed in the next chapter, apoptotic cell uptake causes M2 activation, and hence macrophages of the lymph nodes that handle apoptotic debris, such as the tangible body macrophages, adopt tolerogenic traits [74]. In the context of pathogenic infections, however, this tolerogenic feature of splenic macrophages is unfavorable. Medullary sinus macrophages are tolerogenic to *Toxoplasma* and also allow the replication of certain viruses [69].

4.2 *Apoptotic Cell Uptake: Clearing Neoantigens from the Tissues*

Macrophages remove self-antigens from the tissues by phagocytosis, which is followed by safe disposal of apoptotic cell debris (Fig. 4). Apoptotic cell debris includes apoptotic cell remnants and so-called apoptotic bodies, which are extracellular vesicles emitted during apoptosis [75]. The apoptotic cell debris contains modified self-antigens (neoantigens) generated during cell death, and a potentially harmful cargo of nuclear fragments and other cellular contents.

Macrophages remove self-antigens from the tissues by uptake of apoptotic cells. Apoptotic cell uptake leads to an M2 activation state, which further augments a tolerogenic immune status. Specific macrophages in the spleen and lymph nodes remove apoptotic cells, and ultimately splenic and peripheral lymph nodes have CD11c^{low}, CD40^{low} macrophages, which express IL-10 [42].

Apoptotic cell removal also inhibits the development of inflammation. Heme and iron can induce inflammation, which makes important the safe disposal of aged erythrocytes by splenic macrophages [76]. Rupture of erythrocytes, called hemolysis, also releases heme and iron. Malaria parasites, which grow in erythrocytes,

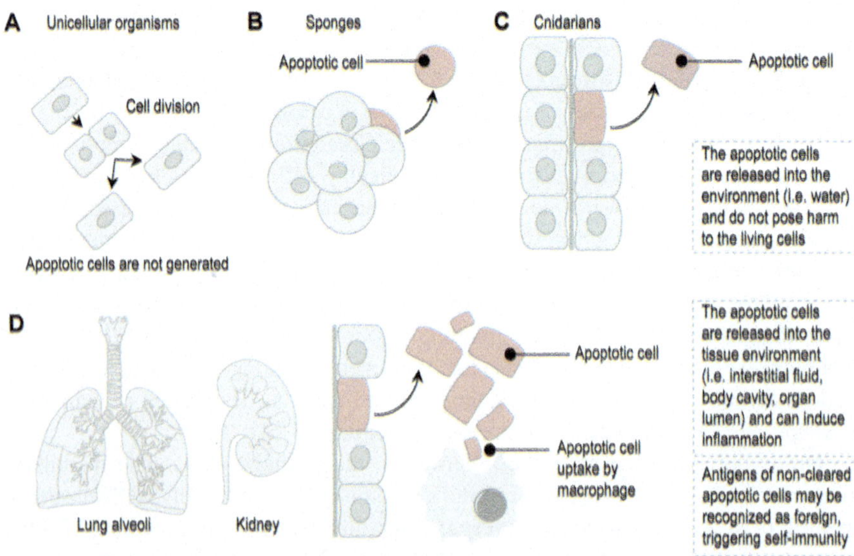

Fig. 4 Immunological impact of apoptotic cell clearance. (**a**) In unicellular organisms, cell divisions mark the end of the life cycle of the individual cell ("reorganizational cell death"). Eventually, no apoptotic cell debris is generated. In colony-forming organisms, such as sponges (**b**), and in tissue-forming organisms, such as cnidarians (**c**), apoptosis is necessary for normal development of the colony or tissues. Apoptotic cells however do not accumulate within the body because they are emitted to the surrounding aqueous environment. (**d**) In animals, the anatomy of some organs does not allow casting away apoptotic cells; apoptotic debris accumulates within the tissues and can cause inflammation or autoimmunity. Macrophages clear the apoptotic remnants

cause hemolysis at the end of their infection cycle. As detailed in chapter "Signal Mechanisms of M2 Macrophage Activation", macrophages recognize aged or malaria-infected erythrocytes and hence avoid the development of inflammation. For example, the expression of CD36, which recognizes malaria-infected erythrocytes, is increased by IL-13 [77]. In addition to tissue-resident macrophages, immature dendritic cells and some other nonprofessional phagocytes, such as fibroblasts and epithelial cells, also participate in apoptotic cell clearance [78]. When macrophages take up apoptotic cells, or apoptotic bodies, they show hallmarks of M2 activation. (This process is detailed in chapter "Signal Mechanisms of M2 Macrophage Activation").

Various cellular contents of apoptotic cells, such as their membrane lipids, are metabolized by macrophages, which eventually reprogram macrophages toward an M2 phenotype. This lipid remodeling of macrophages serves as signals for lipid-sensing nuclear receptors. These nuclear receptors control the expression of gene products responsible for recognition, binding, and internalization of apoptotic debris. For instance, LXRs, PPARs, VDR, RAR, and RXR are known to control genes of the phagocytosis process. Some studies also support the idea that engulfed apoptotic material releases ligands for nuclear receptors, such as retinoids of PPAR and LXR ligands, which hence self-augment apoptotic cell clearance and aid acquisition of M2 activation. This mechanism is detailed in chapter "Mechanisms Which Control the Size of M2 Macrophage-Dominated Tissue Macrophage Niches". Most of the nuclear receptors necessary for apoptotic cell uptake also control various genes of M2 activation. For instance, PPARγ is an auxiliary player in STAT6-mediated signaling.

When macrophages cannot recognize and internalize apoptotic cells, the accumulated apoptotic debris leads to inflammation and can initiate autoimmune disease, such as systemic lupus erythematosus [78].[1] Noncleared apoptotic cells are accumulated in the tissues and undergo secondary necrosis, eventually leading to the release of intracellular contents, such as modified lipids and nucleic acids. These events can trigger inflammation and also cause autoimmunity [78–82]. In addition, inflammation caused by noncleared apoptotic cells can have metabolic impact. In obese adipose tissue the lipid-overloaded adipocytes undergo apoptosis, which attracts monocytes to the adipose tissue [83] and leads to the accumulation of macrophages in the adipose tissue. Macrophages surrounding apoptotic adipocytes can be shifted into M1 activation by lipid derivatives released by the lipid-overloaded adipocytes [83], causing inflammation within the adipose tissue [84]. The inflammatory mediators released by the M1-activated macrophages abrogate insulin signaling and aggravate obesity-associated metabolic impairment, such as adipose tissue hypertrophy and insulin resistance [83]. Macrophage nuclear receptors have potential to abrogate the development of adipose tissue inflammation and insulin resistance [85]. To date it is not known whether increased clearance of the dying

[1]This section is reprinted from: 75. Rőszer, T. (2017) Transcriptional control of apoptotic cell clearance by macrophage nuclear receptors. *Apoptosis* **22**, 284–294.

adipocytes would help reduce adipose tissue inflammation, and whether this process can be augmented by nuclear receptor signaling. However, it is known that endocrine disruptors, which cause obesity through nuclear receptor signaling [86], also affect macrophage phagocytosis and macrophage activation [87, 88].

Phagocytotic uptake and processing of apoptotic cargo trigger M2 activation. Accordingly, insufficient lipid processing of apoptotic cargo has an inflammation-generating effect [89]. Lysosomes are hubs of lipid metabolism [90], and lipid remodeling within the phagosomes and the lysosomes contributes to the synthesis of lipid mediators [91]; this may be interlaced with lipid accumulation within lipid bodies or lipid droplets of macrophages [92]. The exact mechanisms and enzymes responsible for the remodeling of the apoptotic cargo are not yet fully explored. It has been shown that phagosomes containing apoptotic cargo mature faster than phagosomes which contain opsonized cells [93]. There are known mechanisms that may allow release of lipid metabolites from the peroxisomes and activate PPARs [94]; however, it is still unknown how specific lipid metabolites leave the phagosomes and reach the nucleus to evoke nuclear receptor signaling.

In addition to membrane lipids, apoptotic bodies and apoptotic cells contain nucleic acids, which can trigger immune response. DNA fragments, mitochondrial DNA, and various RNA species, including double-stranded microRNAs, constitute the cargo of apoptotic bodies. In systemic lupus erythematosus there is a self-immunity against nuclear fragments, which at least can be explained by deficient clearance of apoptotic debris. Recently it has been shown that, possibly from some deficiency in nucleic acid degradation, immunogenic nucleic acids can be present in apoptotic debris during systemic lupus erythematosus. These nucleic acids can activate the cytoplasmic DNA-sensing pathway (cGAS-STING signaling) and TLR7, TLR8, or TLR9 signaling [95]. These signal pathways induce the expression of interferon-stimulated genes [96].

5 Beneficial and Unfavorable M2 Macrophages

Wound healing involves M2 macrophages. Wound healing is initiated by inflammation, followed by proliferation and remodeling. In brief, M1 macrophages are present during the inflammatory phase of wound healing, and their function is the cleansing of pathogens that may have entered the wounded tissue. Recruited macrophages phagocytose debris in the damaged tissue and also phagocytose apoptotic immune cells. The M1 macrophages are gradually replaced by M2 macrophages, which support wound cell proliferation, capillarization, extracellular matrix synthesis, and resolution of inflammation [97]. When wound healing remains incomplete, the inflammatory phase remains persistent, leading to a chronic wound. In such instances, a therapeutic aim is to increase M2 macrophage abundance in the wounded tissue. However, the lack of the inflammatory phase impairs wound healing. The topic has been reviewed in depth elsewhere, and the main M2 macrophage functions promote tissue repair, neovascularization, and resolution of the initial inflammatory phase of tissue repair [98–100]. However, the very same traits of M2 macrophages are unfavorable in the context of tissue fibrosis, or during tumor tissue development. Tumor-associated macrophages (TAMs) can adopt an M2-like phenotype. Because M2 macrophages support tumor capillarization, tumor growth is supported by M2 activation. Moreover, M2 macrophages are tolerogenic; they block the development of antitumor immune response and eventually permit metastasis. In cancer therapy, inhibition of M2 activation is the aim, with the increase of antitumorigenic, cytotoxic T cell-activating macrophage functions [101] (chapter "Practical Approaches in M2 Macrophage Biology: Analysis, Pharmacology and Didactical Interpretation of M2 Macrophage Functions").

References

1. Jawetz E, Melnick JL, Adelberg EA. Review of medical microbiology. Los Altos: Lange Medical Publications; 1972.
2. McKay DM. The immune response to and immunomodulation by Hymenolepis diminuta. Parasitology. 2009;137:385–94.
3. Bien J, Salamatin R, Sulima A, Savijoki K, Bruce Conn D, Nareaho A, Mlocicki D. Mass spectrometry analysis of the excretory-secretory (E-S) products of the model cestode Hymenolepis diminut a reveals their immunogenic properties and the presence of new E-S proteins in cestodes. Acta Parasitol. 2016;61:429–42.
4. Johnston MJG, Wang A, Catarino MED, Ball L, Phan VC, MacDonald JA, McKay DM. Extracts of the rat tapeworm, Hymenolepis diminuta, suppress macrophage activation in vitro and alleviate chemically induced colitis in mice. Infect Immun. 2010;78:1364–75.
5. Zawistowska-Deniziak A, Basałaj K, Strojny B, Młocicki D. New data on human macrophages polarization by Hymenolepis diminuta tapeworm—an in vitro study. Front Immunol. 2017;8:148.
6. Smallwood TB, Giacomin PR, Loukas A, Mulvenna JP, Clark RJ, Miles JJ. Helminth immunomodulation in autoimmune disease. Front Immunol. 2017;8:453.
7. Locksley RM. Th2 cells: help for helminths. J Exp Med. 1994;179:1405–7.
8. Finkelman FD, Madden KB, Morris SC, Holmes JM, Boiani N, Katona IM, Maliszewski CR. Anti-cytokine antibodies as carrier proteins. Prolongation of in vivo effects of exogenous cytokines by injection of cytokine-anti-cytokine antibody complexes. J Immunol. 1993;151:1235–44.
9. Nair MG, Guild KJ, Artis D. Novel effector molecules in type 2 inflammation: lessons drawn from helminth infection and allergy. J Immunol. 2006;177:1393–9.
10. Nair MG, Gallagher IJ, Taylor MD, Loke P n, Coulson PS, Wilson RA, Maizels RM, Allen JE. Chitinase and Fizz family members are a generalized feature of nematode infection with selective upregulation of Ym1 and Fizz1 by antigen-presenting cells. Infect Immun. 2005;73:385–94.
11. Taylor MD, Harris A, Nair MG, Maizels RM, Allen JE. F4/80^{+} alternatively activated macrophages control CD4^{+} T cell hyporesponsiveness at sites peripheral to filarial infection. J Immunol. 2006;176:6918–27.
12. Rolot M, Dewals BG. Macrophage activation and functions during helminth infection: recent advances from the laboratory mouse. J Immunol Res. 2018;2018:17.
13. Kreider T, Anthony RM, Urban JF Jr, Gause WC. Alternatively activated macrophages in helminth infections. Curr Opin Immunol. 2007;19:448–53.

14. Anthony RM, Urban JF Jr, Alem F, Hamed HA, Rozo CT, Boucher J-L, Van Rooijen N, Gause WC. Memory T(H)2 cells induce alternatively activated macrophages to mediate protection against nematode parasites. Nat Med. 2006;12:955–60.
15. Ziccardi RJ, Cooper NR. Activation of C1 and the classical pathway of complement. In: Podack ER, editor. Cytolytic lymphocytes and complement: effectors of the immune system. Volume 1. Boca Raton: CRC Press, Inc; 1988.
16. Jenkins SJ, Ruckerl D, Cook PC, Jones LH, Finkelman FD, van Rooijen N, MacDonald AS, Allen JE. Local macrophage proliferation, rather than recruitment from the blood, is a signature of TH2 inflammation. Science. 2011;332:1284–8.
17. Campbell SM, Knipper JA, Ruckerl D, Finlay CM, Logan N, Minutti CM, Mack M, Jenkins SJ, Taylor MD, Allen JE. Myeloid cell recruitment versus local proliferation differentiates susceptibility from resistance to filarial infection. elife. 2018;7:e30947.
18. Minutti CM, Jackson-Jones LH, García-Fojeda B, Knipper JA, Sutherland TE, Logan N, Ringqvist E, Guillamat-Prats R, Ferenbach DA, Artigas A, Stamme C, Chroneos ZC, Zaiss DM, Casals C, Allen JE. Local amplifiers of IL-4Rα-mediated macrophage activation promote repair in lung and liver. Science (New York, NY). 2017;356:1076–80.
19. Hams E, Aviello G, Fallon PG. The schistosoma granuloma: friend or foe? Front Immunol. 2013;4:89–98.
20. Kang SA, Park M-K, Park SK, Choi JH, Lee DI, Song SM, Yu HS. Adoptive transfer of Trichinella spiralis-activated macrophages can ameliorate both Th1- and Th2-activated inflammation in murine models. Sci Rep. 2019;9:6547–6547.
21. Wasylnka JA, Hissen AHT, Wan ANC, Moore MM. Intracellular and extracellular growth of Aspergillus fumigatus. Med Mycol. 2005;43:S27–30.
22. Naderer T, McConville MJ. The Leishmania–macrophage interaction: a metabolic perspective. Cell Microbiol. 2008;10:301–8.
23. Bailey H, Bishop WJ. Leishman-Donovan bodies and donovaniasis; sir William Boog Leishman, 1865–1926; Charles Donovan, 1863–1951. Br J Vener Dis. 1959;35:8–9.
24. Pieters J. Mycobacterium tuberculosis and the macrophage: maintaining a balance. Cell Host Microbe. 2008;3:399–407.
25. Jordao L, Bleck CK, Mayorga L, Griffiths G, Anes E. On the killing of mycobacteria by macrophages. Cell Microbiol. 2008;10:529–48.
26. Kahnert A, Seiler P, Stein M, Bandermann S, Hahnke K, Mollenkopf H, Kaufmann SHE. Alternative activation deprives macrophages of a coordinated defense program to Mycobacterium tuberculosis. Eur J Immunol. 2006;36:631–47.
27. Weinberg JB, Volkheimer AD, Rubach MP, Florence SM, Mukemba JP, Kalingonji AR, Langelier C, Chen Y, Bush M, Yeo TW, Granger DL, Anstey NM, Mwaikambo ED. Monocyte polarization in children with falciparum malaria: relationship to nitric oxide insufficiency and disease severity. Sci Rep. 2016;6:29151.
28. Bhatia S, Fei M, Yarlagadda M, Qi Z, Akira S, Saijo S, Iwakura Y, van Rooijen N, Gibson GA, St Croix CM, Ray A, Ray P. Rapid host defense against Aspergillus fumigatus involves alveolar macrophages with a predominance of alternatively activated phenotype. PLoS One. 2011;6:e15943–e15943.
29. El Kasmi KC, Qualls JE, Pesce JT, Smith AM, Thompson RW, Henao-Tamayo M, Basaraba RJ, König T, Schleicher U, Koo M-S, Kaplan G, Fitzgerald KA, Tuomanen EI, Orme IM, Kanneganti T-D, Bogdan C, Wynn TA, Murray PJ. Toll-like receptor-induced arginase 1 in macrophages thwarts effective immunity against intracellular pathogens. Nat Immunol. 2008;9:1399–406.
30. Roy S, Schmeier S, Kaczkowski B, Arner E, Alam T, Ozturk M, Tamgue O, Parihar SP, Kawaji H, Itoh M, Lassmann T, Carninci P, Hayashizaki Y, Forrest ARR, Guler R, Bajic VB, Brombacher F, Suzuki H. Transcriptional landscape of Mycobacterium tuberculosis infection in macrophages. Sci Rep. 2018;8:6758.
31. Zou T, Garifulin O, Berland R, Boyartchuk VL. Listeria monocytogenes infection induces prosurvival metabolic signaling in macrophages. Infect Immun. 2011;79:1526–35.

32. Pascual-Garcia M, Carbo JM, Leon T, Matalonga J, Out R, Van Berkel T, Sarrias MR, Lozano F, Celada A, Valledor AF. Liver X receptors inhibit macrophage proliferation through downregulation of cyclins D1 and B1 and cyclin-dependent kinases 2 and 4. J Immunol. 2011;186:4656–67.
33. A-González N, Castrillo A. Liver X receptors as regulators of macrophage inflammatory and metabolic pathways. Biochim Biophys Acta (BBA) – Mol Basis Dis. 2011;1812:982–94.
34. Muxel SM, Aoki JI, Fernandes JCR, Laranjeira-Silva MF, Zampieri RA, Acuña SM, Müller KE, Vanderlinde RH, Floeter-Winter LM. Arginine and polyamines fate in Leishmania infection. Front Microbiol. 2018;8:2682–2682.
35. Tomiotto-Pellissier F, da Silva Bortoleti BT, Assolini JP, Gonçalves MD, Carloto ACM, Miranda-Sapla MM, Conchon-Costa I, Bordignon J, Pavanelli WR. Macrophage polarization in Leishmaniasis: broadening horizons. Front Immunol. 2018;9:2529–2529.
36. Jensen KDC, Wang Y, Wojno EDT, Shastri AJ, Hu K, Cornel L, Boedec E, Ong Y-C, Chien Y-H, Hunter CA, Boothroyd JC, Saeij JPJ. Toxoplasma polymorphic effectors determine macrophage polarization and intestinal inflammation. Cell Host Microbe. 2011;9:472–83.
37. Xu S, Shinohara ML. Tissue-resident macrophages in fungal infections. Front Immunol. 2017;8:1798–1798.
38. Blériot C, Dupuis T, Jouvion G, Eberl G, Disson O, Lecuit M. Liver-resident macrophage necroptosis orchestrates type 1 microbicidal inflammation and Type-2-mediated tissue repair during bacterial infection. Immunity. 2015;42:145–58.
39. Prihoda J, Tanaka A, de Paula WBM, Allen JF, Tirichine L, Bowler C. Chloroplast-mitochondria cross-talk in diatoms. J Exp Bot. 2012;63:1543–57.
40. Kröger N, Poulsen N. Diatoms—from cell wall biogenesis to nanotechnology. Annu Rev Genet. 2008;42:83–107.
41. Allen AE, Dupont CL, Oborník M, Horák A, Nunes-Nesi A, McCrow JP, Zheng H, Johnson DA, Hu H, Fernie AR, Bowler C. Evolution and metabolic significance of the urea cycle in photosynthetic diatoms. Nature. 2011;473:203–7.
42. Zheng G, Zhong S, Geng Y, Munirathinam G, Cha I, Reardon C, Getz GS, van Rooijen N, Kang Y, Wang B, Chen A. Dexamethasone promotes tolerance in vivo by enriching CD11cloCD40lo tolerogenic macrophages. Eur J Immunol. 2013;43:219–27.
43. Fiorentino DF, Bond MW, Mosmann TR. Two types of mouse T helper cell. IV. Th2 clones secrete a factor that inhibits cytokine production by Th1 clones. J Exp Med. 1989;170:2081–95.
44. Fiorentino DF, Zlotnik A, Mosmann TR, Howard M, O'Garra A. IL-10 inhibits cytokine production by activated macrophages. J Immunol. 1991;147:3815–22.
45. Shouval DS, Biswas A, Goettel JA, McCann K, Conaway E, Redhu NS, Mascanfroni ID, Al Adham Z, Lavoie S, Ibourk M, Nguyen DD, Samsom JN, Escher JC, Somech R, Weiss B, Beier R, Conklin LS, Ebens CL, Santos FGMS, Ferreira AR, Sherlock M, Bhan AK, Müller W, Mora JR, Quintana FJ, Klein C, Muise AM, Horwitz BH, Snapper SB. Interleukin-10 receptor signaling in innate immune cells regulates mucosal immune tolerance and anti-inflammatory macrophage function. Immunity. 2014;40:706–19.
46. Singh V, Kumar M, San Yeoh B, Xiao X, Saha P, Kennett MJ, Vijay-Kumar M. Inhibition of interleukin-10 signaling induces microbiota-dependent chronic colitis in apolipoprotein E deficient mice. Inflamm Bowel Dis. 2016;22:841–52.
47. Mruk DD, Cheng CY. The mammalian blood-testis barrier: its biology and regulation. Endocr Rev. 2015;36:564–91.
48. Wang M, Fijak M, Hossain H, Markmann M, Nüsing RM, Lochnit G, Hartmann MF, Wudy SA, Zhang L, Gu H, Konrad L, Chakraborty T, Meinhardt A, Bhushan S. Characterization of the micro-environment of the testis that shapes the phenotype and function of testicular macrophages. J Immunol. 2017;198(11):4327–40.
49. Bhushan S, Meinhardt A. The macrophages in testis function. J Reprod Immunol. 2017;119:107–12.

50. Mossadegh-Keller N, Gentek R, Gimenez G, Bigot S, Mailfert S, Sieweke MH. Developmental origin and maintenance of distinct testicular macrophage populations. J Exp Med. 2017;214(10):2829–41.
51. Röszer T. Understanding the biology of self-renewing macrophages. Cell. 2018;7:103.
52. Monika F, Andreas M. The testis in immune privilege. Immunol Rev. 2006;213:66–81.
53. Winnall WR, Muir JA, Hedger MP. Rat resident testicular macrophages have an alternatively activated phenotype and constitutively produce interleukin-10 in vitro. J Leukoc Biol. 2011;90:133–43.
54. Winnall WR, Hedger MP. Phenotypic and functional heterogeneity of the testicular macrophage population: a new regulatory model. J Reprod Immunol. 2013;97:147–58.
55. Meinhardt A, Bacher M, Metz C, Bucala R, Wreford N, Lan H, Atkins R, Hedger M. Local regulation of macrophage subsets in the adult rat testis: examination of the roles of the seminiferous tubules, testosterone, and macrophage-migration inhibitory factor1. Biol Reprod. 1998;59:371–8.
56. Jones CJP, Choudhury RH, Aplin JD. Functional changes in Hofbauer cell glycobiology during human pregnancy. Placenta. 2015;36:1130–7.
57. Reyes L, Wolfe B, Golos T. Hofbauer cells: placental macrophages of fetal origin. In: Kloc M, editor. Macrophages: origin, functions and biointervention. Cham: Springer; 2017. p. 45–60.
58. Kammerer U, Eggert AO, Kapp M, McLellan AD, Geijtenbeek TB, Dietl J, van Kooyk Y, Kampgen E. Unique appearance of proliferating antigen-presenting cells expressing DC-SIGN (CD209) in the decidua of early human pregnancy. Am J Pathol. 2003;162:887–96.
59. Wang H, He M, Hou Y, Chen S, Zhang X, Zhang M, Ji X. Role of decidual CD14(+) macrophages in the homeostasis of maternal-fetal interface and the differentiation capacity of the cells during pregnancy and parturition. Placenta. 2016;38:76–83.
60. Ning F, Liu H, Lash GE. The role of decidual macrophages during normal and pathological pregnancy. Am J Reprod Immunol. 2016;75:298–309.
61. Tang Z, Abrahams VM, Mor G, Guller S. Placental Hofbauer cells and complications of pregnancy. Ann N Y Acad Sci. 2011;1221:103–8.
62. Svensson-Arvelund J, Mehta RB, Lindau R, Mirrasekhian E, Rodriguez-Martinez H, Berg G, Lash GE, Jenmalm MC, Ernerudh J. The human fetal placenta promotes tolerance against the semiallogeneic fetus by inducing regulatory T cells and homeostatic M2 macrophages. J Immunol. 2015;194:1534–44.
63. Johnson EL, Chakraborty R. Placental Hofbauer cells limit HIV-1 replication and potentially offset mother to child transmission (MTCT) by induction of immunoregulatory cytokines. Retrovirology. 2012;9:101–101.
64. Simoni MK, Jurado KA, Abrahams VM, Fikrig E, Guller S. Zika virus infection of Hofbauer cells. Am J Reprod Immunol. 2017;77 https://doi.org/10.1111/aji.12613.
65. Tang Z, Niven-Fairchild T, Tadesse S, Norwitz ER, Buhimschi CS, Buhimschi IA, Guller S. Glucocorticoids enhance CD163 expression in placental Hofbauer cells. Endocrinology. 2013;154:471–82.
66. Schliefsteiner C, Peinhaupt M, Kopp S, Lögl J, Lang-Olip I, Hiden U, Heinemann A, Desoye G, Wadsack C. Human placental Hofbauer cells maintain an anti-inflammatory M2 phenotype despite the presence of gestational diabetes mellitus. Front Immunol. 2017;8:888–888.
67. Schliefsteiner C, Ibesich S, Wadsack C. Placental Hofbauer cell polarization resists inflammatory cues in vitro. Int J Mol Sci. 2020;21:736.
68. Kim SY, Romero R, Tarca AL, Bhatti G, Kim CJ, Lee J, Elsey A, Than NG, Chaiworapongsa T, Hassan SS, Kang GH, Kim J-S. Methylome of fetal and maternal monocytes and macrophages at the feto-maternal Interface. Am J Reprod Immunol. 2012;68:8–27.
69. Gray EE, Cyster JG. Lymph node macrophages. J Innate Immun. 2012;4:424–36.
70. Hibbs JB Jr, Chapman HA Jr, Weinberg JB. The macrophage as an antineoplastic surveillance cell: biological perspectives. J Reticuloendothel Soc. 1978;24:549–70.
71. Komohara Y, Ohnishi K, Kuratsu J, Takeya M. Possible involvement of the M2 anti-inflammatory macrophage phenotype in growth of human gliomas. J Pathol. 2008;216:15–24.

72. Klinman DM. Immunotherapeutic uses of CpG oligodeoxynucleotides. Nat Rev Immunol. 2004;4:249–59.
73. Hu J, Xu J, Li M, Zhang Y, Yi H, Chen J, Dong L, Zhang J, Huang Z. Targeting lymph node sinus macrophages to inhibit lymph node metastasis. Mol Ther Nucleic Acids. 2019;16:650–62.
74. Kamperdijk EWA, van Nieuwkerk EBJ, Verdaasdonk MAM, Hoefsmit ECM. Macrophages in different compartments of the non-neoplastic lymph node. In: Grundmann E, Vollmer E, editors. Reaction patterns of the lymph node: part 1 cell types and functions. Berlin/Heidelberg: Springer; 1990. p. 219–45.
75. Röszer T. Transcriptional control of apoptotic cell clearance by macrophage nuclear receptors. Apoptosis. 2017;22:284–94.
76. Alam MZ, Devalaraja S, Haldar M. The heme connection: linking erythrocytes and macrophage biology. Front Immunol. 2017;8:33–33.
77. Berry A, Balard P, Coste A, Olagnier D, Lagane C, Authier H, Benoit-Vical F, Lepert JC, Seguela JP, Magnaval JF, Chambon P, Metzger D, Desvergne B, Wahli W, Auwerx J, Pipy B. IL-13 induces expression of CD36 in human monocytes through PPARgamma activation. Eur J Immunol. 2007;37:1642–52.
78. Penberthy KK, Ravichandran KS. Apoptotic cell recognition receptors and scavenger receptors. Immunol Rev. 2016;269:44–59.
79. Yun JH, Henson PM, Tuder RM. Phagocytic clearance of apoptotic cells: role in lung disease. Expert Rev Respir Med. 2008;2:753–65.
80. Mukundan L, Odegaard JI, Morel CR, Heredia JE, Mwangi JW, Ricardo-Gonzalez RR, Goh YP, Eagle AR, Dunn SE, Awakuni JU, Nguyen KD, Steinman L, Michie SA, Chawla A. PPAR-delta senses and orchestrates clearance of apoptotic cells to promote tolerance. Nat Med. 2009;15:1266–72.
81. Gonzalez NA, Bensinger SJ, Hong C, Beceiro S, Bradley MN, Zelcer N, Deniz J, Ramirez C, Diaz M, Gallardo G, de Galarreta CR, Salazar J, Lopez F, Edwards P, Parks J, Andujar M, Tontonoz P, Castrillo A. Apoptotic cells promote their own clearance and immune tolerance through activation of the nuclear receptor LXR. Immunity. 2009;31:245–58.
82. Röszer T, Menendez-Gutierrez MP, Cedenilla M, Ricote M. Retinoid X receptors in macrophage biology. Trends Endocrinol Metab. 2013;24(9):460–8.
83. Rosen ED, Spiegelman BM. What we talk about when we talk about fat. Cell. 2014;156:20–44.
84. Glass CK, Olefsky JM. Inflammation and lipid signaling in the etiology of insulin resistance. Cell Metab. 2012;15:635–45.
85. Fuentes L, Röszer T, Ricote M. Inflammatory mediators and insulin resistance in obesity: role of nuclear receptor signaling in macrophages. Mediat Inflamm. 2010;2010:219583.
86. Grun F, Blumberg B. Endocrine disrupters as obesogens. Mol Cell Endocrinol. 2009;304:19–29.
87. Couleau N, Falla J, Beillerot A, Battaglia E, D'Innocenzo M, Plancon S, Laval-Gilly P, Bennasroune A. Effects of endocrine disruptor compounds, alone or in combination, on human macrophage-like THP-1 cell response. PLoS One. 2015;10:e0131428.
88. Camarca A, Gianfrani C, Ariemma F, Cimmino I, Bruzzese D, Scerbo R, Picascia S, D'Esposito V, Beguinot F, Formisano P, Valentino R. Human peripheral blood mononuclear cell function and dendritic cell differentiation are affected by bisphenol-A exposure. PLoS One. 2016;11:e0161122.
89. Aflaki E, Borger DK, Grey RJ, Kirby M, Anderson S, Lopez G, Sidransky E. Efferocytosis is impaired in Gaucher macrophages. Haematologica. 2017;102:656–65.
90. Thelen AM, Zoncu R. Emerging roles for the lysosome in lipid metabolism. Trends Cell Biol. 2017;27:833–50.
91. Yu H, Dilbaz S, Coßmann J, Hoang AC, Diedrich V, Herwig A, Harauma A, Hoshi Y, Moriguchi T, Landgraf K, Körner A, Lucas C, Brodesser S, Balogh L, Thuróczy J, Karemore G, Kuefner MS, Park EA, Rapp C, Travers JB, Röszer T. Breast milk alkylglycerols sustain beige adipocytes through adipose tissue macrophages. J Clin Invest. 2019;129:2485–99.

92. Melo RCN, Dvorak AM. Lipid body–phagosome interaction in macrophages during infectious diseases: host defense or pathogen survival strategy? PLoS Pathog. 2012;8:e1002729.
93. Erwig L-P, McPhilips KA, Wynes MW, Ivetic A, Ridley AJ, Henson PM. Differential regulation of phagosome maturation in macrophages and dendritic cells mediated by Rho GTPases and ezrin-radixin-moesin (ERM) proteins. Proc Natl Acad Sci U S A. 2006;103:12825–30.
94. Lodhi IJ, Yin L, Jensen-Urstad APL, Funai K, Coleman T, Baird JH, El Ramahi MK, Razani B, Song H, Fu-Hsu F, Turk J, Semenkovich CF. Inhibiting adipose tissue lipogenesis reprograms thermogenesis and PPARγ activation to decrease diet-induced obesity. Cell Metab. 2012;16:189–201.
95. Li T, Chen ZJ. The cGAS–cGAMP–STING pathway connects DNA damage to inflammation, senescence, and cancer. J Exp Med. 2018;215:1287–99.
96. Kato Y, Park J, Takamatsu H, Konaka H, Aoki W, Aburaya S, Ueda M, Nishide M, Koyama S, Hayama Y, Kinehara Y, Hirano T, Shima Y, Narazaki M, Kumanogoh A. Apoptosis-derived membrane vesicles drive the cGAS-STING pathway and enhance type I IFN production in systemic lupus erythematosus. Ann Rheum Dis. 2018;77:1507–15.
97. Krzyszczyk P, Schloss R, Palmer A, Berthiaume F. The role of macrophages in acute and chronic wound healing and interventions to promote pro-wound healing phenotypes. Front Physiol. 2018;9:419–419.
98. Röszer T. Inflammation as death or life signal in diabetic fracture healing. Inflamm Res. 2011;60:3–10.
99. Ferrante CJ, Leibovich SJ. Regulation of macrophage polarization and wound healing. Adv Wound Care (New Rochelle). 2012;1:10–6.
100. Novak ML, Koh TJ. Macrophage phenotypes during tissue repair. J Leukoc Biol. 2013;93:875–81.
101. Wang W, Marinis JM, Beal AM, Savadkar S, Wu Y, Khan M, Taunk PS, Wu N, Su W, Wu J, Ahsan A, Kurz E, Chen T, Yaboh I, Li F, Gutierrez J, Diskin B, Hundeyin M, Reilly M, Lich JD, Harris PA, Mahajan MK, Thorpe JH, Nassau P, Mosley JE, Leinwand J, Kochen Rossi JA, Mishra A, Aykut B, Glacken M, Ochi A, Verma N, Kim JI, Vasudevaraja V, Adeegbe D, Almonte C, Bagdatlioglu E, Cohen DJ, Wong K-K, Bertin J, Miller G. RIP1 kinase drives macrophage-mediated adaptive immune tolerance in pancreatic cancer. Cancer Cell. 2018;34:757–774.e757.

M2 Macrophages in the Integument and in the Musculoskeletal System

Keywords Skin · Dermal macrophage · Osteoclast · Bone marrow · Macrophage · Bone · Cartilage · Skeletal muscle · Arthritis · Tendonitis · Muscle aging · Muscle regeneration · Fibrosis · Osteoporosis · Arthritis · Allergy · Dermatitis · Macrophage · Langerhans cell

1 Introduction

When macrophages settle in tissues, they become tissue-resident macrophages in response to tissue-specific cues. As a result, each tissue has its own unique macrophage population, which has often distinctive traits and functions (Graphic Abstract 2). This chapter and those following review the functions governed by M2 macrophages in the major organ systems of the human body. We overview the tissue niches occupied by macrophages, the microenvironment in which they reside (e.g., cytokine milieu, exposure to signals of the surrounding cells), and the mutual interactions between macrophages and other elements of the tissues, such as cells and extracellular matrix components. The integument is the primary barrier of the body wherein M2 macrophages are associated with the skin and are called dermal macrophages. The dermal macrophages are members of immune-privileged compartments around the hair follicles and the nails, and interact with melanocytes of the epidermis. The musculoskeletal system or locomotor apparatus constructed of the integration of bones, joints, skeletal muscles, tendons, ligaments, and motor and sensory components of the nervous system. The musculoskeletal system allows voluntary body movement, although it also serves specific human traits such as manipulative tasks with extraordinary accuracy, gestural communication, and speech. The muscles also integrate intermediary metabolism; the bones maintain mineral homeostasis and host the hematopoietic compartment. M2 macrophages are mainly involved in the growth and repair processes of the skeletal elements, muscles, and tendons.

T. Röszer, *The M2 Macrophage*, Progress in Inflammation Research 86,
https://doi.org/10.1007/978-3-030-50480-9_7

2 M2 Features of the Dermal Macrophages

2.1 Perifollicular Macrophages

The skin is composed of the multilayered cornified epithelium or epidermis; the dermis, which contains hair follicles, glands, and sense organs; and subcutaneous tissues, such as connective tissue and white adipose tissue (Fig. 1). The epithelial layer is infiltrated by migratory myeloid cells, the Langerhans cells. These cells are antigen-presenting sentinel cells that develop from yolk sac and fetal liver myeloid progenitors, are maintained by local proliferation, and can migrate to skin-draining lymph nodes to present antigens and shape immune response [1–3]. These cells have dendrite-like protrusions; hence, they were first described as dendritic cells by Paul Langerhans in 1868. Later, they were considered as resident skin macrophages. Recently, these cells are defined as dendritic cells [1], rather than macrophages, and an extensive literature is available about the ontogeny, classification, and function of dendritic cells, including Langerhans cells [4, 5]. Some M2 macrophage markers, however, such as CD206 and CLEC10a, can be expressed by various dendritic cell subsets [5, 6]. For instance, in contact dermatitis, the CD206$^+$ myeloid cells that appear in the skin are immature dendritic cells rather than Langerhans cells or M2 macrophages [6].

The physiologically relevant skin-resident macrophage population is associated with the dermis, especially with the hair follicles, and these macrophages show M2

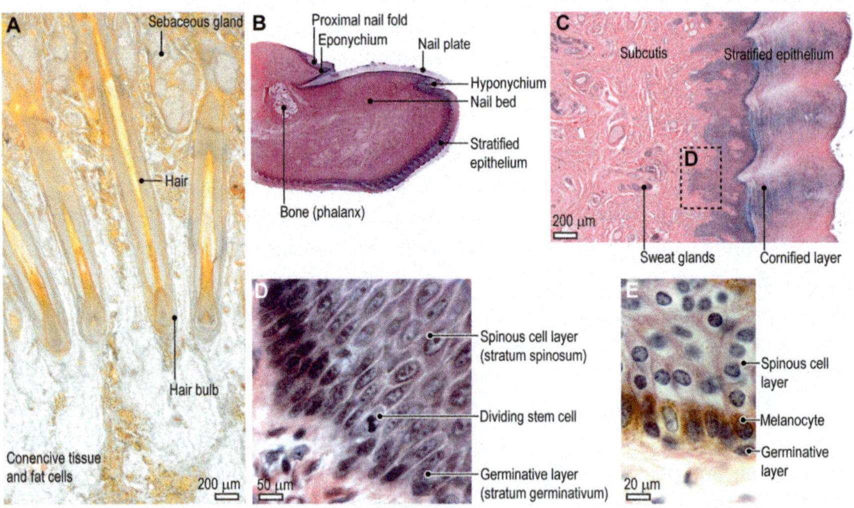

Fig. 1 Skin compartments in which dermal macrophages reside. (**a**, **b**) Immune-privileged sites of the skin: hair follicles and nail compartment. Perifollicular macrophages are associated with hair bulbs and have specific functions with the hair follicle cycle. (**c–e**) Histology of human dermis. Dermal macrophages interact with melanocytes, positioned in the germinative layer. (Images by author)

features [7]. The subcutaneous adipose tissue also contains macrophages; their function is addressed in chapter "M2 Macrophages in the Metabolic Organs and in the Neuroendocrine System". In the mouse, the dermal macrophages express CD301/mMGL (galactose-/*N*-acetylgalactosamine-specific lectin), CD163, dectin-1, coagulation factor XIIIa, and CD16, all of which are M2 activation markers in the human [7–9]. The expression of some of these markers is IL-4/IL-4R independent (CD301, dectin-1) [7], while others are bona fide IL-4- or IL-10-stimulated genes (CD163, CD16, coagulation factor XIIIa) [8, 10]. In addition to cytokine signaling, some skin-specific factors have been recently identified as determinants of macrophage activation in the dermis. In response to high sodium intake, dermal macrophages suffer osmotic stress, which induces the expression of inflammatory cytokines [9]. In contrast, a set of endogenous factors produced within the skin inhibits inflammatory cytokine expression in macrophages: these signals include lipid mediators and alpha melanocyte-stimulating hormone (α-MSH) [11–14].

Most dermal macrophages are associated with the hair follicles (Fig. 1a) and hence are called perifollicular macrophages. These macrophages are part of an immune cell niche surrounding the hair follicles [9]. The perifollicular macrophages are $CD68^+$ and $MHC\text{-}II^+$ in human, $F4/80^+$ in mouse; express CD14, CD163, and CD206; and are enriched in the connective tissue sheath around the bulb of the hair follicles [9, 15]. Their role was first described as collagen degradation during the catagen phase of the follicle cycle [16]: catagen is a transition period between hair growth and the terminal stage of the hair. In mouse, perifollicular macrophage number shows oscillation during the hair follicle cycle: the follicle has more macrophages during active hair growth (the anagen phase) and contains less macrophages after quiescence of the hair follicle (the telogen phase, which produces a fully keratinized hair). The perifollicular macrophages contribute to the activation of skin epithelial stem cells through Wnt signaling [17], and express growth factors such as transforming growth factor-β (TGFβ), hepatocyte growth factor (HGF), insulin-like growth factor-1 (IGF1), and basic fibroblast growth factor (FGF2). Perifollicular macrophages also sustain local iron supply of the hair follicle, as suggested by a recent study [18]. Inactivation of the iron exporter ferroportin in macrophages causes tissue iron deficiency and decreases hair follicle growth [18]. All these traits of the perifollicular macrophages promote hair growth.

When the hair follicle is damaged, macrophages are recruited and polarized to M2 activation in the mouse skin. The perifollicular macrophages in the regenerating hair follicle are $CD11b^+$, $CD45^+$, and $F4/80^+$, express $CD206^+$, and lack CD86. The regenerating hair follicle also expresses M2 macrophage markers *Arg1* and *Ym1*, as well as *Il4* [19]. At specific stages of hair follicle regeneration, the perifollicular macrophages change from $CX_3CR1^{low/med}$ to a CX_3CR1^{high} phenotype, express CCR2 and tumor necrosis factor-α (TNFα), and are $Ly6C^{low}$ [15, 20, 21], resembling alternative monocyte-derived macrophages [22]. A recent study showed that macrophages that express TREM2 (triggering receptor expressed on myeloid cells 2) are enriched within the resting hair follicle, and they produce oncostatin M, which inhibits hair growth. Oncostatin M is an IL-6 family cytokine that maintains the quiescence of hair follicle stem cells. Perifollicular macrophages hence function in maintaining the quiescence of the hair follicle stem cells and support hair regeneration after damage [23]. TREM-2 prevents macrophage apoptosis [24] and is associated with M2 macrophage activation: its expression level is increased by IL-4 in a STAT6-independent manner, and reduced by lipopolysaccharide (LPS) and interferon-γ (IFNγ) [25]. TREM-2 inhibits cytokine production induced by Toll-like receptor (TLR) activation in macrophages [25]. TREM-1 is overexpressed in human M1 macrophages, but TREM-2 is downregulated along with M2 markers [26]. The perifollicular macrophages hence have distinct functions in the hair follicle cycle, and their "M2-ness" is thought to be important for both hair follicle stem cell quiescence and hair growth. The function of perifollicular macrophages is the subject of studies as this book is being completed [15].

The hair follicles and the nails are immune-privileged environment within the skin (Fig. 1b). Both the hair follicle and the nail epithelia lack MHC-Ia antigen expression, and the hair bulb has a cyclic immunosuppressed state associated with the anagen stage [27]. Probably the scalp hair follicles have a prolonged immunosuppressed state, and the collapse of this immune-privileged milieu causes a specific autoimmune disease, alopecia areata [27, 28]. Patients with alopecia areata have elevation of serum TNFα [29], and elevated serum and skin lesion level of macrophage migration inhibitory factor [30]: both are M1 macrophage markers. Antiinflammatory intervention, for instance, the use of topical glucocorticoids, alleviates the symptoms of alopecia areata [28]. Inflammatory mediators induce alopecia in mouse; for example, inflammatory lipid mediators of the breast milk can cause skin inflammation and hair loss in infant mice [31]. Accordingly, various forms of alopecia are linked to macrophages [32]. For instance, perifollicular macrophages show a different cell-surface expression pattern in frontal fibrosing alopecia (FFA) and lichen planopilaris (LPP): LPP has a higher number of $CD68^+$ macrophages, reduced CD86, and increased CD163 and IL-4 compared with FFA samples. Lack of macrophage iron export also leads to alopecia in mice [18].

The hair follicles are connected to dermal glands, the sebaceous glands (Fig. 1a). These glands are composed of lipid-laden cells that undergo apoptosis and discharge their intracellular contents to the hair follicle by holocrine secretion: this eventually generates a fatty layer, called sebum, on the surface of the epithelium that protects the skin from dehydration. Holocrine secretion means that the gland cells undergo apoptosis and are cast away, instead of emitting their intracellular contents by exocytosis (i.e., by merocrine secretion) or by releasing cell particles (i.e., by extracellular vesicles, or apocrine secretion). Squalene is capable of neutralizing reactive oxygen species generated by UV irradiation in the skin [33]. The sebaceous gland lipids are immunoregulatory, and the composition of sebum has specific changes in various skin diseases, such as acne or psoriasis [14]. Some physiological constituents of the sebum inhibit M1 activation of macrophages [14].

The nail epithelium has constant immunosuppression, and the proximal nail matrix expresses antiinflammatory and immunosuppressive proteins; the surrounding of the nail-generating tissue has few immune cells compared to other skin regions [34, 35]. Fungal infections of the skin often become persistent in the nails, causing onychomycosis. Macrophages, upon infection with *Trichophyton rubrum*, a common fungal pathogen causing onychomycosis, produce IL-10 and fail to polarize toward the M1 state. Moreover, the fungal conidia reduce macrophage viability, and the conidia grow into hyphae inside the macrophages, eventually rupturing them [36].

2.2 *Macrophage–Melanocyte Interaction*

Dermal macrophages also interact with melanocytes (Fig. 1c–e). Melanocytes are melanin-accumulating cells of the epidermis that are attached to the basal membrane of the stratified epithelium and protect the germinative layer from UV radiation of sunlight (Fig. 1e). Melanocytes are hence critical for protecting the skin tissues from UV light [37]. Melanocytes proliferate and produce melanin in response to UV radiation, and also extend their dendrite-like processes. The latter mechanism allows the melanin-filled cytoplasm of the melanocyte to project into the epithelial layer, eventually shielding the epithelial cells from UV light. One report from 1993 shows that macrophages release a protein that increases dendrite extension of the melanocytes [38]. The nature of this protein has remained unidentified, but the elicited effects mimic the response of melanocytes to α-MSH [11]. It is known that murine macrophage RAW264.7 cells express proopiomelanocortin, the precursor molecule of α-MSH [12]. The melanocortin system has antiinflammatory effects in macrophages and inhibits M1 activation [12, 13]. More recently it was shown that UV radiation-induced proliferation of melanocytes is dependent on $Ly6c^{low}$, $MHC\text{-}II^{high}$ macrophages. These macrophages are recruited to the tissues in a CCR2-independent manner, contrasting with the extravasation process of inflammatory monocytes [39]. IL-10 may provide pro-survival cues to melanocytes and be applied in the treatment of vitiligo and other depigmenting disorders [40]. Local inhibition of M1 activation is important to maintain skin melanin. For instance, vitiligo is an acquired disorder in which depigmented macules result from mostly autoimmune loss of melanocytes. The melanocyte remnants are mostly phagocytosed by $CD68^+$ macrophages, and the clearance of debris by macrophages is essential to complete repigmentation [41] (Fig. 2a).

2.3 *M2 Macrophages in the Mammary Gland*

The skin has a specific modification of its glands that forms the mammary glands in the female body. The mammary glands, modified sweat glands, secrete breast milk during lactation. Protein and carbohydrate components of the breast milk are secreted by exocytosis (i.e., by merocrine secretion), whereas the lipid components are emitted by apocrine secretion. The apocrine secretion releases a large lipid vesicle, enveloped in the cell membrane and the endoplasmic reticule membranes. Maternal membrane lipids are hence emitted into the breast milk and deliver lipid

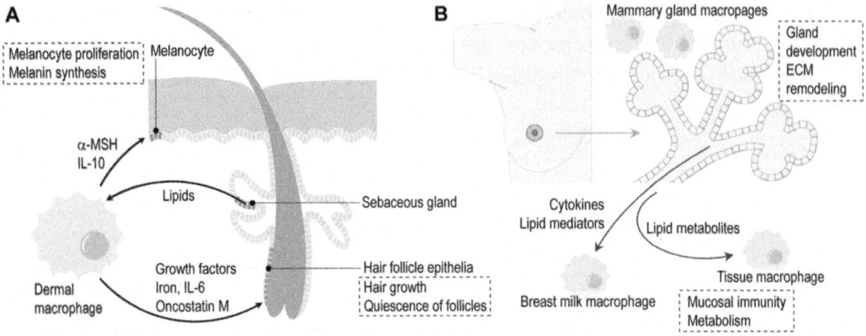

Fig. 2 Interactions between dermal macrophages and skin compartments. (**a**) Dermal macrophages are members of specific niches of the skin: they interact with the hair follicles and the melanocytes via cytokines, hormones, and lipid metabolites. (**b**) Mammary gland macrophages are necessary for the morphogenesis of the gland tubules and acini. Macrophages are emitted into the breast milk, with various mediators that can reach tissue-resident macrophages of the infant. The postpartum mammary gland is rich in M2 macrophages, which are needed for extracellular matrix (ECM) remodeling of the breast

mediators to the infant [42]. These maternal signals are decisive for the metabolic and immune development of the infant [42]. Importantly, at least one breast milk-specific lipid family, the alkylglycerols (AKGs), are metabolized by macrophages of the infant adipose tissue, producing lipid mediators that sustain a heat-generating function of the adipose tissue [42]. The breast gland has its own tissue-resident macrophage pool, which is necessary for development of the mammary gland [43]. During lactation these macrophages are emitted into the breast milk [44], where they have immune functions: they neutralize pathogens, deliver mediators, and increase infant immunity [42, 45]. Breast milk has a wide range of immune molecules, immunoglobulins, cytokines, inflammatory, and antiinflammatory mediators [46, 47]; these signals ensure immunization of the infant. In infancy the intestine is colonized by microbiota that are mainly obtained from the breast milk [48], which poses a risk for an immune response of the infant toward the microbiota and can cause necrotizing enterocolitis. Establishing a tolerogenic macrophage response toward the microbiota seems to be facilitated by breast milk signals [49]. Accordingly, breast milk components have an M2 polarizing effect on some macrophages [50], and breast milk protects the infant from necrotizing enterocolitis [49]. The underlying mechanism requires neutralization of inflammatory mediators of the breast milk. For instance, platelet-activating factor (PAF) is an inflammatory lipid mediator present in the breast milk. It is thought that PAF strengthens the infant's mucosal immunity; however, excess PAF can cause inflammation [51]. The breast milk and the infant intestine can eliminate excess PAF by the PAF-acetyl hydrolase (PAFAH, also called PLA2G7) enzyme [52, 53]. Interestingly, the PAFAH content of breast milk is increased by macrophages [51]. Moreover, breast milk AKGs are noninflammatory precursors of PAF, and they are safely transported to the adipose tissue through the lymphatic circulation of the infant [42]. When AKGs reach the adipose tissue, they are metabolized into PAF locally [42], without posing harm to the infant's intestinal mucosa.

The mammary gland macrophage population shows M2 features after pregnancy (Fig. 2b). As a result of the lipid-rich breast milk production, adipocyte size is strongly reduced during lactation, called breast involution [54]. This process is associated with remodeling of the extracellular matrix of the breast stroma. The postpartum breast macrophages are mainly M2 activated: they express arginase-1, MRC1, and the breast tissue is rich in Th2 cytokines and macrophage chemotactic factors. The main function of M2 macrophages is remodeling of the extracellular matrix of the breast [55]. However, the Th2/M2 immune milieu can increase the tolerance of cancer cells in the postpartum breast, which may explain the temporal increase of breast cancer risk after pregnancy [56].

3 M2 Macrophages in the Skeletal System

3.1 Macrophage Types of the Skeletal System

The human skeleton is built from bones and cartilages. The bones are arranged in a structure that forms a scaffold for the muscular system. Bones are interconnected to form joints and allow movements of the body parts. Some joints have special importance in the human posture, bipedalism, walking, and manual abilities, and their dysfunction causes disabling diseases (Fig. 3). These functionally important joints are complemented with cartilages: articular cartilages cover the interconnected bone heads, a cartilaginous meniscus supports the knee joint, and special forms of cartilage interconnect the spine, the ribs, the sternum, and the symphysis of the pelvic bones: all are essential for human posture and motion. Inflammation of the joints, erosion of the articular cartilage, and loss of bone mass cause rheumatoid arthritis, osteoarthritis, osteoporosis, and temporomandibular disorders, and lead to chronic joint pain, impaired mobility, loss of manual abilities, chronic inflammation, and fragility fractures (Fig. 4) [57, 58].

Macrophages are present both in the bones and in the cartilages. The bone marrow has resident macrophages, and the bone itself has highly specialized macrophages, the so-called osteoclasts, or bone-resorbing cells [62]. Bone marrow-resident macrophages support red blood cell generation and reside in the erythroblastic islands: these are nursing cells that are in contact with red blood cell progenitors, the erythroblasts. Because these bone marrow-resident macrophages are positioned in the centers of the erythroblastic islands, they are also called central macrophages: they express CD11b, VCAM-1, F4/80, CD163, and CD169 [63]. Moreover, DARC (atypical chemokine receptor 1)-expressing bone marrow macrophages maintain the quiescence of long-term hematopoietic stem cells [64]. Osteoclast precursors are not polarized, although they have a T cell suppressor effect [65]. Osteoclasts are highly specialized macrophages that enzymatically lyse the bone extracellular

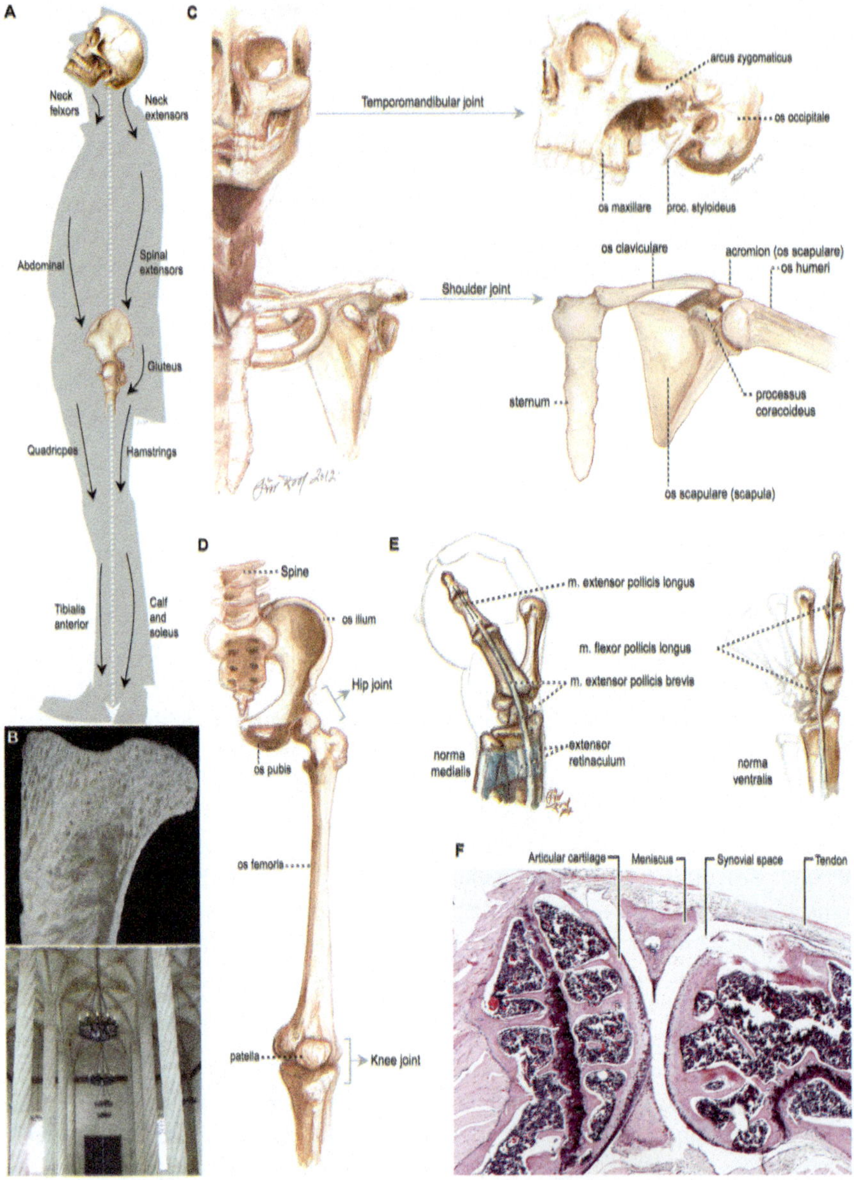

Fig. 3 The human skeletal system and the relevant joints. (**a**) The human posture allows bipedalism as a result of the interplay between the skeletal scaffold and large antigravity muscle groups. (**b**) The bones are built from trabeculae, which are arranged similarly to the arches of stone buildings. (**c–e**) Joints relevant in human physiology. (**f**) Histology of the knee joint. (Drawings (**a**, **c–e**) and photographs (**b**, **f**) by author)

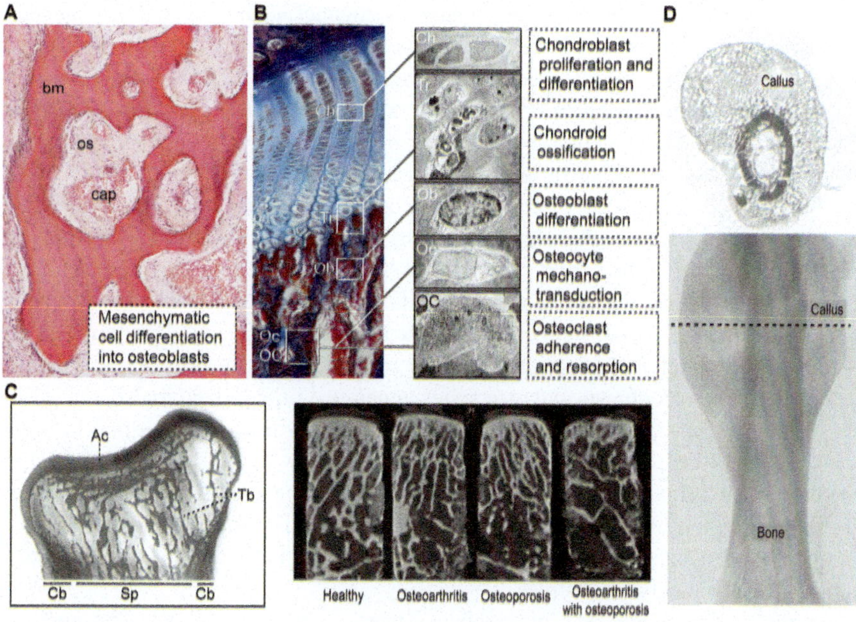

Fig. 4 Bone development, bone loss, and bone repair. (**a**) Cross section of human calvarial bone shows desmal osteogenesis. *bm* bone matrix, *cap* capillary, *os* osteoid (newly formed bone). (**b**) Endochondral ossification occurs in the epiphysis plates of long bones. Longitudinal section of human knee. *Ch* chondrocyte, *Tr* transition zone, *Ob* osteoblast, *Oc* osteocyte, *OC* osteoclast. (Images by author; reprinted from [59]. OC image is reprinted with permission from [60].) (**c**) Longitudinal section of human humerus showing compact (*Cb*) and spongy (*Sp*) bone compartments. Mechanical strength mainly relies on the amount and thickness of bone trabeculae (*Tb*) of the spongy bone and the thickness of the compact bone. *Ac* articular cartilage (image by author, reprinted from [59]). Bone diseases can deteriorate bone structure by reducing trabeculae number and thickness. (Micro-computed tomography images of rabbit bones, reprinted with permission [61]). (**d**) Regenerating bone fracture. Micro-computed tomography image of mouse femur. *Dotted line* indicates site of fracture; cross-sectional reconstruction shown on *top*. (Image by author; reprinted from [59])

matrix, hence allowing its constant remodeling. Because the bone matrix is mineralized and contains calcium and phosphate crystals, the osteoclasts also release calcium and phosphate into the blood circulation. Inflammatory signaling is essential for bone development and regeneration [66], and inflammatory joint diseases are associated with increased osteoclast number and activity (reviewed elsewhere [59, 67, 68]). As the osteoclast has a highly specialized nature, is does not fall into the M1/M2 classification of macrophage activation.

The cartilage also has its specific macrophage type, the chondroclast [69]. Chondroclasts phagocytose apoptotic cartilage cells and glycosaminoglycans during transition of cartilage to bone [69, 70]. Similar to osteoclasts, they do not fit into the M1 or M2 category, although their apoptotic cell-clearing function vests them with M2-like traitd.

The development of bone tissue is called osteogenesis. One form of osteogenesis takes place in a connective tissue cast, in which mesenchymal stem cells give rise to bone cell precursors, the osteoprogenitors, which eventually differentiate into bone-forming cells, the osteoblasts. Osteoblasts mature into osteocytes: this process is called desmal or intramembranous ossification and typically appears in the flat bones of the skull. Another form of ossification that begins with a cartilage cast of the bone is called chondral or cartilaginous ossification. Typically, it is seen in the development of long bones and the spine. Once the cartilage template is formed, its cells undergo apoptosis and leave behind an extracellular matrix scaffold. Chondroclasts clear the apoptotic debris [70]. Capillaries invade the apoptotic region, and mesenchymal stem cells colonize the scaffold. Eventually, similar to the events of the desmal ossification, the stem cells differentiate into osteoblasts and osteocytes. The osteoblasts and osteocytes generate a mineralized matrix that gives the mechanical properties of the bone. During the lifetime the bone matrix has a constant turnover: it is deposited and resorbed, along with the dynamic renewal of the osteoblasts. Bone renewal or turnover maintains bone strength and also provides for the adaptation of bone to changing mechanical needs: this latter process is bone remodeling. Physical activity remodels the microanatomy of the bone and adapts it to the needs of mechanical stress.

In the mature bone, M1 macrophages cause inflammatory diseases, such as osteoarthritis, joint cartilage inflammation, and bone erosion [71]. The synovial fluid in arthritis contains M1 macrophages and relatively fewer M2 macrophages [72, 73]. In this setting, inflammation impairs the structure of the cartilage surrounding the joints, causing painful conditions such as temporomandibular joint inflammation and arthritis. Antiinflammatory therapy aims to restore the physiological macrophage activation state, hence reducing total macrophage number and curbing M1 activation in the joint cartilage and the joint synovial fluid [59, 74, 75]. The impact of M1 macrophages in inflammatory joint diseases, such as rheumatoid arthritis, and the hurdle of developing specific antiinflammatory drugs to reduce M1 macrophages in the joints, have been extensively studied and reviewed [57, 59, 71, 74, 76–80].

3.2 M2 Macrophages and Bone Repair

When bone is broken as a result of traumatic injury, the bone-healing process is initiated. Fracture healing recapitulates the main elements of physiological bone development, including cartilage development, osteogenesis, and bone remodeling. Most studies addressing the mechanisms of bone development actually studied bone fracture healing. The initial phase of bone healing is called the inflammatory phase, and M1 macrophages are indispensable for this early priming period of fracture healing [66]. The impact of inflammatory signaling in bone healing has been extensively reviewed [81], and diseases that are associated with chronic inflammation cause osteoporosis and impede bone healing [82–84]. M2 macrophages appear after the inflammatory phase and are necessary for cartilage cast development and osteogenesis. M2 macrophages function in the development of the cartilage cast of bones. Mesenchymal stem cells and M2 macrophages synergistically increase cartilage formation in vitro by increased glycosaminoglycan synthesis [85]. In the endochondral ossification phase of bone fracture healing, when the cartilage callus is being replaced by bone, M2 macrophages are present. Induction of M2 macrophages through IL-4 and IL-13 significantly enhances bone formation in mouse [81]. Also, the cartilage callus seems to favor M2 macrophage polarization [86]. Signals that inhibit M1 activation can improve the ossification phase but also can inhibit the early priming phase of bone healing [66, 83].

4 M2 Macrophages in the Skeletal Muscles and Tendons

The skeletal muscles contain resident macrophages. Under physiological conditions, the M2 macrophages improve insulin sensitivity of the skeletal muscles [87, 88]. Moreover, endurance exercise training increases $CD206^+$, $CD163^+$ macrophage number in the muscles, whereas resistance training does not affect it [89]. Muscle fiber hypertrophy and satellite cell number are correlated with M2 macrophage number [89]. Satellite cells are stem cell reserves of the skeletal muscle. In muscle damage these cells serve for regeneration of the tissue, and M2 macrophages may promote this process [90, 91]. However, aging is also associated with an increase of skeletal muscle M2 macrophage number, which is considered unfavorable because of the increase of fibrosis within the muscle [92, 93]. Muscle fibers express neuronal NO synthase (nNOS) [59, 94]. The release of NO from the sarcolemmal membrane is necessary for proper perfusion of the working muscles [94]. The complex role of NO in the muscle is reviewed elsewhere [59]. In brief, NO is a vasodilator signal molecule in the circulatory system, and when skeletal muscle contracts, the locally produced NO dilates the supporting blood vessels [95]. This effect overcomes the hypoxia caused by the compression of the vessel walls by the contracting muscle bulk. Expression of a muscle-specific nNOS transgene in mice prevents age-related increases in M2 macrophages and also age-related muscle fibrosis [92]. Whether resistance exercise also increases M2 macrophage activation in the muscle is dependent on the age of the individual [96].

The shoulder joint is formed by a complex arrangement of muscles and tendons around the head of the humerus bone (Fig. 5). It allows very flexible motion of the arms in concert with the chest and back musculature. The movement possibilities of the shoulder are versatile, allowing a forceful and precise throwing of objects toward a goal with the support of precise manual abilities. The adaptive success of the modern human largely relies in these extraordinary manual skills. This trait has allowed development of the technical environment of human societies, eventually leading to an extreme dependence on technology in the contemporary era. Acquisition of manual abilities thus has great importance in hominin evolution [97, 98]. Functioning of the shoulder joint hence is key for everyday human life, and damage or pain arising

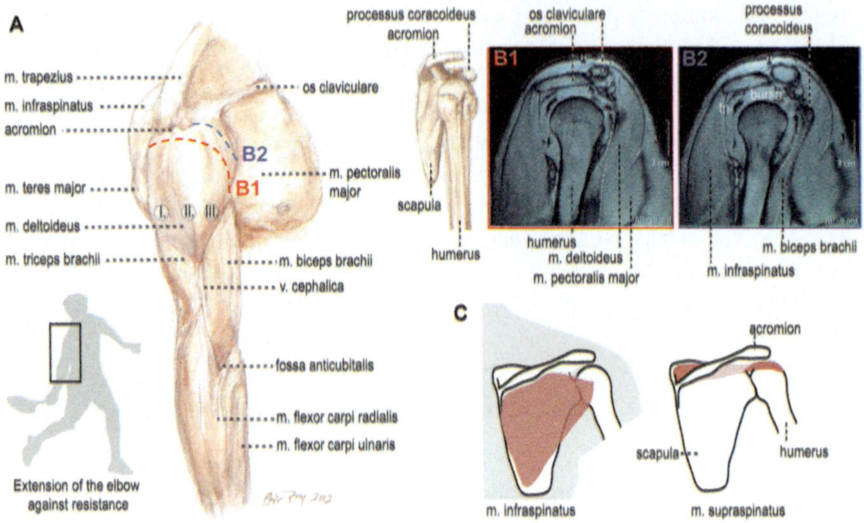

Fig. 5 Inflammation of the shoulder joint causes chronic pain and manual disability. (**a**) The shoulder is formed by a complex arrangement of muscles and tendons around the upper limb bones. The main muscle groups of the shoulder are exemplified in the motion of a tennis player. *I*, *II*, *III*, posterior, medial, and anterior parts of deltoid muscle, respectively. *Planes* **b1** and **b2** are also shown in computed tomography images in panels **b1** and **b2**. *tn* tendon of infraspinatus muscle. (**c**) Schematic representation of the infraspinatus and supraspinatus muscles, key components of the rotator cuff. (Drawings **a**, **c** by author; **b1**, **b2** with courtesy, see Acknowledgments)

from the shoulder joint impairs forearm mobility, manual skills, and quality of life. The shoulder joint can be damaged by acute injury, causing dislocation, fracture, and tendon and nerve injury. Moreover, chronic shoulder pain and immobility, which can develop from tendon damage and nerve compression in the shoulder joint, is among the most frequent musculoskeletal diseases today. Chronic rotator cuff tears, a form of tendon damage, are the main cause of chronic shoulder pain (Fig. 5).

A tear of the rotator cuff, most frequently on the tendon of the supraspinatus muscle (Fig. 5b, c), leads to atrophy of the muscle fibers, accumulation of fat within the muscle fibers, and fibrosis and infiltration of macrophages among the fibers. The stem cells differentiate into fat cells [99]; the infiltrated macrophages are polarized toward the M2 state and express CD206. For instance, in mouse, denervation or cutting the supraspinatus tendon doubles the number of M2 macrophages in the affected muscle: CD206-expressing macrophage count increases from 30–40% to 60–70% [100]. Complete loss of monocyte-derived macrophages, however, worsens muscle atrophy. In the early stage of tendon injury, M1 macrophages may prevail, leading to the concept that pain is a consequence of tendon inflammation, and the term tendonitis also reflects the inflammation of tendons in this setting [101]. Most recently however, the term tendinopathy is used, instead of tendonitis, which does reflect the more complex role of immune cells in tendon-related diseases [101].

M1 traits of macrophages have various roles in the tendons. There is a cytokine communication between the tendon cells, the tenocytes, and the macrophages. Macrophages differentiated from human peripheral blood monocytes induce the expression of human leukocyte antigen (HLA) and IL-6 in human tenocytes, isolated from a ruptured supraspinatus tendon. The cytokine signal from macrophages may include IFNγ, TNFα, and IL-1β. In turn, tenocytes stimulate proinflammatory polarization of macrophages, and macrophage–tenocyte co-cultures release more IL-6, IL-8, and MCP-1 than tenocyte cultures alone [102]. The inflammatory signature as a hallmark of early disease stage is mainly governed by interferon and NFκB signaling [103]. Interestingly, there is an increased and prolonged inflammation in the regeneration of the slow-twitch soleus muscle, whereas the regeneration of the fast-twitch muscle extensor digitorum longus has a less prominent inflammatory phase [104] (Fig. 6a). Inflammation is associated with the early repair phase of bone fracture as well, and certain inflammatory pathways are indispensable for proper healing process, as well as normal musculoskeletal physiology [59, 66]. The inflammatory cytokine wave promotes the recruitment of macrophages after injury [105], allowing macrophages to participate in the tissue reaction to damage. Shoulder injury can also include peripheral nerve damage, and the inflammatory cytokine IL-1β helps nerve regeneration by promoting Schwann cell proliferation, de-differentiation, and axon outgrowth [106, 107]. Moreover, hypoxia around the nerve injury triggers HIF1α activation in macrophages, eventually leading to vascular endothelial growth factor-A (VEGFA) release. VEGFA is a signal for capillary growth and nerve regeneration, and VEGFA-expressing macrophages are considered M2 (by some classification M2d) macrophages. Macrophages in the damaged muscle also release IL-10 and TGFβ, which aid regeneration and resolution of inflammation [108]. Accordingly, advanced-stage rotator cuff tear is associated with

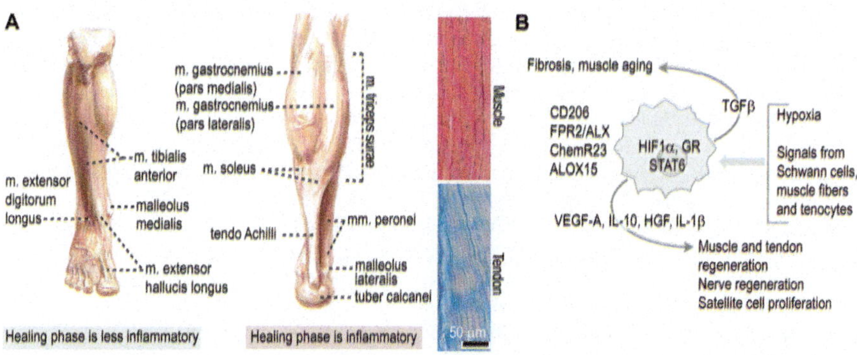

Fig. 6 Macrophages have site-specific functions in the muscles and tendons. (**a**) Antigravity muscles of the human leg. The regeneration of the fast-twitch muscle extensor digitorum longus has a less prominent inflammatory phase; inflammation is prolonged in the regeneration of the slow-twitch soleus muscle. (**b**) Skeletal muscle-associated macrophages can either cause fibrosis and promote aging of the muscle, or can improve its functions and allow the regeneration of muscles and tendons. Macrophage functions are shaped by local cues, such as signals from the neighboring cells. (Drawings and images by author)

STAT6 and glucocorticoid receptor activation [103]. The early disease stage also has its own pro-resolving factors, such as formylpeptide receptor (FPR2/ALX) and the resolving receptor ChemR23, and the relief from symptoms is associated with *CD206* and *ALOX15* (encoding a lipoxygenase) mRNA expression [103]. CD206-expressing macrophages produce hepatocyte growth factor in injured muscle, which may aid muscle fiber regeneration [109].

Similar to the interplay between macrophages, Schwann cells, and tenocytes, a satellite cell–macrophage interplay occurs in the muscle (Fig. 6b). Satellite cells are stem cells that reside within the skeletal muscles and have the potential to generate new muscle. Macrophages and conditioned media of macrophages induce proliferation of satellite cells in vitro [110]. Mesenchymal stromal cells, such as the satellite cells, can release extracellular vesicles which elicit M2 activation in macrophages. This extracellular vesicle-induced macrophage activation seems to improve tendon regeneration and reduce inflammation [111].

References

1. Merad M, Ginhoux F, Collin M. Origin, homeostasis and function of Langerhans cells and other langerin-expressing dendritic cells. Nat Rev Immunol. 2008;8:935–47.
2. Wang Y, Szretter KJ, Vermi W, Gilfillan S, Rossini C, Cella M, Barrow AD, Diamond MS, Colonna M. IL-34 is a tissue-restricted ligand of CSF1R required for the development of Langerhans cells and microglia. Nat Immunol. 2012;13:753–60.
3. Perdiguero EG, Klapproth K, Schulz C, Busch K, Azzoni E, Crozet L, Garner H, Trouillet C, de Bruijn MF, Geissmann F, Rodewald HR. Tissue-resident macrophages originate from yolk-sac-derived erythro-myeloid progenitors. Nature. 2015;518(7540):547–51.
4. Doebel T, Voisin B, Nagao K. Langerhans cells: the macrophage in dendritic cell clothing. Trends Immunol. 2017;38:817–28.
5. Collin M, Bigley V. Human dendritic cell subsets: an update. Immunology. 2018;154:3–20.
6. Wollenberg A, Oppel T, Schottdorf E-M, Günther S, Moderer M, Mommaas M. Expression and function of the mannose receptor CD206 on epidermal dendritic cells in inflammatory skin diseases. J Invest Dermatol. 2002;118:327–34.
7. Dupasquier M, Stoitzner P, Wan H, Cerqueira D, van Oudenaren A, Voerman JS, Denda-Nagai K, Irimura T, Raes G, Romani N, Leenen PJ. The dermal microenvironment induces the expression of the alternative activation marker CD301/mMGL in mononuclear phagocytes, independent of IL-4/IL-13 signaling. J Leukoc Biol. 2006;80:838–49.
8. Törőcsik D, Bárdos H, Nagy L, Ádány R. Identification of factor XIII-A as a marker of alternative macrophage activation. Cell Mol Life Sci CMLS. 2005;62:2132–9.
9. Yanez DA, Lacher RK, Vidyarthi A, Colegio OR. The role of macrophages in skin homeostasis. Pflugers Arch. 2017;469:455–63.
10. Mendoza-Coronel E, Ortega E. Macrophage polarization modulates FcγR- and CD13-mediated phagocytosis and reactive oxygen species production, independently of receptor membrane expression. Front Immunol. 2017;8:303–303.
11. Swope V, Abdel-Malek Z. Significance of the melanocortin 1 and endothelin B receptors in melanocyte homeostasis and prevention of sun-induced genotoxicity. Front Genet. 2016;7:146.
12. Star RA, Rajora N, Huang J, Stock RC, Catania A, Lipton JM. Evidence of autocrine modulation of macrophage nitric oxide synthase by alpha-melanocyte-stimulating hormone. Proc Natl Acad Sci USA. 1995;92:8016–20.

13. Taylor AW. The immunomodulating neuropeptide alpha-melanocyte-stimulating hormone (alpha-MSH) suppresses LPS-stimulated TLR4 with IRAK-M in macrophages. J Neuroimmunol. 2005;162:43–50.
14. Lovaszi M, Mattii M, Eyerich K, Gacsi A, Csanyi E, Kovacs D, Ruhl R, Szegedi A, Kemeny L, Stahle M, Zouboulis CC, Eyerich S, Torocsik D. Sebum lipids influence macrophage polarization and activation. Br J Dermatol. 2017;177:1671–82.
15. Muneeb F, Hardman JA, Paus R. Hair growth control by innate immunocytes: perifollicular macrophages revisited. Exp Dermatol. 2019;28:425–31.
16. Parakkal PF. Role of macrophages in collagen resorption during hair growth cycle. J Ultrastruct Res. 1969;29:210–7.
17. Castellana D, Paus R, Perez-Moreno M. Macrophages contribute to the cyclic activation of adult hair follicle stem cells. PLoS Biol. 2014;12:e1002002–e1002002.
18. Recalcati S, Gammella E, Buratti P, Doni A, Anselmo A, Locati M, Cairo G. Macrophage ferroportin is essential for stromal cell proliferation in wound healing. Haematologica. 2019;104:47–58.
19. Chu S-Y, Chou C-H, Huang H-D, Yen M-H, Hong H-C, Chao P-H, Wang Y-H, Chen P-Y, Nian S-X, Chen Y-R, Liou L-Y, Liu Y-C, Chen H-M, Lin F-M, Chang Y-T, Chen C-C, Lee OK. Mechanical stretch induces hair regeneration through the alternative activation of macrophages. Nat Commun. 2019;10:1524.
20. Rahmani W, Liu Y, Rosin NL, Kline A, Raharjo E, Yoon J, Stratton JA, Sinha S, Biernaskie J. Macrophages promote wound-induced hair follicle regeneration in a CX3CR1- and TGF-β1-dependent manner. J Invest Dermatol. 2018;138:2111–22.
21. Kasuya A, Ito T, Tokura Y. M2 macrophages promote wound-induced hair neogenesis. J Dermatol Sci. 2018;91:250–5.
22. Dal-Secco D, Wang J, Zeng Z, Kolaczkowska E, Wong CHY, Petri B, Ransohoff RM, Charo IF, Jenne CN, Kubes P. A dynamic spectrum of monocytes arising from the in situ reprogramming of $CCR2^{+}$ monocytes at a site of sterile injury. J Exp Med. 2015;212:447–56.
23. Wang ECE, Dai Z, Ferrante AW, Drake CG, Christiano AM. A subset of $TREM2^{+}$ dermal macrophages secretes oncostatin M to maintain hair follicle stem cell quiescence and inhibit hair growth. Cell Stem Cell. 2019;24:654–669.e656.
24. Wu K, Byers DE, Jin X, Agapov E, Alexander-Brett J, Patel AC, Cella M, Gilfilan S, Colonna M, Kober DL, Brett TJ, Holtzman MJ. TREM-2 promotes macrophage survival and lung disease after respiratory viral infection. J Exp Med. 2015;212:681–97.
25. Turnbull IR, Gilfillan S, Cella M, Aoshi T, Miller M, Piccio L, Hernandez M, Colonna M. Cutting edge: TREM-2 attenuates macrophage activation. J Immunol. 2006;177:3520–4.
26. Subramanian S, Pallati PK, Sharma P, Agrawal DK, Nandipati KC. TREM-1 associated macrophage polarization plays a significant role in inducing insulin resistance in obese population. J Transl Med. 2017;15:85–85.
27. Ito T, Meyer K, Ito N, Paus R. Immune privilege and the skin. In: Nickoloff BJ, Nestle FO, editors. Dermatologic Immunity. Curr Dir Autoimmun (Basel, Karger). 2008;10:27–52.
28. Kam Lun EH, Alexander KCL. Alopecia areata. Recent Patents Inflamm Allergy Drug Discov. 2011;5:98–107.
29. Kasumagic-Halilovic E, Prohic A, Cavaljuga S. Tumor necrosis factor-alpha in patients with alopecia areata. Indian J Dermatol. 2011;56:494–6.
30. Salem SA, Asaad MK, Elsayed SB, Sehsah HM. Evaluation of macrophage migration inhibitory factor (MIF) levels in serum and lesional skin of patients with alopecia areata. Int J Dermatol. 2016;55:1357–61.
31. Wan Y, Saghatelian A, Chong LW, Zhang CL, Cravatt BF, Evans RM. Maternal PPAR gamma protects nursing neonates by suppressing the production of inflammatory milk. Genes Dev. 2007;21:1895–908.
32. Skoutelis A, Freinkel RK, Kaufman DS, Leibovich SJ. Angiogenic activity is defective in monocytes from patients with alopecia universalis. J Invest Dermatol. 1990;95:139–43.

33. Lovászi M, Szegedi A, Zouboulis CC, Törőcsik D. Sebaceous-immunobiology is orchestrated by sebum lipids. Dermatoendocrinology. 2017;9:e1375636–e1375636.
34. Mescher AL. Macrophages and fibroblasts during inflammation and tissue repair in models of organ regeneration. Regeneration (Oxf). 2017;4:39–53.
35. Ito T, Ito N, Saathoff M, Stampachiacchiere B, Bettermann A, Bulfone-Paus S, Takigawa M, Nickoloff BJ, Paus R. Immunology of the human nail apparatus: the nail matrix is a site of relative immune privilege. J Invest Dermatol. 2005;125:1139–48.
36. Campos MRM, Russo M, Gomes E, Almeida SR. Stimulation, inhibition and death of macrophages infected with *Trichophyton rubrum*. Microbes Infect. 2006;8:372–9.
37. Yaar M, Park H-Y. Melanocytes: a window into the nervous system. J Invest Dermatol. 2012;132:835–45.
38. Suzuki S, Takayuki K, Takeuchi T. Macrophages release melanocyte dendrite extension factor in response to ultra-violet ray. In Vitro Cell Dev Biol Anim. 1993;29A:419–26.
39. Handoko HY, Rodero MP, Boyle GM, Ferguson B, Engwerda C, Hill G, Konrad Muller H, Khosrotehrani K, Walker GJ. UVB-induced melanocyte proliferation in neonatal mice driven by CCR2-independent recruitment of Ly6clow MHCIIhi macrophages. J Invest Dermatol. 2013;133:1803–12.
40. Zhou J, Ling J, Song J, Wang Y, Feng B, Ping F. Interleukin 10 protects primary melanocyte by activation of Stat-3 and PI3K/Akt/NF-κB signaling pathways. Cytokine. 2016;83:275–81.
41. Oiso N, Tanemura A, Kotobuki Y, Kimura M, Katayama I, Kawada A. Role of macrophage infiltration in successful repigmentation in a new periphery-spreading vitiligo lesion in a male Japanese patient. J Dermatol. 2013;40:915–8.
42. Yu H, Dilbaz S, Coßmann J, Hoang AC, Diedrich V, Herwig A, Harauma A, Hoshi Y, Moriguchi T, Landgraf K, Körner A, Lucas C, Brodesser S, Balogh L, Thuróczy J, Karemore G, Kuefner MS, Park EA, Rapp C, Travers JB, Röszer T. Breast milk alkylglycerols sustain beige adipocytes through adipose tissue macrophages. J Clin Invest. 2019;129:2485–99.
43. Gouon-Evans V, Lin EY, Pollard JW. Requirement of macrophages and eosinophils and their cytokines/chemokines for mammary gland development. Breast Cancer Res BCR. 2002;4:155–64.
44. Ichikawa M, Sugita M, Takahashi M, Satomi M, Takeshita T, Araki T, Takahashi H. Breast milk macrophages spontaneously produce granulocyte-macrophage colony-stimulating factor and differentiate into dendritic cells in the presence of exogenous interleukin-4 alone. Immunology. 2003;108:189–95.
45. Maschmann J, Goelz R, Witzel S, Strittmatter U, Steinmassl M, Jahn G, Hamprecht K. Characterization of human breast milk leukocytes and their potential role in cytomegalovirus transmission to newborns. Neonatology. 2015;107:213–9.
46. Garofalo R. Cytokines in human milk. J Pediatr. 2010;156:S36–40.
47. Bardanzellu F, Fanos V, Reali A. "Omics" in human colostrum and mature milk: looking to old data with new eyes. Nutrients. 2017;9(8):843.
48. Stewart CJ, Ajami NJ, O'Brien JL, Hutchinson DS, Smith DP, Wong MC, Ross MC, Lloyd RE, Doddapaneni H, Metcalf GA, Muzny D, Gibbs RA, Vatanen T, Huttenhower C, Xavier RJ, Rewers M, Hagopian W, Toppari J, Ziegler A-G, She J-X, Akolkar B, Lernmark A, Hyoty H, Vehik K, Krischer JP, Petrosino JF. Temporal development of the gut microbiome in early childhood from the TEDDY study. Nature. 2018;562:583–8.
49. He Y, Lawlor NT, Newburg DS. Human milk components modulate Toll-like receptor-mediated inflammation. Adv Nutr. 2016;7:102–11.
50. Panahipour L, Kochergina E, Kreissl A, Haiden N, Gruber R. Milk modulates macrophage polarization in vitro. Cytokine X. 2019;1:100009.
51. Du Y, Yang M, Wei W, Huynh HD, Herz J, Saghatelian A, Wan Y. Macrophage VLDL receptor promotes PAFAH secretion in mother's milk and suppresses systemic inflammation in nursing neonates. Nat Commun. 2012;3:1008.
52. Furukawa M, Narahara H, Yasuda K, Johnston JM. Presence of platelet-activating factor-acetyl hydrolase in milk. J Lipid Res. 1993;34:1603–9.

53. Lu J, Pierce M, Franklin A, Jilling T, Stafforini DM, Caplan M. Dual roles of endogenous platelet-activating factor acetylhydrolase in a murine model of necrotizing enterocolitis. Pediatr Res. 2010;68:225–30.
54. Wang QA, Song A, Chen W, Schwalie PC, Zhang F, Vishvanath L, Jiang L, Ye R, Shao M, Tao C, Gupta RK, Deplancke B, Scherer PE. Reversible de-differentiation of mature white adipocytes into preadipocyte-like precursors during lactation. Cell Metab. 2018;28:282–288.e283.
55. O'Brien J, Lyons T, Monks J, Lucia MS, Wilson RS, Hines L, Man Y-G, Borges V, Schedin P. Alternatively activated macrophages and collagen remodeling characterize the postpartum involuting mammary gland across species. Am J Pathol. 2010;176:1241–55.
56. Lambe M, Hsieh C-C, Trichopoulos D, Ekbom A, Pavia M, Adami H-O. Transient increase in the risk of breast cancer after giving birth. N Engl J Med. 1994;331:5–9.
57. Henrotin YE, Bruckner P, Pujol JP. The role of reactive oxygen species in homeostasis and degradation of cartilage. Osteoarthr Cartilage. 2003;11:747–55.
58. Hofkens W, Storm G, van den Berg W, van Lent P. Inhibition of M1 macrophage activation in favour of M2 differentiation by liposomal targeting of glucocorticoids to the synovial lining during experimental arthritis. Ann Rheum Dis. 2011;70:A40–A40.
59. Rőszer T. Nitric oxide signaling and nitrosative stress in the musculoskeletal system. In: Laher I, editor. Systems biology of free radicals and antioxidants. Berlin/Heidelberg: Springer; 2014. p. 2895–926.
60. Melhus G, Brorson SH, Baekkevold ES, Andersson G, Jemtland R, Olstad OK, Reinholt FP. Gene expression and distribution of key bone turnover markers in the callus of estrogen-deficient, vitamin D-depleted rats. Calcif Tissue Int. 2010;87:77–89.
61. Bellido M, Lugo L, Roman-Blas JA, Castaneda S, Caeiro JR, Dapia S, Calvo E, Largo R, Herrero-Beaumont G. Subchondral bone microstructural damage by increased remodelling aggravates experimental osteoarthritis preceded by osteoporosis. Arthritis Res Ther. 2010;12:R152.
62. Dey A, Allen J, Hankey-Giblin PA. Ontogeny and polarization of macrophages in inflammation: blood monocytes versus tissue macrophages. Front Immunol. 2015;5:683–683.
63. Seu KG, Papoin J, Fessler R, Hom J, Huang G, Mohandas N, Blanc L, Kalfa TA. Unraveling macrophage heterogeneity in erythroblastic islands. Front Immunol. 2017;8:1140–1140.
64. Hur J, Choi JI, Lee H, Nham P, Kim TW, Chae CW, Yun JY, Kang JA, Kang J, Lee SE, Yoon CH, Boo K, Ham S, Roh TY, Jun JK, Lee H, Baek SH, Kim HS. CD82/KAI1 maintains the dormancy of long-term hematopoietic stem cells through interaction with DARC-expressing macrophages. Cell Stem Cell. 2016;18:508–21.
65. Charles JF, Hsu L-Y, Niemi EC, Weiss A, Aliprantis AO, Nakamura MC. Inflammatory arthritis increases mouse osteoclast precursors with myeloid suppressor function. J Clin Invest. 2012;122:4592–605.
66. Röszer T. Inflammation as death or life signal in diabetic fracture healing. Inflamm Res. 2011;60:3–10.
67. Sato K, Takayanagi H. Osteoclasts, rheumatoid arthritis, and osteoimmunology. Curr Opin Rheumatol. 2006;18:419–26.
68. McHugh J. Precursors of 'bad' osteoclasts identified in arthritis. Nat Rev Rheumatol. 2020;16:63–63.
69. Lewinson D, Silbermann M. Chondroclasts and endothelial cells collaborate in the process of cartilage resorption. Anat Rec. 1992;233:504–14.
70. Wlodarski K, Brodzikowska A, Kuzaka B. Are chondroclasts and osteoclasts identical? Folia Biol. 2014;62:143–7.
71. Martel-Pelletier J, Barr AJ, Cicuttini FM, Conaghan PG, Cooper C, Goldring MB, Goldring SR, Jones G, Teichtahl AJ, Pelletier J-P. Osteoarthritis. Nat Rev Dis Primers. 2016;2:16072.
72. Liu B, Zhang M, Zhao J, Zheng M, Yang H. Imbalance of M1/M2 macrophages is linked to severity level of knee osteoarthritis. Exp Ther Med. 2018;16:5009–14.

73. Wood MJ, Leckenby A, Reynolds G, Spiering R, Pratt AG, Rankin KS, Isaacs JD, Haniffa MA, Milling S, Hilkens CMU. Macrophage proliferation distinguishes 2 subgroups of knee osteoarthritis patients. JCI Insight. 2019;44(2):e125325.
74. Kou X-X, Li C-S, He D-Q, Wang X-D, Hao T, Meng Z, Zhou Y-H, Gan Y-H. Estradiol promotes M1-like macrophage activation through cadherin-11 to aggravate temporomandibular joint inflammation in rats. J Immunol. 2015;194:2810–8.
75. Zhao L, Xiao E, He L, Duan D, He Y, Chen S, Zhang Y, Gan Y. Reducing macrophage numbers alleviates temporomandibular joint ankylosis. Cell Tissue Res. 2020;379(3):521–36.
76. Laria A, Lurati A, Marrazza M, Mazzocchi D, Re KA, Scarpellini M. The macrophages in rheumatic diseases. J Inflamm Res. 2016;9:1–11.
77. Culemann S, Grüneboom A, Krönke G. Chapter three: Origin and function of synovial macrophage subsets during inflammatory joint disease. In: Alt F, editor. Advances in immunology, vol. 143. New York: Academic; 2019. p. 75–98.
78. Abramson SB. Nitric oxide in inflammation and pain associated with osteoarthritis. Arthritis Res Ther. 2008;10(suppl 2):S2.
79. Hong HS, Son Y. Substance P ameliorates collagen II-induced arthritis in mice via suppression of the inflammatory response. Biochem Biophys Res Commun. 2014;453:179–84.
80. Menendez-Gutierrez MP, Röszer T, Fuentes L, Nunez V, Escolano A, Redondo JM, De Clerck N, Metzger D, Valledor AF, Ricote M. Retinoid X receptors orchestrate osteoclast differentiation and postnatal bone remodeling. J Clin Invest. 2015;125(2):809–23.
81. Schlundt C, El Khassawna T, Serra A, Dienelt A, Wendler S, Schell H, van Rooijen N, Radbruch A, Lucius R, Hartmann S, Duda GN, Schmidt-Bleek K. Macrophages in bone fracture healing: their essential role in endochondral ossification. Bone. 2018;106:78–89.
82. Hamann C, Kirschner S, Gunther KP, Hofbauer LC. Bone, sweet bone—osteoporotic fractures in diabetes mellitus. Nat Rev Endocrinol. 2012;8:297–305.
83. Röszer T, Jozsa T, Kiss-Toth ED, De Clerck N, Balogh L. Leptin receptor deficient diabetic (db/db) mice are compromised in postnatal bone regeneration. Cell Tissue Res. 2014;356:195–206.
84. Kojima H, Kim J, Chan L. Emerging roles of hematopoietic cells in the pathobiology of diabetic complications. Trends Endocrinol Metab. 2014;25:178–87.
85. Sesia SB, Duhr R, Medeiros da Cunha C, Wolf F, Padovan E, Spagnoli GC, Martin I, Barbero A. M2-macrophages modulate the cartilage-forming capacity of human bone marrow-derived mesenchymal stem/progenitor cell. Osteoarthr Cartilage. 2014;22:S147–8.
86. Huang J, Zhou Y, Wang Y, Cai X, Wang Y. The mutual effects between macrophages and cartilage templates in the process of subcutaneous endochondral bone formation. RSC Adv. 2018;8:23679–87.
87. Hevener AL, Olefsky JM, Reichart D, Nguyen MT, Bandyopadyhay G, Leung HY, Watt MJ, Benner C, Febbraio MA, Nguyen AK, Folian B, Subramaniam S, Gonzalez FJ, Glass CK, Ricote M. Macrophage PPAR gamma is required for normal skeletal muscle and hepatic insulin sensitivity and full antidiabetic effects of thiazolidinediones. J Clin Invest. 2007;117:1658–69.
88. Odegaard JI, Ricardo-Gonzalez RR, Goforth MH, Morel CR, Subramanian V, Mukundan L, Red Eagle A, Vats D, Brombacher F, Ferrante AW, Chawla A. Macrophage-specific PPARgamma controls alternative activation and improves insulin resistance. Nature. 2007;447:1116–20.
89. Walton RG, Kosmac K, Mula J, Fry CS, Peck BD, Groshong JS, Finlin BS, Zhu B, Kern PA, Peterson CA. Human skeletal muscle macrophages increase following cycle training and are associated with adaptations that may facilitate growth. Sci Rep. 2019;9:969–969.
90. Rigamonti E, Zordan P, Sciorati C, Rovere-Querini P, Brunelli S. Macrophage plasticity in skeletal muscle repair. Biomed Res Int. 2014;2014:560629.
91. Ruffell D, Mourkioti F, Gambardella A, Kirstetter P, Lopez RG, Rosenthal N, Nerlov C. A CREB-C/EBPbeta cascade induces M2 macrophage-specific gene expression and promotes muscle injury repair. Proc Natl Acad Sci USA. 2009;106:17475–80.

92. Wang Y, Wehling-Henricks M, Samengo G, Tidball JG. Increases of M2a macrophages and fibrosis in aging muscle are influenced by bone marrow aging and negatively regulated by muscle-derived nitric oxide. Aging Cell. 2015;14:678–88.
93. Wehling-Henricks M, Jordan M, Gotoh T, Grody W, Roos K, Tidball J. Arginine metabolism by macrophages promotes cardiac and muscle fibrosis in mdx muscular dystrophy. PLoS One. 2010;5:e10763.
94. Percival JM, Anderson KN, Huang P, Adams ME, Froehner SC. Golgi and sarcolemmal neuronal NOS differentially regulate contraction-induced fatigue and vasoconstriction in exercising mouse skeletal muscle. J Clin Invest. 2010;120:816–26.
95. Rőszer T. Harboring of NOS to the cell membrane. In: The biology of subcellular nitric oxide. Dordrecht: Springer; 2012. p. 105–32.
96. Przybyla B, Gurley C, Harvey JF, Bearden E, Kortebein P, Evans WJ, Sullivan DH, Peterson CA, Dennis RA. Aging alters macrophage properties in human skeletal muscle both at rest and in response to acute resistance exercise. Exp Gerontol. 2006;41:320–7.
97. Susman RL. Hand function and tool behavior in early hominids. J Hum Evol. 1998;35:23–46.
98. Marzke MW, Marzke RF. Evolution of the human hand: approaches to acquiring, analysing and interpreting the anatomical evidence. J Anat. 2000;197(t 1):121–40.
99. Schubert MF, Noah AC, Bedi A, Gumucio JP, Mendias CL. Reduced myogenic and increased adipogenic differentiation capacity of rotator cuff muscle stem cells. JBJS. 2019;101:228–38.
100. Krieger JR, Tellier LE, Ollukaren MT, Temenoff JS, Botchwey EA. Quantitative analysis of immune cell subset infiltration of supraspinatus muscle after severe rotator cuff injury. Regen Eng Transl Med. 2017;3:82–93.
101. Abraham AC, Shah SA, Thomopoulos S. Targeting inflammation in rotator cuff tendon degeneration and repair. Tech Shoulder Elbow Surg. 2017;18:84–90.
102. Stolk M, Klatte-Schulz F, Schmock A, Minkwitz S, Wildemann B, Seifert M. New insights into tenocyte–immune cell interplay in an in vitro model of inflammation. Sci Rep. 2017;7:9801.
103. Dakin SG, Martinez FO, Yapp C, Wells G, Oppermann U, Dean BJF, Smith RDJ, Wheway K, Watkins B, Roche L, Carr AJ. Inflammation activation and resolution in human tendon disease. Sci Transl Med. 2015;7:311ra173–311ra173.
104. Zimowska M, Kasprzycka P, Bocian K, Delaney K, Jung P, Kuchcinska K, Kaczmarska K, Gladysz D, Streminska W, Ciemerych MA. Inflammatory response during slow- and fast-twitch muscle regeneration. Muscle Nerve. 2017;55:400–9.
105. Fregnan F, Muratori L, Simões AR, Giacobini-Robecchi MG, Raimondo S. Role of inflammatory cytokines in peripheral nerve injury. Neural Regen Res. 2012;7:2259–66.
106. Temporin K, Tanaka H, Kuroda Y, Okada K, Yachi K, Moritomo H, Murase T, Yoshikawa H. Interleukin-1 beta promotes sensory nerve regeneration after sciatic nerve injury. Neurosci Lett. 2008;440:130–3.
107. Chen G, Luo X, Wang W, Wang Y, Zhu F, Wang W. Interleukin-1β promotes Schwann cells de-differentiation in Wallerian degeneration via the c-JUN/AP-1 pathway. Front Cell Neurosci. 2019;13:304.
108. Liu P, Peng J, Han G-H, Ding X, Wei S, Gao G, Huang K, Chang F, Wang Y. Role of macrophages in peripheral nerve injury and repair. Neural Regen Res. 2019;14:1335–42.
109. Sakaguchi S, Shono J, Suzuki T, Sawano S, Anderson JE, Do MK, Ohtsubo H, Mizunoya W, Sato Y, Nakamura M, Furuse M, Yamada K, Ikeuchi Y, Tatsumi R. Implication of anti-inflammatory macrophages in regenerative moto-neuritogenesis: promotion of myoblast migration and neural chemorepellent semaphorin 3A expression in injured muscle. Int J Biochem Cell Biol. 2014;54:272–85.
110. Cantini M, Massimino ML, Bruson A, Catani C, Dallalibera L, Carraro U. Macrophages regulate proliferation and differentiation of satellite cells. Biochem Biophys Res Commun. 1994;202:1688–96.
111. Chamberlain CS, Clements AEB, Kink JA, Choi U, Baer GS, Halanski MA, Hematti P, Vanderby R. Extracellular vesicle-educated macrophages promote early Achilles tendon healing. Stem Cells. 2019;37:652–62.

M2 Macrophages in the Circulatory, Respiratory, and Excretory Organs

Keywords Lung · Alveolar macrophage · Lung cancer · Bronchitis · Allergy · Heart · Atherosclerosis · Cardiac repair · Fibrosis · Myocardium · Adventitia · Macrophage · Angiogenesis · Tumor

1 Introduction

This chapter reviews M2 macrophage functions in three organ systems that complement each other and have synergistic functions: the circulatory, respiratory, and excretory systems. These organ systems evolved when the complexity of the body plan made it necessary to develop fluid compartments around the cells. In the human body these fluid compartments are the interstitial fluid, the lymph, the blood, and other special fluid compartments such as the cerebrospinal fluid, the aqueous humor in the eye, and the endolymph in the inner ear. The circulatory system generates, channels, and keeps in motion these fluid components, and the respiratory system ensures oxygenation and proper carbon dioxide tension of the fluid compartments. The excretory organs maintain volume and ionic composition, and clear potentially harmful materials, such as metabolic end products. Macrophages are present in all these three organ systems: they are enriched in the vessel walls, within the heart, in the respiratory tract, and in the lung alveoli, kidney, and lower urinary tract.

2 M2 Macrophages in the Circulatory System

2.1 M2 Macrophages in the Vasculature

Early development of macrophage progenitors is associated with the development of the vasculature [1]. Yolk sac progenitors of macrophages are spherical, mobile, CX_3CR1^+, $CD45^+$, and Kit^- cells. They spread within the developing vascular network, either by rolling on the endothelia or traveling with the blood flow [1] (also

T. Röszer, *The M2 Macrophage*, Progress in Inflammation Research 86,
https://doi.org/10.1007/978-3-030-50480-9_8

see chapter "Mechanisms Which Control the Size of M2 Macrophage-Dominated Tissue Macrophage Niches"). As the vasculature grows into the developing organs between embryonic days 9 and 12.5, with peak on embryonic day 10.5 in mouse, macrophage progenitors settle in tissues. They change their shape into a dendrite-bearing form and begin to self-renew and further colonize the organs [1] (also see chapter "Mechanisms Which Control the Size of M2 Macrophage-Dominated Tissue Macrophage Niches"). There are intra-aortic hematopoietic cell clusters in the caudal part of the E10.5 mouse embryo [2], and the adventitia of the aorta retains embryonic macrophage progenitors after birth. This adventitial niche generates macrophages locally. The macrophage colony-forming unit of the aorta contains Ki67$^+$, Lin$^-$, c-Kit$^+$, CD135$^-$, CD115$^+$, CX_3CR1^+, Ly6C$^+$, and CD11b$^-$ macrophages [3]. The adventitia surrounds the external surface of the vessels and consists of a single layer of epithelial cells, with connective tissue beneath (Fig. 1). The adventitial macrophages are M2 macrophages and express CD206 and CD136 [4]. Macrophages of the adventitia can interact with other cells, such as fibroblasts, myofibroblasts, dendritic cells, and endothelial progenitors. These interactions determine local inflammatory status in the adventitia and affect vessel growth within the wall of large arteries [5, 6]. These vessels perfuse thick vessel walls and are called vasa vasorum, which means vessels of the vessels. In atherosclerosis these vessels may grow into a deeper region of vessel wall, which is called the tunica

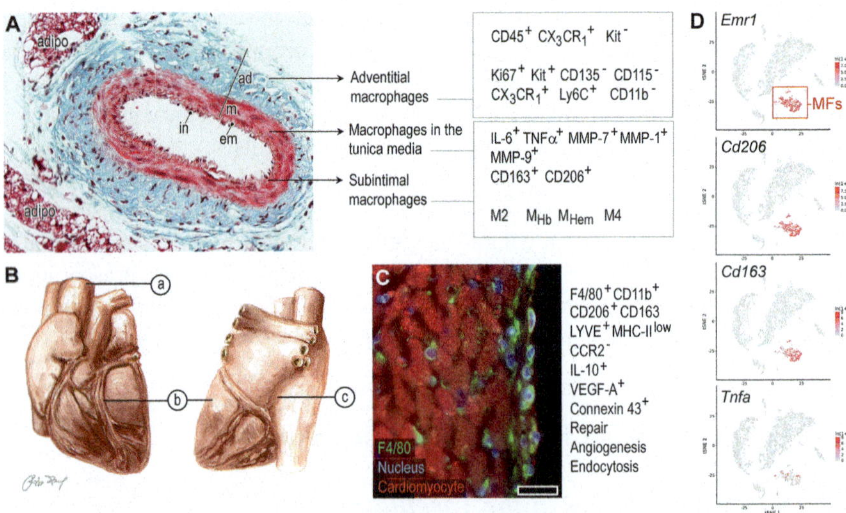

Fig. 1 Macrophages associated with the vessel wall and the heart. (**a**) Histology of a blood vessel, trichrome staining. *ad* adventitia with blue-stained collagen fibers, *m* muscle cells, *em* elastic membrane of the tunica media, *in* intima, *adipo* adipose tissue surrounding the vessel. The adventitia contains resident macrophages that locally replenish. The tunica media and the tunica adventitia are infiltrated by macrophages in atherosclerosis and vascular wall inflammation. (**b**) Heart-associated macrophages are confined to the aortic wall (*a*), the heart muscle (*b*), and the pacemaker region (*c*). (**c**) Heart-associated macrophages in the mouse heart. (Images and drawings by author.) (**d**) Single-cell sequencing data of heart and vessel wall resident macrophages of the mouse. (Retrieved from the Tabula Muris Consortium [37])

media; this eventually disrupts the immune privilege of the tunica media and leads to increased leukocyte infiltration and inflammation. In the media, the immune cells are mainly localized around the vasa vasorum [7]. Adventitial macrophages hence have a role in vascular inflammation, and recently it has been shown that vascular inflammation may be initiated in the adventitia and progress in an outside-in manner [5, 6]. In vascular inflammation and in atherosclerosis the subintimal space, the space beneath the endothelial lining of the vessels, is rich in macrophages (Fig. 1). The atherosclerotic plaque becomes filled with lipid-laden macrophages, called foam cells [4, 8]. These cells have a major role in vascular inflammation and atherosclerotic plaque development. The activation state of plaque macrophages in the progression, stabilization, rupture, and regression of atherosclerosis has been extensively reviewed. There are many opposing views about the assignment of plaque macrophages to M1 or M2 functional categories [9, 10]. In brief, M1 macrophages, often identical with foam cells, promote atherosclerosis and plaque rupture, while various antiinflammatory and pro-fibrotic, hence M2 macrophage subsets, inhibit lesion progression, surround intraplaque hemorrhages, and may stabilize the plaque [9, 11, 12]. There are atherosclerosis plaque-specific macrophage types [11, 13, 14]. Proinflammatory macrophages of the atherosclerotic plaque are considered M1 if they secrete tumor necrosis factor-α (TNFα), interleukin (IL)-6, and IL-12; M4 if they secrete IL-6, TNFα, and metalloproteinase 7 (MMP-7). The M1 macrophages develop in response to TLR4 activation, whereas the similarly pro-inflammatory M4 macrophages develop in response to CXCL4. The antiinflammatory and antioxidant macrophages of the atherosclerotic plaque are considered M2 macrophages, called $M_{(Hb)}$ macrophages and $M_{(hem)}$ macrophages, respectively. $M_{(Hb)}$ and the $M_{(hem)}$ macrophages develop in response to hemoglobin, or haptoglobin and hem [12], without inducing lipid accumulation and foam cell differentiation [15]. Moreover, a pro-atherogenic role of M2 macrophages is also known [16], and in some regions of the plaque, such as in the fibrous cap, both M1 and M2 macrophages are equally present [4]. Because M2 macrophages secrete metalloproteinases (MMPs), they can remodel the extracellular matrix of the tunica media [17]. Accumulation of M2 macrophages in the aortic arch causes fibrosis and the loss of elastin content of the vessel wall, eventually worsening the vascular effect of angiotensin II-induced hypertension [18]. This profibrotic role of M2 macrophages has some beneficial effects in atherosclerotic plaques, where it may increase plaque stability, thus avoiding plaque rupture [19].

Vein walls also contain macrophages, and varicose veins have increased macrophage and monocyte content [20]. Vein wrapping causes IL-4 synthesis in the vein walls, which increases M2 macrophage activation, and this is thought to improve neuropathy of the sciatic nerve [21]. In deep venous thrombosis M2 macrophages impede resolution of the thrombus, and the inhibition or loss of p53 inhibits M2 activation and improves thrombus resolution [22].

In a recently described trait, M2 macrophages can form capillary structures, which phenomenon is called vascular mimicry. In the tumor environment, $CD163^+$, MSR1 (macrophage scavenger receptor-1)$^+$, and $CD206^+$ macrophages may fuse and form tubular, capillary-like channels [23]. As mentioned before, M2-activating signals also increase macrophage fusion (e.g., development of Langhans cells; see chapter "What Is an M2 Macrophage? Historical Overview of the Macrophage Polarization Model. The Th1/Th2 and M1/M2 Paradigm, the Arginine Fork"), which might be the cause of the fusion and vascular mimicry of tumor-associated macrophages (TAMs). M2 macrophages produce angiogenic growth factors, such as basic fibroblast growth factor, insulin-like growth factor-1, chemokine (C-C) motif ligand 2, placental growth factor, and vascular growth factor-A (VEGFA) [24]. In normal tissue development, and in wound healing, M2 macrophages support vascularization and blood supply of the regenerating tissue. However, the same effect of TAMs supports the growth of tumor tissue [25, 26].

2.2 M2 Macrophages in the Heart

The heart contains resident macrophages. These macrophages are scattered between heart muscle cells, arranged in the vicinity of coronary arteries, are present in the distal atrioventricular node, and in ischemic heart disease macrophage number increases in the diseased heart muscle [27, 28] (Fig. 1). Cardiac macrophages develop from embryonic macrophage progenitors and self-renew, replenishing the cardiac macrophage niche without the need of monocytes [29, 30]. With aging there is a varying degree of monocyte infiltration in the heart, and monocytes infiltrate the damaged tissues following cardiac infarct [29]. However, monocyte-derived macrophages and resident cardiac macrophages have distinct features, and they are thought to have distinct roles in cardiac repair [31, 32]. Resident cardiac macrophages show M2 features and are $CD206^+F4/80^+CD11b^+$. $TIMD4^+$ lymphatic vessel endothelial hyaluronan receptor 1 $(LYVE1)^+$ $MHC\text{-}II^{lo}CCR2^-$ cardiac resident macrophages express repair-associated genes [31] and genes associated with angiogenesis and endocytosis [33]. In general, resident cardiac macrophages are considered as a cardioprotective role of resident cardiac macrophages. In disease settings the main function of cardiac M2 macrophages is cardiac repair: their number increases following cardiac infarct, they express antiinflammatory and repair genes such as *Il10* and *Vegfa*, and they are essential for fibroblast-mediated cardiac repair [34]. However, profibrotic macrophages can cause cardiac fibrosis and contribute to heart failure [35].

Another role for cardiac resident macrophages is related to the pacemaker activity of the heart. Macrophages in the atrioventricular node express the gap junction component connexin 43, which is thought to couple them to cardiomyocytes depolarized in synchrony with cardiomyocytes [28]. Connexin 43 however does not necessarily form gap junctions; it may also be present in the mitochondria [36].

3 M2 Macrophages in the Respiratory System

The upper respiratory tract is a gateway for airborne infections; hence, it has strong immunosurveillance. The inhaled air passes through the nasal cavity and the pharynx, which have mucosal-associated lymphoid tissues (MALTs), the tonsils. The tonsils are arranged in Waldeyer's lymphatic ring,[1] and each tonsil is placed at relevant gateways: at the palatine, at the entrance of the auditory tube, at the root of the tongue, and on the wall of the pharynx. Macrophages, as well as dendritic cells, are constituents of the tonsils [38], and they present antigens to lymphocytes to initiate immune response against pathogens [39]. The larynx and the bronchi also have associated immune tissue, which is called bronchial-associated lymphoid tissue (BALT). Such as all lymphoid organs, tonsils also have a high apoptosis rate from the elimination of nonreactive and autoreactive lymphocytes. The apoptotic cell debris is cleared by tonsil macrophages [40, 41]. Tonsil macrophages express the M2 marker CD163$^+$, and have a high phagocytosis rate; however, they are prone to be activated toward an inflammatory state by lipopolysaccharide (LPS) [42]. They also express receptors for IL-1 and IL-36, which are inducers of inflammatory cytokine secretion [43]. Tonsils protect the airways from infections, so little function can be found for pathogen-tolerating macrophages in the tonsils. This consideration suggests that tonsil macrophages have a dual role: they safely dispose apoptotic cells, which maintains an M2 activation state, and simultaneously they respond to pathogens, IL-1, and IL-36 with Th1 cytokine secretion. As detailed in chapter "What Is an M2 Macrophage? Historical Overview of the Macrophage Polarization Model. The Th1/Th2 and M1/M2 Paradigm, the Arginine Fork", the macrophage phenotype induced by IL-1 receptor ligands and LPS is considered to be M2b macrophage activation by some authors.

Recurrent infections can lead to the hypertrophy of the tonsils, which condition mainly affects children and adolescents [44]. Recurrent tonsillar inflammation and tonsil hypertrophy are both Th1 inflammatory responses; however, in tonsillar hypertrophy an antiinflammatory cytokine IL-37 (formerly called IL-1 family member 7) has increased expression [45]. IL-37 is known to inhibit Notch1 and NFκB signaling, and oxidized low-density lipoprotein-induced M1 activation, and increases CD206, IL-10, and arginase-1 expression [46, 47]. Tonsils can also enlarge in allergic reactions, eventually, as a result of Th2 immune response; this suggests an M2 macrophage activation in tonsillar hypertrophy, which may intend to resolve the chronic Th1 inflammation. Interestingly, recurrent tonsil inflammation, a Th1 condition, is negatively associated with allergic rhinitis, which is a Th2 condition [44]. Also, tonsil-derived mesenchymal stem cells significantly attenuate allergic symptoms [48].

The larynx is a crossroads of the airway and the esophagus. The laryngeal muscles are necessary for swallowing food and allowing airflow into the bronchi. It is important that solid particles or fluids do not enter the bronchi, because that would

[1] Named after Wilhelm von Waldeyer-Hartz (1836–1921).

occlude the lower respiratory tract. The laryngeal muscles also allow speech, which is a secondary function of the respiratory system. Interestingly, some laryngeal muscles are enriched in M2 macrophages: the posterior cricoarytenoideus muscle, which is needed for whispering [49]; the cricopharyngeus muscle, necessary for swallowing [50]; and the genioglossus muscle, which elevates the larynx [51]. The specific function of these macrophages is unknown. It is though that because M2 macrophages act in aging of the skeletal muscles, thinning and death of striated muscle fibers may occur more frequently in the larynx and pharynx than in other parts of the body in the elderly [51]. It is plausible to assume that M2 macrophages of the laryngeal muscles contribute to aging-associated change of voice.

The vocal folds also contain macrophages and affect voice quality. Macrophages are arranged in the superficial layer of the vocal fold, where they can sense stimuli from airway pathogens. LPS induces an M2-like, IL-10-producing macrophage activation; mechanical injury leads to an early *Tnfa*-expressing, NOS2$^+$, CD163$^+$ macrophage activation, followed by CD206$^+$, *Tgfb1*-expressing macrophage phenotype [52, 53]. Vocal fold fibroblasts secrete both inflammatory cytokines and antiinflammatory mediators, and they can induce the activation of vocal fold macrophages. Fibroblasts from scarring in the vocal cord mitigate inflammatory macrophage activation. In contrast, vocal fold fibroblasts from polypoid tissue of the vocal cord increase LPS-induced M1 activation of macrophages [54].

The lower respiratory tract is rich in macrophages, which constitute about 70% of the immune cells (Fig. 2). The bronchi and the lungs contain three macrophage niches: BALT, alveolar macrophages, and interstitial or pulmonary tissue macrophages. By origin, these macrophages can be resident or infiltrating and can develop from embryonic or adult hematopoiesis [57].

During embryonic development of the lungs, macrophages are abundant in the developing branches of the alveoli (Fig. 2). In postnatal lung development, M2 macrophages are necessary for the development of alveoli [55]. In the adult lung, however, M2 macrophages represent a minor fraction of macrophages. CD163$^+$ macrophages are situated at the peripheral site of BALT, near arteries and the bronchus and in the interstitium of the lungs; and M2 macrophages comprise a CD43$^-$-infiltrating macrophage population in alveoli in rat lung [58, 59].

The M2 macrophage population is expanding in lung cancer, lung fibrosis, allergy, and asthma. In lung cancer the M2 macrophages are present in the tumor stroma and infiltrate into tumor islets, and overall an increased M2 macrophage number is a sign of poor prognosis in lung cancer [60]. Mice that lack IL-4Rα are more resistant to lung tumor development than their wild-type counterparts, and conversely, deficiency in IFNγ receptor promotes tumor development [61]. M2 macrophages are tolerogenic and secrete immunosuppressing cytokines, hence allowing development of cancerous cells. Moreover, M2 macrophages of the lung promote epithelial to mesenchyme transition and eventual invasion of cancer cells by activating the ERK1/2 signaling pathway [62]. M2 macrophages also contribute to resistance to chemotherapy, so that it is a therapeutic challenge to impede M2 macrophage activation in cancer. One approach is the pharmacological blockage of M2-activating signal pathways. Examples include gefitinib, imatinib, and

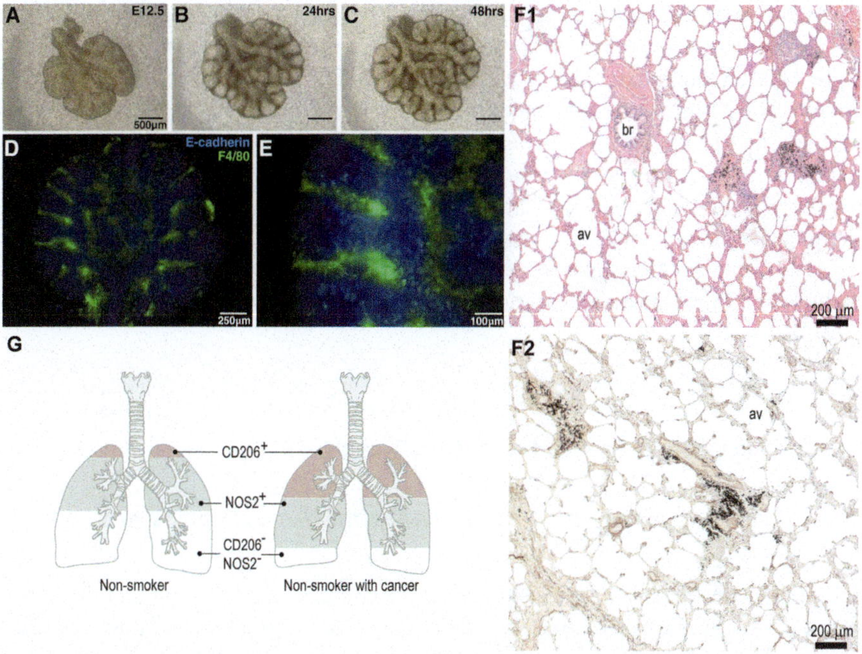

Fig. 2 Macrophages in the lower respiratory tract. (**a-e**) Macrophages in the developing mouse lung. (Reprinted from [55].) Lung explants of a 12.5-day-old mouse embryo show the development of bronchiole branches and the connected alveoli. Macrophages are enriched between the developing branches. (**d**, **e**) Immunofluorescence labelling of the lung epithelium (anti-E-cadherin; blue) and macrophages (anti-F4/80; green). (**f1**) Hematoxylin and eosin-stained section of human lung showing a bronchiole (*br*) and numerous alveoli (*av*). (**f2**) The same lung specimen with Van Gieson staining of elastic fibers. Carbon-laden macrophages are visible in *black* in both panels. (Images by author.) (**g**) Proportions of naïve (CD206⁻, NOS2⁻), M1 (NOS2⁺), and M2 (CD206⁺) macrophages in the human lung, using values from [56]

resveratrol, which inhibit lung cancer growth by impeding M2 activation through inhibiting STAT3 and STAT6 signaling [63–65]. The second strategy is to skew macrophage polarization toward the M1 state, and several possible mechanisms have been identified already, ranging from lipid metabolites, synthetic ligands, Th1 cytokines, and activators of Th1 cells in the tumor environment [66–68]. For example, inhibition of galectin-3 increases M1 polarization of macrophages and promotes CD8⁺ T-cell infiltration [69]. Galectin-3 is a lectin that is highly expressed within the tumor microenvironment of aggressive cancers [69] and is known to induce M2 macrophage activation [70]. A galectin-3 inhibitor molecule (GB1107) has been shown to reduce human and mouse lung adenocarcinoma growth and metastasis [69]. The third strategy to reduce M2 macrophage number is the targeted elimination of M2 macrophages. A recent study shows that a derivative of the honey bee venom melittin binds preferentially to M2-like CD206⁺ macrophages and induces their apoptosis [71].

M2 alveolar macrophages are necessary for the resolution of fungal and helminth infections. However, excess M2 macrophage activation causes fibrosis, myofibroblast differentiation, allergy, and asthma [72]. Allergen exposure activates lung epithelial cells, which triggers macrophage responses: M1 and M2 activation of resident and recruited macrophages. It has also been suggested that polyamines produced by macrophage arginase-1 may attract and activate mast cells, thus promoting airway inflammation [73]. The M2 marker FIZZ1 is abundant in the bronchoalveolar lavage fluid in allergic airway inflammation in the mouse, where it causes vascular inflammation, exhibits chemotactic and fibrogenic properties, induces myofibroblast differentiation, and recruits bone marrow-derived cells [74–78].

4 Macrophages in the Excretory Organs

4.1 M2 Macrophages in the Kidney and the Urinary Tract

The kidney develops from mesenchyme of the nephrogonotome region of the mesoderm [79] (Fig. 3). The earliest developmental stage of the kidney is the so-called pronephros, which remains the functional kidney in adult cephalochordates and in some chordates, such as the hagfish *Myxine glutinosa*. In teleost fishes, the pronephros develops into the head kidney, which is a lymphoid organ with immune and endocrine functions [80]. The teleost head kidney is rich in macrophages, among which are found M2-like cells (see in chapter "Evolution of M2 Macrophage Functions in Chordates"). In most vertebrates the pronephros is succeeded by the mesonephros, which develops into the kidney in fishes and amphibia. In amniotes (i.e., reptiles, birds, and mammals), the mesonephros atrophies during embryonic development, and its tubular system transforms into the Wolffian duct, which further develops into the male reproductive organs in mammals. The functional kidney develops from the so-called metanephros in amniotes. In the mouse embryo, the metanephros is rich in macrophages, which are derived from the yolk sac, and are present in the embryo from embryonic day (E)9.5 onward. The first wave of kidney macrophages are enriched at the caudal region of the Wolffian duct at E9.5–E10.5. Early macrophages arrange alongside anterior nephrogenic cells; later, they appear in the cortical nephrogenic zone, and are arranged around newly forming cortical interstitial blood vessels. Throughout renal organogenesis, most kidney macrophages are F4/80$^+$, CD206$^+$, and express genes that have functions in blood vessel morphogenesis. The proportion of CD206$^+$ cells decreases in later kidney development whereas the amount of galectin-3 expressing myeloid cells increases. The function of macrophages is to phagocytose components of the renal vascular system and support vessel development and endothelial cross-connections in the developing kidney [2]. The absence of macrophages reduces nephrogenic cell clearance and delays ureteric bud development.

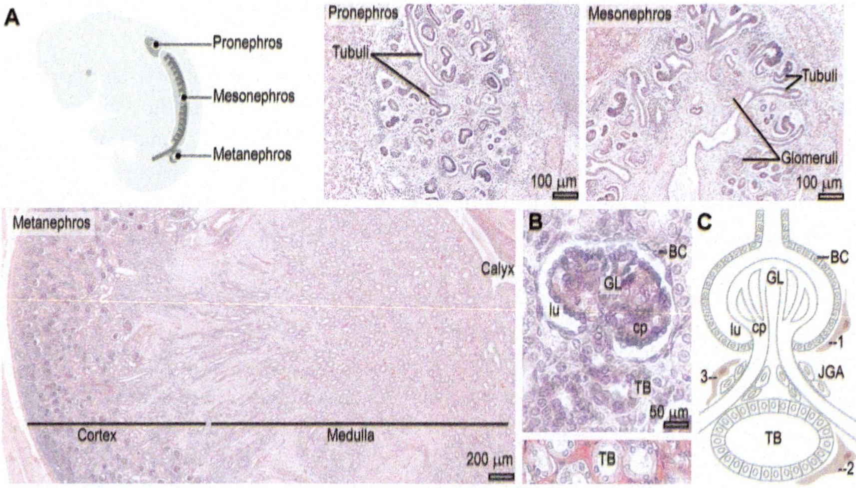

Fig. 3 Compartments of the kidney and the niche for macrophages. (**a**) Embryonic kidney development: localization of the pronephros, the mesonephros, and the metanephros. Tissue sections of human pronephros and metanephros. The pronephros has so-called aglomerular nephrons, which are built up from nephron tubuli, without a capillary glomerulus. The mesonephros develops into the metanephros, which has functional glomeruli connected to the nephron tubuli. The metanephros is divided into the cortex, containing glomeruli within Bowman capsules, and the medulla, containing nephron tubuli and collective ducts. The urine is collected in the calyx. (**b**) Cross section of a mature nephron. *GL* capillary glomerule, *cp* capillary, *BC* Bowman capsule, *lu* lumen of capsule, *TB* tubuli. *Top panel* shows cortical part of the nephron; *bottom panel* shows corresponding medullary region. (**c**) Scheme of the nephron glomeruli and tubuli, with cells of the JGA. Macrophages are associated with the interstitium, adjacent to the Bowman capsule [1], the nephron tubules [2], and the JGA [3]. (Images by author)

In the adult kidney, there are long-lived resident macrophages and monocyte-derived transient macrophages [81]. In renal inflammation, macrophage recruitment and proliferation are increased [82]. Macrophages constitute the majority of the renal medullary interstitial cell population. Macrophages are also associated with cortical distal and proximal nephron tubules and Bowman's capsule [83]. Macrophages are also present in the juxtaglomerular complex or juxtaglomerular apparatus (JGA) (Fig. 3). JGA is the endocrine hotspot of the kidney, and it regulates blood volume and blood pressure through the renin-angiotensin-aldosterone system (RAAS) [84]. Macrophages also express renin [85]. However, it is still unknown whether JGA-resident macrophages would affect hormone secretion from the JGA. What has been explored in depth is the effect of RAAS on macrophages. RAAS elicits both inflammatory and pro-resolving macrophage traits, which have impact in hypertension, vascular diseases, and chronic kidney disease. For instance, angiotensin II promotes M1 polarization [86], and the blockade of angiotensin receptor 2 increases M2 macrophage number in the kidney and reduces glomerulonephritis [87]. However, angiotensin II-induced aortic fibrosis is caused by excess

M2 activation [18], and the lack of angiotensin receptor 1a worsens kidney inflammation in obese mice [88].

The homeostatic function of kidney macrophages is the removal of apoptotic cells. In the absence of safe disposal of apoptotic debris, an autoimmune reaction may arise that destroys the kidney glomeruli and leads to inflammation [89, 90]. Necrotic cell loss in the kidney activates an inflammatory macrophage activation. The underlying mechanism is the activation of cytoplasmic DNA sensor AIM2 by DNA fragments of the necrotic cells [91]. Macrophages accumulate in various forms of kidney inflammation: diabetic glomerulonephritis, obesity-induced kidney inflammation, interstitial kidney inflammation, acute kidney inflammation, chronic kidney disease, and obstructive nephropathy [92]. These diseases are associated with increased macrophage number, from monocyte infiltration and local macrophage proliferation and M1 activation [93, 94]. It is thought that M2 macrophages help the resolution of inflammation in such inflammatory settings [95]. However, accumulation of M2 macrophages in the kidney interstitium is associated with diabetic nephropathy [90], and excess M2 macrophages cause kidney fibrosis [93, 96].

The urinary tract mucosa contains resident macrophages, which are important for protection from pathogens that may invade the urinary tract from the urethral opening [97]. M2 macrophages in the urinary tract are, however, associated with malignancies, such as remission and poor prognosis of bladder cancer [98], by the tolerogenic, IL-10-producing trait of M2 macrophages, which impairs antitumor immunity [97]. An immunotherapy option of bladder cancer treatment is the installation of an attenuated form of *Mycobacterium bovis*, bacillus Calmette–Guérin (BCG), in the bladder, which is thought to stimulate antitumor immunity and probably also M1 macrophage activation. M2 macrophages impede the effectiveness of BCG therapy [97, 98]. As detailed in chapter "Signal Mechanisms of M2 Macrophage Activation", STAT6 signaling is one of the M2-activating signal mechanisms; it is inactivated by a ubiquitin ligase, so-called GRAIL or RNF128. Reduced expression of RNF128 sustains STAT6 signaling [99] and is associated with poor prognosis of urinary tract malignancies [100].

A recent report showed that urethral M2 macrophages, which highly express IL-1 receptor, CD206, and IL-4Rα, but not CD163, serve as reservoirs of HIV-1. When these macrophages are maintained in vitro, they are capable of releasing HIV-1 virions [101]. The inner lining of the foreskin contains macrophages that express high amounts of CD4 and CD-SIGN, which may also be a route of HIV-1 transmission. Circumcision may reduce the transmission risk of HIV-1 [102].

4.2 Secondary Excretory Organs and M2 Macrophages

In addition to the kidney, humans have secondary excretory organs, such as the sweat glands and the lachrymal glands. The sweat glands of the skin allow perspiration and ultimately control body temperature. Perspiration also causes water and electrolyte loss. The sweat glands also express an antibacterial peptide, dermcidin

[103]. Dermcidin is further processed by proteases in the sweat, resulting in antimicrobial peptides. Dermcidin also attenuates LPS-induced macrophage activation in vitro [104]. The mammary gland, which is a specialized sweat gland, also contains a macrophage population. The function of these macrophages is detailed in chapter “M2 Macrophages in the Integument and in the Musculoskeletal System”. Macrophages are also present in the lacrimal glands. Th2-dominant diseases of the lacrimal glands are associated with M2 macrophages and fibrosis: autoimmune dry eyes [105], severe dry eyes [106], IgG4-related dacryoadenitis, and sialoadenitis (Mikulicz’s disease) [107]. IL-33 produced by M2 macrophages might contribute to the aberrant activation of the Th2 immune response [108].

References

1. Stremmel C, Schuchert R, Wagner F, Thaler R, Weinberger T, Pick R, Mass E, Ishikawa-Ankerhold HC, Margraf A, Hutter S, Vagnozzi R, Klapproth S, Frampton J, Yona S, Scheiermann C, Molkentin JD, Jeschke U, Moser M, Sperandio M, Massberg S, Geissmann F, Schulz C. Yolk sac macrophage progenitors traffic to the embryo during defined stages of development. Nat Commun. 2018;9:75.
2. Munro DAD, Wineberg Y, Tarnick J, Vink CS, Li Z, Pridans C, Dzierzak E, Kalisky T, Hohenstein P, Davies JA. Macrophages restrict the nephrogenic field and promote endothelial connections during kidney development. eLife. 2019;8:e43271.
3. Psaltis PJ, Puranik AS, Spoon DB, Chue CD, Hoffman SJ, Witt TA, Delacroix S, Kleppe LS, Mueske CS, Pan S, Gulati R, Simari RD. Characterization of a resident population of adventitial macrophage progenitor cells in postnatal vasculature. Circ Res. 2014;115:364–75.
4. Stöger JL, Gijbels MJJ, van der Velden S, Manca M, van der Loos CM, Biessen EAL, Daemen MJAP, Lutgens E, de Winther MPJ. Distribution of macrophage polarization markers in human atherosclerosis. Atherosclerosis. 2012;225:461–8.
5. Maiellaro K, Taylor WR. The role of the adventitia in vascular inflammation. Cardiovasc Res. 2007;75:640–8.
6. Stenmark KR, Yeager ME, El Kasmi KC, Nozik-Grayck E, Gerasimovskaya EV, Li M, Riddle SR, Frid MG. The adventitia: essential regulator of vascular wall structure and function. Annu Rev Physiol. 2013;75:23–47.
7. Zorc-Pleskovič R, Pleskovič A, Vraspir-Porenta O, Zorc M, Milutinović A. Immune cells and vasa vasorum in the tunica media of atherosclerotic coronary arteries. Bosn J Basic Med Sci. 2018;18:240–5.
8. Chistiakov DA, Bobryshev YV, Orekhov AN. Macrophage-mediated cholesterol handling in atherosclerosis. J Cell Mol Med. 2016;20:17–28.
9. Barrett TJ. Macrophages in atherosclerosis regression. Arterioscler Thromb Vasc Biol. 2020;40:20–33.
10. Vinchi F, Muckenthaler M, Costa da Silva M, Balla G, Balla J, Jeney V. Atherogenesis and iron: from epidemiology to cellular level. Front Pharmacol. 2014;5:94.
11. Colin S, Chinetti-Gbaguidi G, Staels B. Macrophage phenotypes in atherosclerosis. Immunol Rev. 2014;262:153–66.
12. Bobryshev YV, Ivanova EA, Chistiakov DA, Nikiforov NG, Orekhov AN. Macrophages and their role in atherosclerosis: pathophysiology and transcriptome analysis. Biomed Res Int. 2016;2016:9582430–9582430.
13. Moore KJ, Sheedy FJ, Fisher EA. Macrophages in atherosclerosis: a dynamic balance. Nat Rev Immunol. 2013;13:709–21.

14. Chinetti-Gbaguidi G, Colin S, Staels B. Macrophage subsets in atherosclerosis. Nat Rev Cardiol. 2014;12:10.
15. Finn AV, Nakano M, Polavarapu R, Karmali V, Saeed O, Zhao X, Yazdani S, Otsuka F, Davis T, Habib A, Narula J, Kolodgie FD, Virmani R. Hemoglobin directs macrophage differentiation and prevents foam cell formation in human atherosclerotic plaques. J Am Coll Cardiol. 2012;59:166–77.
16. Guo L, Akahori H, Harari E, Smith SL, Polavarapu R, Karmali V, Otsuka F, Gannon RL, Braumann RE, Dickinson MH, Gupta A, Jenkins AL, Lipinski MJ, Kim J, Chhour P, de Vries PS, Jinnouchi H, Kutys R, Mori H, Kutyna MD, Torii S, Sakamoto A, Choi CU, Cheng Q, Grove ML, Sawan MA, Zhang Y, Cao Y, Kolodgie FD, Cormode DP, Arking DE, Boerwinkle E, Morrison AC, Erdmann J, Sotoodehnia N, Virmani R, Finn AV. $CD163^+$ macrophages promote angiogenesis and vascular permeability accompanied by inflammation in atherosclerosis. J Clin Invest. 2018;128:1106–24.
17. Yamashita A, Shoji K, Tsuruda T, Furukoji E, Takahashi M, Nishihira K, Tamura S, Asada Y. Medial and adventitial macrophages are associated with expansive atherosclerotic remodeling in rabbit femoral artery. Histol Histopathol. 2008;23:127–36.
18. Moore JP, Vinh A, Tuck KL, Sakkal S, Krishnan SM, Chan CT, Lieu M, Samuel CS, Diep H, Kemp-Harper BK, Tare M, Ricardo SD, Guzik TJ, Sobey CG, Drummond GR. M2 macrophage accumulation in the aortic wall during angiotensin II infusion in mice is associated with fibrosis, elastin loss, and elevated blood pressure. Am J Phys Heart Circ Phys. 2015;309:H906–17.
19. Medbury HJ, James V, Ngo J, Hitos K, Wang Y, Harris DC, Fletcher JP. Differing association of macrophage subsets with atherosclerotic plaque stability. Int Angiol. 2013;32:74–84.
20. Sayer GL, Smith PDC. Immunocytochemical characterisation of the inflammatory cell infiltrate of varicose veins. Eur J Vasc Endovasc Surg. 2004;28:479–83.
21. Hirosawa N, Uchida K, Kuniyoshi K, Murakami K, Inoue G, Miyagi M, Matsuura Y, Orita S, Inage K, Suzuki T, Takaso M, Ohtori S. Vein wrapping promotes M2 macrophage polarization in a rat chronic constriction injury model. J Orthop Res. 2018;36:2210–7.
22. Mukhopadhyay S, Antalis TM, Nguyen KP, Hoofnagle MH, Sarkar R. Myeloid p53 regulates macrophage polarization and venous thrombus resolution by inflammatory vascular remodeling in mice. Blood. 2017;129:3245–55.
23. Barnett FH, Rosenfeld M, Wood M, Kiosses WB, Usui Y, Marchetti V, Aguilar E, Friedlander M. Macrophages form functional vascular mimicry channels in vivo. Sci Rep. 2016;6:36659.
24. Jetten N, Verbruggen S, Gijbels MJ, Post MJ, De Winther MP, Donners MM. Anti-inflammatory M2, but not pro-inflammatory M1 macrophages promote angiogenesis in vivo. Angiogenesis. 2014;17:109–18.
25. Pollard JW. Trophic macrophages in development and disease. Nat Rev Immunol. 2009;9:259–70.
26. Sica A, Schioppa T, Mantovani A, Allavena P. Tumour-associated macrophages are a distinct M2 polarised population promoting tumour progression: potential targets of anti-cancer therapy. Eur J Cancer. 2006;42:717–27.
27. Azzawi M, Kan SW, Hillier V, Yonan N, Hutchinson IV, Hasleton PS. The distribution of cardiac macrophages in myocardial ischaemia and cardiomyopathy. Histopathology (Oxf). 2005;46:314–9.
28. Hulsmans M, Clauss S, Xiao L, Aguirre AD, King KR, Hanley A, Hucker WJ, Wülfers EM, Seemann G, Courties G, Iwamoto Y, Sun Y, Savol AJ, Sager HB, Lavine KJ, Fishbein GA, Capen DE, Da Silva N, Miquerol L, Wakimoto H, Seidman CE, Seidman JG, Sadreyev RI, Naxerova K, Mitchell RN, Brown D, Libby P, Weissleder R, Swirski FK, Kohl P, Vinegoni C, Milan DJ, Ellinor PT, Nahrendorf M. Macrophages facilitate electrical conduction in the heart. Cell. 2017;169:510–522.e520.
29. Molawi K, Wolf Y, Kandalla PK, Favret J, Hagemeyer N, Frenzel K, Pinto AR, Klapproth K, Henri S, Malissen B, Rodewald HR, Rosenthal NA, Bajenoff M, Prinz M, Jung S, Sieweke

MH. Progressive replacement of embryo-derived cardiac macrophages with age. J Exp Med. 2014;211:2151–8.

30. Ensan S, Li A, Besla R, Degousee N, Cosme J, Roufaiel M, Shikatani EA, El-Maklizi M, Williams JW, Robins L, Li C, Lewis B, Yun TJ, Lee JS, Wieghofer P, Khattar R, Farrokhi K, Byrne J, Ouzounian M, Zavitz CC, Levy GA, Bauer CM, Libby P, Husain M, Swirski FK, Cheong C, Prinz M, Hilgendorf I, Randolph GJ, Epelman S, Gramolini AO, Cybulsky MI, Rubin BB, Robbins CS. Self-renewing resident arterial macrophages arise from embryonic CX3CR1(+) precursors and circulating monocytes immediately after birth. Nat Immunol. 2016;17:159–68.
31. Dick SA, Macklin JA, Nejat S, Momen A, Clemente-Casares X, Althagafi MG, Chen J, Kantores C, Hosseinzadeh S, Aronoff L, Wong A, Zaman R, Barbu I, Besla R, Lavine KJ, Razani B, Ginhoux F, Husain M, Cybulsky MI, Robbins CS, Epelman S. Self-renewing resident cardiac macrophages limit adverse remodeling following myocardial infarction. Nat Immunol. 2019;20:29–39.
32. Bajpai G, Bredemeyer A, Li W, Zaitsev K, Koenig AL, Lokshina I, Mohan J, Ivey B, Hsiao H-M, Weinheimer C, Kovacs A, Epelman S, Artyomov M, Kreisel D, Lavine KJ. Tissue resident $CCR2^-$ and $CCR2^+$ cardiac macrophages differentially orchestrate monocyte recruitment and fate specification following myocardial injury. Circ Res. 2019;124:263–78.
33. Wang Z, Koenig AL, Lavine KJ, Apte RS. Macrophage plasticity and function in the eye and heart. Trends Immunol. 2019;40:825–41.
34. Shiraishi M, Shintani Y, Shintani Y, Ishida H, Saba R, Yamaguchi A, Adachi H, Yashiro K, Suzuki K. Alternatively activated macrophages determine repair of the infarcted adult murine heart. J Clin Invest. 2016;126:2151–66.
35. Lim GB. Macrophages promote cardiac fibrosis and diastolic dysfunction. Nat Rev Cardiol. 2018;15:196–7.
36. Kim S-N, Kwon H-J, Im S-W, Son Y-H, Akindehin S, Jung Y-S, Lee SJ, Rhyu IJ, Kim IY, Seong J-K, Lee J, Yoo H-C, Granneman JG, Lee Y-H. Connexin 43 is required for the maintenance of mitochondrial integrity in brown adipose tissue. Sci Rep. 2017;7:7159.
37. Tabula Muris Consortium; Overall coordination; Logistical coordination; Organ collection and processing; Library preparation and sequencing; Computational data analysis; Cell type annotation; Writing group; Supplemental text writing group; Principal investigators. Single-cell transcriptomics of 20 mouse organs creates a Tabula Muris. Nature. 2018;562:367–72.
38. Yamamoto Y, Okato S, Takahashi H, Takeda K, Magari S. Distribution and morphology of macrophages in palatine tonsils. Acta Otolaryngol. 1988;105:83–95.
39. Scadding GK. Immunology of the tonsil: a review. J R Soc Med. 1990;83:104–7.
40. Schrijvers DM, Meyer GRYD, Kockx MM, Herman AG, Martinet W. Phagocytosis of apoptotic cells by macrophages is impaired in atherosclerosis. Arterioscler Thromb Vasc Biol. 2005;25:1256–61.
41. Schrijvers DM, De Meyer GRY, Kockx MM, Herman AG, Martinet W. Comparison of apoptosis detection markers combined with macrophage immunostaining to study phagocytosis of apoptotic cells in situ. Biomark Insights. 2007;1:193–200.
42. Soldevila F, Edwards JC, Graham SP, Stevens LM, Crudgington B, Crooke HR, Werling D, Steinbach F. Characterization of the myeloid cell populations' resident in the porcine palatine tonsil. Front Immunol. 2018;9:1800.
43. Dietrich D, Martin P, Flacher V, Sun Y, Jarrossay D, Brembilla N, Mueller C, Arnett HA, Palmer G, Towne J, Gabay C. Interleukin-36 potently stimulates human M2 macrophages, Langerhans cells and keratinocytes to produce pro-inflammatory cytokines. Cytokine. 2016;84:88–98.
44. Ameli F, Brocchetti F, Tosca MA, Schiavetti I, Ciprandi G. Tonsil volume and allergic rhinitis in children. Allergy Rhinol (Providence). 2014;5:137–42.
45. Mikola E, Elenius V, Saarinen M, Palomares O, Waris M, Turunen R, Puhakka T, Ivaska L, Rückert B, Aab A, Vahlberg T, Vuorinen T, Allander T, Camargo CA, Akdis M, Akdis CA,

Jartti T. Tonsillar cytokine expression between patients with tonsillar hypertrophy and recurrent tonsillitis. Clin Transl Allergy. 2018;8:22.
46. Huang J, Hou FL, Zhang AY, Li ZL. Protective effect of the polarity of macrophages regulated by IL-37 on atherosclerosis. Genet Mol Res. 2016;15 https://doi.org/10.4238/gmr.15027616.
47. Zeng Q, Ma Q, Dong W, Zhou P, Zong S, Su S, Yang X, Li J, Lai W, Xu D. GW28-e0637 Interleukin 37 suppresses the M1 polarization of macrophages through Notch1 and Nuclear factor kappa B pathways. J Am Coll Cardiol. 2017;70:C22.
48. Li H, Tian Y, Xie L, Liu X, Huang Z, Su W. Mesenchymal stem cells in allergic diseases: current status. Allergol Int. 2020;69:35–45.
49. Tsunoda K, Niimi S, Hirose H. The roles of the posterior cricoarytenoid and thyropharyngeus muscles in whispered speech. Folia Phoniatr Logo. 1994;46:139–51.
50. Curtis DJ, Cruess DF, Berg T. The cricopharyngeal muscle: a videorecording review. Am J Roentgenol. 1984;142:497–500.
51. Rhee S, Yamamoto M, Kitamura K, Masaaki K, Katori Y, Murakami G, Abe S-I. Macrophage density in pharyngeal and laryngeal muscles greatly exceeds that in other striated muscles: an immunohistochemical study using elderly human cadavers. Anat Cell Biol. 2016;49:177–83.
52. King SN, Guille J, Thibeault SL. Characterization of the leukocyte response in acute vocal fold injury. PLoS One. 2015;10:e0139260.
53. Kaba S, Nakamura R, Yamashita M, Katsuno T, Suzuki R, Tateya I, Kishimoto Y, Omori K. Alterations in macrophage polarization in injured murine vocal folds. Laryngoscope. 2019;129:E135–42.
54. King SN, Chen F, Jetté ME, Thibeault SL. Vocal fold fibroblasts immunoregulate activated macrophage phenotype. Cytokine. 2013;61:228–36.
55. Jones CV, Williams TM, Walker KA, Dickinson H, Sakkal S, Rumballe BA, Little MH, Jenkin G, Ricardo SD. M2 macrophage polarisation is associated with alveolar formation during postnatal lung development. Respir Res. 2013;14:41.
56. Bazzan E, Tinè M, Radu C, Turato G, Mutti E, Simioni P, Cazzuffi R, Baraldo S, Saetta M, Cosio MG. Phenotypes of alveolar macrophages in normal human lung and their polarization with disease. Eur Respir J. 2014;44:1472.
57. Röszer T. Understanding the biology of self-renewing macrophages. Cells. 2018;7(8):103.
58. Van Der Brugge-Gamelkoorn GJ, Dijkstra CD, Sminia T. Characterization of pulmonary macrophages and bronchus-associated lymphoid tissue (BALT) macrophages in the rat. An enzyme-cytochemical and immunocytochemical study. Immunobiology. 1985;169:553–62.
59. Venosa A, Malaviya R, Choi H, Gow AJ, Laskin JD, Laskin DL. Characterization of distinct macrophage subpopulations during nitrogen mustard-induced lung injury and fibrosis. Am J Respir Cell Mol Biol. 2016;54:436–46.
60. Cao L, Che X, Qiu X, Li Z, Yang B, Wang S, Hou K, Fan Y, Qu X, Liu Y. M2 macrophage infiltration into tumor islets leads to poor prognosis in non-small-cell lung cancer. Cancer Manag Res. 2019;11:6125–38.
61. Redente EF, Dwyer-Nield LD, Barrett BS, Riches DWH, Malkinson AM. Lung tumor growth is stimulated in IFN-$\gamma^{-/-}$ mice and inhibited in IL-4R$\alpha^{-/-}$ mice. Anticancer Res. 2009;29:5095–101.
62. Guo Z, Song J, Hao J, Zhao H, Du X, Li E, Kuang Y, Yang F, Wang W, Deng J, Wang Q. M2 macrophages promote NSCLC metastasis by upregulating CRYAB. Cell Death Dis. 2019;10:377.
63. Sun L, Chen B, Jiang R, Li J, Wang B. Resveratrol inhibits lung cancer growth by suppressing M2-like polarization of tumor associated macrophages. Cell Immunol. 2017;311:86–93.
64. Yao Z, Zhang J, Zhang B, Liang G, Chen X, Yao F, Xu X, Wu H, He Q, Ding L, Yang B. Imatinib prevents lung cancer metastasis by inhibiting M2-like polarization of macrophages. Pharmacol Res. 2018;133:121–31.
65. Tariq M, Zhang J-Q, Liang G-K, He Q-J, Ding L, Yang B. Gefitinib inhibits M2-like polarization of tumor-associated macrophages in Lewis lung cancer by targeting the STAT6 signaling pathway. Acta Pharmacol Sin. 2017;38:1501–11.

66. Ren F, Fan M, Mei J, Wu Y, Liu C, Pu Q, You Z, Liu L. Interferon-γ and celecoxib inhibit lung-tumor growth through modulating M2/M1 macrophage ratio in the tumor microenvironment. Drug Des Dev Ther. 2014;8:1527–38.
67. Poh AR, Ernst M. Targeting macrophages in cancer: from bench to bedside. Front Oncol. 2018;8:49.
68. Liu M, Luo F, Ding C, Albeituni S, Hu X, Ma Y, Cai Y, McNally L, Sanders MA, Jain D, Kloecker G, Bousamra M, Zhang H-G, Higashi RM, Lane AN, Fan TW-M, Yan J. Dectin-1 activation by a natural product β-glucan converts immunosuppressive macrophages into an M1-like phenotype. J Immunol. 2015;195:5055–65.
69. Vuong L, Kouverianou E, Rooney CM, McHugh BJ, Howie SEM, Gregory CD, Forbes SJ, Henderson NC, Zetterberg FR, Nilsson UJ, Leffler H, Ford P, Pedersen A, Gravelle L, Tantawi S, Schambye H, Sethi T, MacKinnon AC. An orally active galectin-3 antagonist inhibits lung adenocarcinoma growth and augments response to PD-L1 blockade. Cancer Res. 2019;79:1480–92.
70. MacKinnon AC, Farnworth SL, Hodkinson PS, Henderson NC, Atkinson KM, Leffler H, Nilsson UJ, Haslett C, Forbes SJ, Sethi T. Regulation of alternative macrophage activation by galectin-3. J Immunol. 2008;180:2650–8.
71. Lee C, Jeong H, Bae Y, Shin K, Kang S, Kim H, Oh J, Bae H. Targeting of M2-like tumor-associated macrophages with a melittin-based pro-apoptotic peptide. J Immunother Cancer. 2019;7:147.
72. Hou J, Shi J, Chen L, Lv Z, Chen X, Cao H, Xiang Z, Han X. M2 macrophages promote myofibroblast differentiation of LR-MSCs and are associated with pulmonary fibrogenesis. Cell Commun Signal. 2018;16:89–89.
73. Nair MG, Guild KJ, Artis D. Novel effector molecules in type 2 inflammation: lessons drawn from helminth infection and allergy. J Immunol. 2006;177:1393–9.
74. Liu T, Yu H, Ullenbruch M, Jin H, Ito T, Wu Z, Liu J, Phan SH. The in vivo fibrotic role of FIZZ1 in pulmonary fibrosis. PLoS One. 2014;9:e88362.
75. Yamaji-Kegan K, Su Q, Angelini DJ, Myers AC, Cheadle C, Johns RA. Hypoxia-induced mitogenic factor (HIMF/FIZZ1/RELMalpha) increases lung inflammation and activates pulmonary microvascular endothelial cells via an IL-4-dependent mechanism. J Immunol. 2010;185:5539–48.
76. Holcomb IN, Kabakoff RC, Chan B, Baker TW, Gurney A, Henzel W, Nelson C, Lowman HB, Wright BD, Skelton NJ, Frantz GD, Tumas DB, Peale FV Jr, Shelton DL, Hebert CC. FIZZ1, a novel cysteine-rich secreted protein associated with pulmonary inflammation, defines a new gene family. EMBO J. 2000;19:4046–55.
77. Raes G, De Baetselier P, Noel W, Beschin A, Brombacher F, Hassanzadeh Gh G. Differential expression of FIZZ1 and Ym1 in alternatively versus classically activated macrophages. J Leukoc Biol. 2002;71:597–602.
78. Yamaji-Kegan K, Takimoto E, Zhang A, Weiner NC, Meuchel LW, Berger AE, Cheadle C, Johns RA. Hypoxia-induced mitogenic factor (FIZZ1/RELMalpha) induces endothelial cell apoptosis and subsequent interleukin-4-dependent pulmonary hypertension. Am J Physiol Lung Cell Mol Physiol. 2014;306:L1090–103.
79. Smyth IM, Cullen-McEwen LA, Caruana G, Black MJ, Bertram JF. Development of the kidney: morphology and mechanisms. In: Polin RA, Abman SH, Rowitch DH, Benitz WE, Fox WW, editors. Fetal and neonatal physiology. 5th ed. Philadelphia: Elsevier; 2017. p. 953–964.e954.
80. Geven EJW, Klaren PHM. The teleost head kidney: integrating thyroid and immune signalling. Dev Comp Immunol. 2017;66:73–83.
81. Puranik AS, Leaf IA, Jensen MA, Hedayat AF, Saad A, Kim K-W, Saadalla AM, Woollard JR, Kashyap S, Textor SC, Grande JP, Lerman A, Simari RD, Randolph GJ, Duffield JS, Lerman LO. Kidney-resident macrophages promote a proangiogenic environment in the normal and chronically ischemic mouse kidney. Sci Rep. 2018;8:13948.

82. Lenda DM, Kikawada E, Stanley ER, Kelley VR. Reduced macrophage recruitment, proliferation, and activation in colony-stimulating factor-1-deficient mice results in decreased tubular apoptosis during renal inflammation. J Immunol. 2003;170:3254–62.
83. Hume DA, Gordon S. Mononuclear phagocyte system of the mouse defined by immunohistochemical localization of antigen F4/80. Identification of resident macrophages in renal medullary and cortical interstitium and the juxtaglomerular complex. J Exp Med. 1983;157:1704–9.
84. Röszer T, Ricote M. PPARs in the renal regulation of systemic blood pressure. PPAR Res. 2010;2010:698730.
85. Iwai N, Inagami T, Ohmichi N, Kinoshita M. Renin is expressed in rat macrophage/monocyte cells. Hypertension. 1996;27:399–403.
86. Qi GM, Jia LX, Li YL, Li HH, Du J. Adiponectin suppresses angiotensin II-induced inflammation and cardiac fibrosis through activation of macrophage autophagy. Endocrinology. 2014;155:2254–65.
87. Aki K, Shimizu A, Masuda Y, Kuwahara N, Arai T, Ishikawa A, Fujita E, Mii A, Natori Y, Fukunaga Y, Fukuda Y. ANG II receptor blockade enhances anti-inflammatory macrophages in anti-glomerular basement membrane glomerulonephritis. Am J Physiol Renal Physiol. 2010;298:F870–82.
88. Ma LJ, Corsa BA, Zhou J, Yang H, Li H, Tang YW, Babaev VR, Major AS, Linton MF, Fazio S, Hunley TE, Kon V, Fogo AB. Angiotensin type 1 receptor modulates macrophage polarization and renal injury in obesity. Am J Physiol Renal Physiol. 2011;300:F1203–13.
89. Poon IK, Lucas CD, Rossi AG, Ravichandran KS. Apoptotic cell clearance: basic biology and therapeutic potential. Nat Rev Immunol. 2014;14:166–80.
90. Zhang X, Yang Y, Zhao Y. Macrophage phenotype and its relationship with renal function in human diabetic nephropathy. PLoS One. 2019;14:e0221991.
91. Komada T, Chung H, Lau A, Platnich JM, Beck PL, Benediktsson H, Duff HJ, Jenne CN, Muruve DA. Macrophage uptake of necrotic cell DNA activates the AIM2 inflammasome to regulate a proinflammatory phenotype in CKD. J Am Soc Nephrol. 2018;29:1165–81.
92. Lo T-H, Tseng K-Y, Tsao W-S, Yang C-Y, Hsieh S-L, Chiu AW-H, Takai T, Mak TW, Tarng D-C, Chen N-J. TREM-1 regulates macrophage polarization in ureteral obstruction. Kidney Int. 2014;86:1174–86.
93. Engel DR, Krause TA, Snelgrove SL, Thiebes S, Hickey MJ, Boor P, Kitching AR, Kurts C. CX3CR1 reduces kidney fibrosis by inhibiting local proliferation of profibrotic macrophages. J Immunol. 2015;194(4):1628–38.
94. Guiteras R, Flaquer M, Cruzado JM. Macrophage in chronic kidney disease. Clin Kidney J. 2016;9:765–71.
95. Geng Y, Zhang L, Fu B, Zhang J, Hong Q, Hu J, Li D, Luo C, Cui S, Zhu F, Chen X. Mesenchymal stem cells ameliorate rhabdomyolysis-induced acute kidney injury via the activation of M2 macrophages. Stem Cell Res Ther. 2014;5:80.
96. Bhatia D, Chung K-P, Nakahira K, Patino E, Rice MC, Torres LK, Muthukumar T, Choi AM, Akchurin OM, Choi ME. Mitophagy-dependent macrophage reprogramming protects against kidney fibrosis. JCI Insight. 2019;4:e132826.
97. Lacerda Mariano L, Ingersoll MA. Bladder resident macrophages: mucosal sentinels. Cell Immunol. 2018;330:136–41.
98. Suriano F, Santini D, Perrone G, Amato M, Vincenzi B, Tonini G, Muda A, Boggia S, Buscarini M, Pantano F. Tumor associated macrophages polarization dictates the efficacy of BCG instillation in non-muscle invasive urothelial bladder cancer. J Exp Clin Cancer Res. 2013;32:87.
99. Waqas SFH, Hoang A, Lin Y, Ampem G, Röszer T, et al. Neuropeptide FF increases M2 activation and self-renewal of adipose tissue macrophages. J Clin Invest. 2017;127:2842–54.
100. Lee Y-Y, Wang C-T, Huang SK-H, Wu W-J, Huang C-N, Li C-C, Chan T-C, Liang P-I, Hsing C-H, Li C-F. Downregulation of RNF128 predicts progression and poor prognosis in patients with urothelial carcinoma of the upper tract and urinary bladder. J Cancer. 2016;7:2187–96.

101. Ganor Y, Real F, Sennepin A, Dutertre C-A, Prevedel L, Xu L, Tudor D, Charmeteau B, Couëdel-Courteille A, Marion S, Zenak A-R, Jourdain J-P, Zhou Z, Schmitt A, Capron C, Eugenin E, Cheynier R, Revol M, Cristofari S, Bomsel M. HIV-1 reservoirs in urethral macrophages of patients under suppressive antiretroviral therapy. Nat Microbiol. 2019;4:633–44.
102. Tobian AAR, Kacker S, Quinn TC. Male circumcision: a globally relevant but under-utilized method for the prevention of HIV and other sexually transmitted infections. Annu Rev Med. 2014;65:293–306.
103. Schittek B, Hipfel R, Sauer B, Bauer J, Kalbacher H, Stevanovic S, Schirle M, Schroeder K, Blin N, Meier F, Rassner G, Garbe C. Dermcidin: a novel human antibiotic peptide secreted by sweat glands. Nat Immunol. 2001;2:1133–7.
104. Wang E, Qiang X, Li J, Zhu S, Wang P. The in vitro immune-modulating properties of a sweat gland-derived antimicrobial peptide dermcidin. Shock. 2016;45:28–32.
105. Zhou D, Chen Y-T, Chen F, Gallup M, Vijmasi T, Bahrami AF, Noble LB, van Rooijen N, McNamara NA. Critical involvement of macrophage infiltration in the development of Sjögren's syndrome-associated dry eye. Am J Pathol. 2012;181:753–60.
106. Inaba T, Ogawa Y, Shinmura S, Tadaki M, Kawakami Y, Tsubota K. A role of macrophages in the lacrimal gland inflammation of patient with chronic graft versus host disease. Invest Ophthalmol Vis Sci. 2008;49:2387–2387.
107. Furukawa S, Moriyama M, Tanaka A, Maehara T, Tsuboi H, Iizuka M, Hayashida J-N, Ohta M, Saeki T, Notohara K, Sumida T, Nakamura S. Preferential M2 macrophages contribute to fibrosis in IgG4-related dacryoadenitis and sialoadenitis, so-called Mikulicz's disease. Clin Immunol. 2015;156:9–18.
108. Furukawa S, Moriyama M, Miyake K, Nakashima H, Tanaka A, Maehara T, Iizuka-Koga M, Tsuboi H, Hayashida J-N, Ishiguro N, Yamauchi M, Sumida T, Nakamura S. Interleukin-33 produced by M2 macrophages and other immune cells contributes to Th2 immune reaction of IgG4-related disease. Sci Rep. 2017;7:42413.

M2 Macrophages in the Metabolic Organs and in the Neuroendocrine System

Keywords Adipose tissue · Obesity · Adipose tissue macrophage · Kupffer cell · Immunometabolism · Diabetes · Fatty liver · Endocrinology · Metabolism · Nervous system · Endocrine organs

1 Introduction

Metabolism is a principle of the living organism. It is a complex network of enzymatic and nonenzymatic biochemical transformation processes to sustain life that produce, degrade, and convert vital materials for the organism, generate energy, and eliminate nonfunctional or toxic materials [1, 2]. The proper function of this complex network requires regulatory mechanisms that integrate and synchronize this plethora of processes [1, 3]. In the simplest cellular organisms, robust factors control metabolism, such as substrate availability and the chemical environment that determine the activity of the metabolic pathways. The growing complexity of metabolic networks, however, requires more sophisticated regulatory mechanisms: the endocrine organs and the nervous system become upstream-acting regulators of the metabolism, and signaling modules and complex interactions evolve between the metabolic organs and the neuroendocrine system. Macrophages reside within the metabolic organs and the neuroendocrine system, where they have direct effects on metabolism. Macrophages aid the development of metabolic organs such as the adipose tissue and the liver, and also shape the function of metabolic cells. This trait of macrophages has mostly been studied in metabolic diseases and in the context of M1 activation. We know how suppression of M1 activation would affect metabolism; however, little is known about the physiological function of M2 macrophages in the metabolic organs and in metabolic regulation.

T. Röszer, *The M2 Macrophage*, Progress in Inflammation Research 86,
https://doi.org/10.1007/978-3-030-50480-9_9

2 Metabolic Impact of Macrophages in Adipose Tissue

In vertebrates, the so-called white adipose tissue (WAT) stores neutral lipids and fatty acids, serving as an energy storage site. WAT has several secondary functions as well: it aids thermal isolation, supports the musculoskeletal system (e.g., in the form of fat pads between muscles or in the foot sole), and metabolizes and generates hormones (e.g., estrogens). In mammals a special form of adipose tissue is also known, the so-called brown adipose tissue (BAT). BAT oxidizes lipids, and its oxidative phosphorylation is uncoupled from ATP synthesis to generate heat. Mammal newborns also have WAT depots that share some features with BAT and are capable of heat generation. This form of adipose tissue is often termed brite (brown-in-white) or beige adipose tissue (BeAT).

Adipose tissue macrophages (ATMs) are resident macrophages of WAT [4–9]. They are already present at birth, and thus populate the BeAT depots of the infant [10]. The first wave of ATMs develops from embryonic macrophage progenitors, and later they are accompanied by monocyte-derived ATMs. However, there is a monocyte-independent replenishment of ATMs in the lean WAT that is increased in the obese WAT, expanding the size of the ATM niche [11, 12]. ATMs comprise approximately 10% of all cells in the WAT under physiological conditions, which increases up to 50% in severe obesity [13]. Interestingly, BAT is scarce in macrophages, and obesity further reduces macrophage number in BAT [11, 14].

ATMs were first identified in obese mouse WAT in the 1960s; however, their presence in human WAT and their role in metabolic diseases was recognized only in the 2000s [15–18]. Obesity causes a dramatic increase in ATM number as the result of monocyte infiltration and increased proliferation of ATMs [19]. ATMs gather around lipid-overloaded and apoptotic fat cells, forming crown-like structures (Fig. 1). The crown-like structures are ATM clusters and, similarly to the Langhans cells, the ATMs fuse to form giant cells or syncytia [4–9, 12, 20–22].

Lean WAT also contains ATMs; however, these are scattered among the fat cells and patrol the tissue (Fig. 1). In obesity, ATMs become prevalent in the obese WAT and scavenge remnants of dying fat cells (Fig. 1). Functions of ATMs have been explored in the context of obesity and in obesity-associated metabolic diseases. Excess development of WAT leads to obesity: a complex and multifactorial disease involving genetic and environmental factors, including impaired neuroendocrine signaling, overnutrition, and sedentary lifestyle [23] (Fig. 2). It is estimated that 38% of the world's adult population will be overweight and another 20% will be obese by 2030 [23]. Obesity increases the prevalence of chronic incurable diseases, such as insulin resistance (IR) and type 2 diabetes mellitus (T2DM). Today, 9% of the world's population suffers from diabetes, and more than 90% of newly diagnosed diabetic patients have T2DM. Indeed, T2DM is one of the fastest growing noncommunicable diseases and is projected to be the seventh leading global cause of death within the next decade [24–26]. Obesity is therefore a significant health problem and a major medical challenge for our society [27]. The obese WAT generates a local inflammatory environment from the secretion of chemotactic and

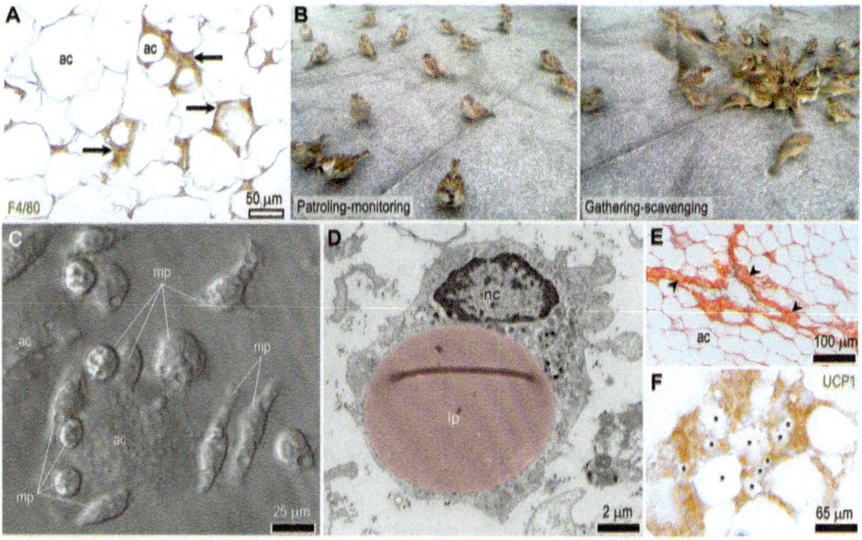

Fig. 1 Adipose tissue macrophages. (**a**) Obese white adipose tissue (WAT) in mouse. Adipose tissue macrophages (ATMs) are labeled with an antibody against the mouse macrophage marker F4/80 antigen. ATMs form crown-like structures (*arrows*) around adipocytes (*ac*). (**b**) ATM functions in lean and obese fat depots. *Left:* ATMs (such as these 'sparrows') are scattered among the fat cells in the lean state. ATMs have an immune surveillance function by patrolling in the tissue and monitoring antigens and apoptotic cells. *Right:* When the same sparrows find food, they gather and scavenge it. Similarly, ATMs gather around lipid-overloaded, dying fat cells and scavenge their materials. (**c**) Apoptotic fat cells (*ac*) and macrophages (*mp*) in vitro. Such as in the adipose tissue, macrophages form clusters around the apoptotic fat cells and clear their remnants and lipid molecules. (**d**) Transmission electron microscopy of a mouse ATM engulfing a large lipid droplet. *nc* nucleus, *lp* lipid droplet. For better visibility the lipid droplet is labeled with pseudo-color. (**e**) Obese WAT in human: collagen fibers (*arrowheads*) are stained with picrosirius red. Excessive collagen production leads to fibrosis. (**f**) Beige adipose tissue in an infant mouse. The fat cells contain multiple lipid vesicles (*asterisks*). The cells were labeled with an antibody against UCP1. (Images by author)

inflammatory mediators by the adipocytes. The release of modified lipid species, which together attract monocytes, increases local macrophage proliferation and bone marrow monocyte development, and promotes monocyte infiltration and M1 activation of ATMs [7, 20, 28, 29].

Single-cell sequencing data, and the growing body of studies using flow cytometry, in macrophage-specific knockout models show there are various ATM subsets that show varying degree of M1 and M2 traits [30–32]. Depending on the status of obesity, the ATM population may be dominated by M2 ATMs, or a mixture of M1 and M2 ATMs, and eventually can be dominated by M1, inflammatory ATMs [33–35]. There are also depot-specific differences between ATM pools [36]. The locally produced inflammatory mediators impair insulin signaling in the WAT, and when released into the circulation, the inflammatory mediators can cause liver, muscle, and pancreatic islet damage. Eventually, obesity can lead to insulin resistance, type

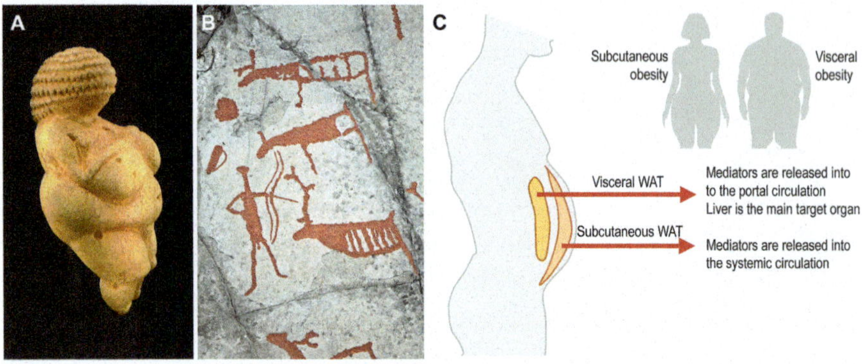

Fig. 2 Physiological adipose tissue development and types of obesity. (**a**) WAT has endocrine functions and is especially important for maintaining estrogen levels. Prehistoric artwork exaggerated subcutaneous obesity as a hallmark of fertility. (**b**) WAT accumulation during resource-rich periods, and utilization of stored fat in resource limitation, was considered an advantage in human evolution. Lipid- and carbohydrate-rich nutrition was hence physiologically impactful in prehistoric times, as depicted in this hunting scene in a cave painting. Fat storage needed some degree of insulin resistance. This feature of human WAT, however, can lead to obesity and metabolic diseases when nutrition is not limiting. (**c**) Two major types of obesity: subcutaneous and visceral. (**a**, **b** with courtesy, see Acknowledgments)

2 diabetes, type 1 diabetes, cardiovascular disease, and fatty liver disease. Inflammatory ATM functions are responsible for some of the etiology and pathogenesis of these diseases [5–7, 17, 20, 37–39].

3 What Is the Function of the M2 ATMs?

Because M1 ATMs of obese WAT are harmful to the metabolism, the assumption was prevalent in the medical literature of the 2010s that lean WAT should be rich in M2 ATMs. Skewing ATMs toward the M2 state curbs the unfavorable effects of ATMs [39], and eventually M2 ATMs were supposed to be metabolically beneficial [38]. Considerable effort was made to reprogram M1 ATMs toward an M2 state: nuclear receptor signaling, lipid metabolites, TLR signaling, adipokines, chemotactic factors and cytokines, gut-derived microbial products, content of exosomes released by neighboring cells, epigenetic modifications, vitamins, and hormones have been identified as crucial regulators of M1 ATM activation [28, 40–46].

However, and surprisingly, we know little about the exact functions of M2 ATMs to date. The "metabolically good" tag was already glued on the M2 ATM before its actual roles were explored. Recent research has provided a more detailed picture of the M2 ATMs. M2 activation of ATMs can be elicited by IL-4/STAT6 signaling [22], IL-6/STAT3 signaling [47, 48], PPARγ [49], and IL-10 signaling. Endogenous signals which induce M2 ATM activation include pancreatic neuropeptide FF and cold-stress induced IL-4 synthesis. A plethora of endogenous and exogenous signals have been identified that inhibit M1 ATM activation. However, a curbed M1 activation does not lead to M2 effector functions. The visceral WAT is perfused by blood delivering absorbed nutrients from the gut, also carrying microbial products of the normal microbiome. It is plausible that these microbial products sustain a certain degree of M1 activation of ATMs under physiological conditions, and that maintenance of a compensatory M2 state is necessary to avoid immune response against gut microbiota. ATMs and regulatory T cells produce IL-10 in the lean WAT [50, 51], which is a possible endogenous signal for suppressing ATM activation.

When obesity develops, M2 ATMs can also be prevalent in the WAT such as M1 ATMs, and M2 ATMs can cause adipose tissue fibrosis [52] (Fig. 1). Excess production of extracellular matrix encapsulates the fat cells, limiting their expansion; this eventually limits fat-storing capacity, and causes fat cell death and release of inflammation-provoking lipid species. In this context, M2 ATMs do not provide a metabolic benefit. Nevertheless, ATMs of the lean WAT express M2 markers, and reducing M1 ATM number improves metabolic health and reduces IR. However, M1 traits of ATMs are necessary for WAT physiology, and a transient inflammatory signaling is essential for normal WAT development [53]. Safe disposal of apoptotic cells is another M2 function for ATMs, and obesity-inducing substances may also impair this function [54}. WAT has key functions in metabolism by secreting hormones and adipokines that have direct impact on various organ systems, including reproductive, circulatory, and skeletal systems [38], and it is thought that the endocrine function of macrophages is associated with their M2 state (see chapter "Signal mechanisms of M2 macrophage activation").

ATMs hence have both M2 and M1 traits in the lean WAT, which are equally necessary for WAT physiology. Following are detailed examples of the interplay between M1 and M2 ATM traits.

4 Timing of M1 and M2 Traits of ATMs Guides WAT Development

In infancy, the fat depots are rich in so-called beige adipocytes. These adipocytes have multiple lipid vesicles, in contrast to the white adipocytes which store lipids in a single, large lipid vesicle [55]. The beige adipocytes are multivesicular or multilocular fat cells, meaning that they store fat in multiple vesicles. The beige adipocytes are also rich in mitochondria, and eventually oxidize lipids and generate energy (Fig. 1). Moreover, mitochondrial ATP synthesis may be uncoupled by the presence of mitochondrial uncoupling protein 1 (UCP1), and probably because of the presence of fatty acid metabolites that can also uncouple mitochondrial respiration from ATP synthesis [56]. As a result, beige adipocytes have a high degree of oxidative metabolism, and by mitochondrial uncoupling, they generate heat. These metabolic features allow an active lipid metabolism in the neonate and fulfill the energy demands of early development. The fat depots are mainly subcutaneous in the infant, forming an insulating mantle. The subcutaneous heat-generating beige fat cells eventually contribute to the maintenance of the core body temperature in infancy, when the endogenous body temperature regulation is still immature, and the body is poikilotherm (i.e., depends on the temperature of the environment). The fat depots contain ATMs after birth and metabolize lipid species derived from the breast milk [10]. Eventually, inflammatory mediators, platelet-activating factor and IL-6, are generated by the ATMs, which maintain beige adipocyte development. In the infant, fat depots, hence an M1-like ATM trait, can be recognized, which induces the expression of UCP-1, increases mitochondrial biogenesis, and sustains heat generation. The underlying signaling is an IL-6/JAK/STAT3 pathway, which increases *Ucp1* transcription, mitobiogenesis, and the expression of genes required for mitobiogenesis [10]. A local inflammatory signaling is necessary for early fat development [57].

However, lipid metabolism and fat distribution change during development. For instance, the facial fat pad (Bichat fat pad) is well developed in infants, and it is thought be necessary for the sucking buccal movements during breastfeeding [58]. Later, this fat pad is reduced in size. Subcutaneous fat distribution is gender dependent, as it is determined by sex hormones during puberty. Also, the fat metabolism of an adult is substantially different from an infant's metabolic demands and metabolic performance. Fat intake in infancy is magnitudes higher than in adulthood [59], and this lipid-rich nutrition covers the ATP demands of early growth. It is possible in infancy to afford the waste of lipids to burn them off in the form of heat, instead of ATP. However, as lipid intake (i.e., from breast milk) decreases, the fat depots change their function from fat oxidation and heat generation to fat storage. The transition of multilocular fat cells into unilocular fat cells in human postnatal development was observed early in the twentieth century [60]; today, this process is called beige-to-white transition [61]. During this transition ATMs lose their ability to generate PAF and IL-6 from breast milk AKGs [10]. The underlying mechanism involves increased ATM expression of AKG-metabolizing AKG monooxygenase (AGMO). Increased AGMO level is associated with M2 activation, making plausible that an M1–M2 switch of ATMs allows a transition from fat burning to fat storage [10].

In adulthood it is possible to elicit heat-generating activity of the white adipocytes. The trigger is cold exposure, which induces noradrenaline release from nerves within the WAT, leading to beta-adrenergic receptor stimulation in the white adipocytes [62, 63]. As a result, UCP1-expressing, heat-generating adipocytes appear in the WAT. This "renaissance" of beige adipocyte features leads to the development of a brite (brown-in-white) phenotype of the WAT [63]. The process is also called adipocyte beige transformation or adipocyte browning. It is a recently identified candidate to treat obesity, using IR and T2DM, by increasing cellular energy expenditure. Converting white adipose cells into a beige phenotype increases lipolysis and thermogenesis [55, 64]. Many genes and hormonal signals have been identified to induce adipocyte browning; however, there are gaps in our understanding of the physiological control of this process, and we are far from the use of adipocyte browning as a therapeutic option. The therapeutic potential of browning agents and

WAT browning remains an extremely active area of research in the field of metabolism with the potential to lead to new therapies.

ATMs and mast cells, which are myeloid cells related to macrophages residing in WAT, are involved in the "renaissance" of beige adipocytes [8, 65, 66]. Mast cells release IL-4 in response to cold stress, which stimulates UCP1 expression in adipocytes, promoting WAT browning and ultimately lipolysis and thermogenesis [8]. IL-4 increases the prevalence of M2 ATMs, which has been shown to further increase WAT browning [8, 67]. In addition to M2 polarization, cold stress also augments M2 ATM content via increased ATM proliferation [67]. It was first suggested that ATMs promote adipocyte browning by secreting catecholamines, which serve as hormonal signals for beige adipocyte transformation [68]. However, it has been recently shown by independent studies that ATMs do not produce catecholamines [69, 70]. In BAT, which expresses UCP1 lifelong, ATMs are scarce [11], and they even reduce the level of catecholamines that are released from nerve endings [71].

The Th2 cytokine environment favors beige transition of white adipocytes, although some inflammatory signals act against beige adipogenesis [8, 40, 70, 72]. However, inflammatory signaling, such as IL-6 and STAT3 signaling, is also essential for the development of brown and beige adipose tissue [10, 57, 73]. It is plausible that IL-4 induces beige transformation without the need of M2 ATMs, and cytokines can have distinct effects on ATMs and on adipocyte precursors. It is also possible that the contribution of ATMs to beige cell development is restricted to perinatal life, and that M2 activation diminishes the ability of ATMs to generate beige-inducing mediators [10]. A recent study also suggests that IL-10, a cytokine associated with M2 ATMs of the lean WAT, acts against beige adipogenesis [74].

5 M2 Traits of Liver Macrophages and Skeletal Muscle Macrophages

The fetal liver is colonized by yolk sac progenitors, which proliferate locally and generate fetal monocytes. The fetal liver is hence functioning as an embryonic hematopoietic organ. Fetal monocytes seed developing organs of the embryo and give rise to various tissue-resident macrophage types. Proliferating embryonic macrophages differentiate into liver-specific tissue-resident macrophages, the Kupffer cells, in an MCSF-dependent manner [75]. The Kupffer cells are fixed, or immotile, macrophages abundant in the liver, and they constitute about 80% of the total macrophage count of the body (Fig. 3). In adulthood the Kupffer cells are replenished by long-lived embryonic macrophage progenitors, but monocytes also settle in the differentiation into macrophages [76–79]. The liver macrophage niche composition is sensitive to inflammation and liver damage [80, 81].

The liver, as is the WAT, is exposed to antigens and microbial products, such as a low level of endotoxin. This condition results from the central role of the liver in intermediate metabolism: it receives blood from the portal vein, which delivers the

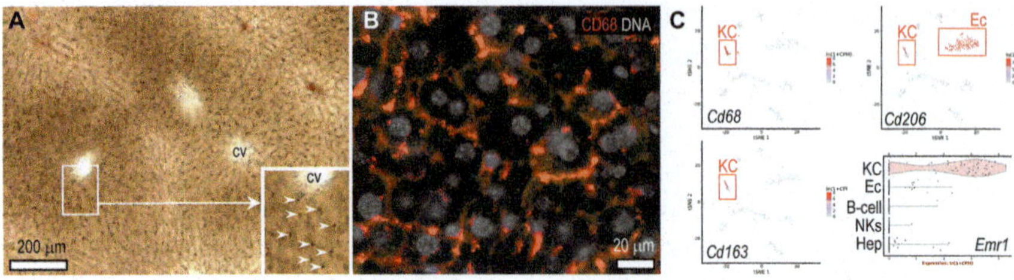

Fig. 3 M2 macrophages in the liver. (**a**) Section of mouse liver: *black*-colored cells are liver macrophages. The macrophages had phagocytosed black-colored carbon particles. *cv* central vein. (**b**) Kupffer cells in mouse liver, labeled with an antibody against CD68. (**c**) Expression of macrophage markers and M2 activation markers by mouse Kupffer cells. (Single-cell sequencing data retrieved from the Tabula Muris Consortium [91])

absorbed nutrients from the gut to the liver. Microbial products entering the portal circulation hence reach the liver, posing the risk of the initiation of an immune response. Moreover, lipids and mediators emitted by the WAT also reach the liver through the portal vein, with potential to trigger inflammation. Not surprisingly, obese WAT increases the risk of liver inflammation, fatty liver disease (steatosis), nonalcoholic steatohepatitis, and liver insulin resistance. As a self-protecting mechanism, Kupffer cells adopt a tolerogenic activation state; hence, they are considered examples of resident M2 macrophages [82]. The tolerogenic Kupffer cells release IL-10, interact with regulatory T cells, and elicit their IL-10 synthesis [82, 83]. M2 activation of Kupffer cells can be induced through IL-4/STAT6 [84], KLF4 [85], and PPARβ/δ [86], and lipopolysaccharide (LPS) increases their IL-10 secretion [83]. Kupffer cells also engulf apoptotic cells, which further augments their antiinflammatory state. M2 Kupffer cells induce apoptosis of M1 Kupffer cells, and liver inflammation is ameliorated by M2 Kupffer cells [85, 87]. M2 Kupffer cells improve the insulin sensitivity of the inflamed liver [86]. Studies to understand the metabolic role of Kupffer cells were inspired by studies on ATMs, leading to the assumption that inhibiting M1 Kupffer cell activation is metabolically beneficial. Various signals, such as adiponectin, serotonin, melatonin, and NR ligands, have been shown to mitigate M1 activation of Kupffer cells [48, 84, 88]. However, M2 traits of Kupffer cells can lead to liver fibrosis. The liver harbors monocyte-derived further macrophages, which appear transiently or settle and proliferate in the liver. The recruitment of $Ly6C^+$ monocytes is initiated by Kupffer cells in response to damage. These macrophages have various roles in liver physiology, largely depending on the disease context. Some of these recruited monocytes differentiate into $Ly6C^+$ macrophages, which release TGFβ. Another subset differentiates into $Ly6C^-$ macrophages, which have a tissue-regenerating property and resolve inflammation [89].

The skeletal muscles also contain resident macrophages, which show both M1 and M2 traits, where M1 traits are often associated with inflammation and muscle dystrophy [90]. M2 macrophages of the skeletal muscle impede the development of IR, and aid muscle and tendon regeneration; however, these also contribute to aging and fibrosis of the muscle (see chapter “M2 macrophages in the integument and in the musculoskeletal system”).

6 M2 Macrophages in the Nervous System and in the Endocrine Organs

Microglia are the resident macrophage type of the central nervous system [92]. These macrophages are yolk sac derived and long lived [93]. Microglia clear apoptotic cells during the development of the nervous system, promote proliferation and development of neuron precursors, and in the adult brain microglia dissolve unnecessary synaptic contacts [92]. The central nervous system is protected from most pathogens by the blood–brain barrier, which also impedes the extravasation of monocytes, making the microglia population ontogenetically homogeneous, that is, it develops from the embryonic hematopoiesis. The central nervous system was long considered as an immune-privileged environment, and under physiological conditions microglia are exposed to TGFβ and IL-10, which impede microglia activation [94–96]. This observation has led to the view that the microglia is in an M2-like state under physiological conditions [97]. However, because of the unique features of the immune system of the brain, microglia show functional traits that are difficult to define by the M1/M2 polarization model [97]. Such as in other organs, M1-like macrophage activation in the nervous system is associated with response to pathogens and injury, and is necessary for early repair, whereas an M2-like activation is needed to mend inflammation and aid repair [98–102]. Prolonged M1-like activation of the microglia, which is called microglia priming, leads to neurodegeneration [92]. However, the most recent progress in the field shows that distinct brain regions have their own, specific microglia populations, where the activation state has different functional significance, and M2-polarizing stimuli have region-specific effects in the brain [92, 102, 103]. Moreover, there is a gender difference in microglia development and the expression of M1/M2 markers [92, 104].

In addition to microglia, resident macrophages have been identified in the neural ganglia, in the sensory organs, peripheral nerves, and in the endocrine organs (Fig. 4) [71, 105, 106]. For instance, the inner ear contains macrophages that express CD68, CD11b, MHC-II, and the microglia marker IBA1 (ionized calcium-binding adapter molecule-1). These cells are motile, shuttle between the endolymph and the endolymphatic sac wall [107, 108], and interact with lymphocytes at the blood–labyrinth barrier. The auditory nerve within the cochlea also contains resident macrophages, which are thought to serve local immune surveillance and aid nerve regeneration. Macrophages of the inner ear and the auditory nerve interact with hair cells and neurons through cell membrane specializations and protrusions [105]. The Pacinian corpuscles also contain resident macrophages, which are present in spaces between glia cell lamellae formed around the sensory nerve ending in the core of the Pacinian corpuscle [109]. When the sensory nerve is injured, the Pacinian

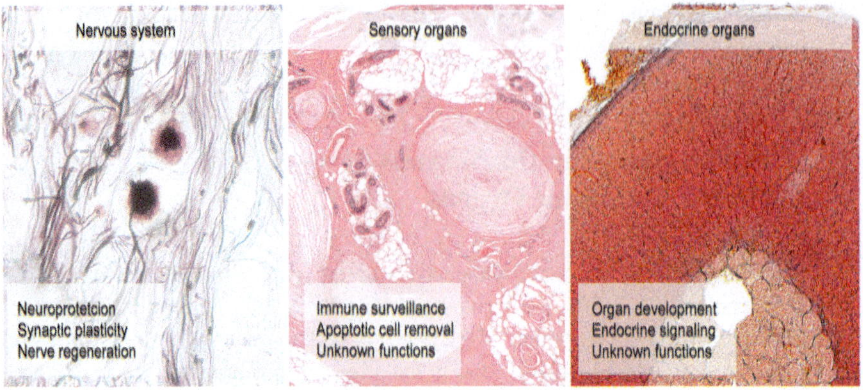

Fig. 4 M2 macrophage functions in the neuroendocrine system. The central nervous system has its own, specific resident macrophage pool, the so-called microglia. M2-like features of the microglia are neuroprotective, contrasting M1 traits which are associated with neurodegeneration and neuroinflammation. Peripheral nerves contain resident macrophages, which may aid nerve regeneration. Macrophages are also present in some sensory organs, such as the inner ear and the Pacinian corpuscles. Similarly, endocrine organs have resident macrophages that are necessary for the normal development of the endocrine cells. Macrophages are also sources of hormones and other endocrine signals. Histology images show a peripheral neural ganglion with two neurons and several nerve fibers (*left*), Pacinian corpuscles in the skin (*middle*), and the adrenal gland (*right*). (Images by author)

corpuscles undergo degeneration. In such case the resident macrophages remove the cell debris of degenerating nonneuronal cells [110].

Macrophages reside in the endocrine organs as well. The pancreatic islet contain macrophages under physiological conditions, and their number increases in autoimmune reaction against the pancreatic islet cells [111]. The islet macrophages show self-renewal and replenish independently from the monocyte supply. M2 markers are expressed by these macrophages, and the lack of macrophages during development impairs islet differentiation and insulin production. Resident, monocyte-derived macrophages are found in the stroma of the pancreas, and these cells also express certain M2 markers. Although it is known that an inflammatory macrophage activation is destructive for the endocrine pancreas, and is associated with diabetes development, the role of resident pancreatic M2-like macrophages is still to be defined [112]. Similarly, resident macrophages are present in the adrenal gland, where they establish contacts with the endocrine cells [113]. The pituitary gland also contains M2 resident macrophages, which are capable of proliferating, and their number is increased in prolactinoma [114]. The function of M2 macrophages in the endocrine organs is still unknown. Endocrine and neuronal signals shape macrophage activation. Neuropeptides such as vasoactive intestinal peptide (VIP), pituitary adenylate cyclase-activating peptide (PACAP), and neuropeptide FF (NPFF), as well as adenosine, capsaicin, acetylcholine, and beta-adrenergic receptor activation, induce M2 activation [22, 115–123]. Macrophages in the adipose tissue join peripheral nerves and eliminate the released noradrenaline [124]. M2 macrophages are constituents of the so-called diffuse or tissue neuroendocrine system, and release neurotransmitters and hormones, such as substance P, neuropeptide Y, VIP, insulin-like growth factor, and adiponectin [86, 115, 125–127]. The tissue endocrine system is a collective term to describe endocrine cells that are not arranged into a specific endocrine organ, but rather appear as dispersed cells within tissue. Examples include the endocrine cells of the gastrointestinal mucosa and scattered endocrine cells in the parenchyma of the pancreas. The endocrine function of the tissue-resident macrophages still should be explored; however, it has already been shown that it has a major role in the regulation of metabolism, tissue development, and repair.

References

1. Berkhout J, Bosdriesz E, Nikerel E, Molenaar D, Ridder DD, Teusink B, Bruggeman FJ. How biochemical constraints of cellular growth shape evolutionary adaptations in metabolism. Genetics. 2013;194:505–12.
2. van Ooijen G, Millar AJ. Non-transcriptional oscillators in circadian timekeeping. Trends Biochem Sci. 2012;37:484–92.
3. Del Vecchio D. Modularity in signaling systems. Phys Biol. 2012;9:045008.
4. Epelman S, Lavine KJ, Randolph GJ. Origin and functions of tissue macrophages. Immunity. 2014;41:21–35.
5. Osborn O, Olefsky JM. The cellular and signaling networks linking the immune system and metabolism in disease. Nat Med. 2012;18:363–74.
6. Lee BC, Lee J. Cellular and molecular players in adipose tissue inflammation in the development of obesity-induced insulin resistance. Biochim Biophys Acta. 2014;1842:446–62.
7. Glass CK, Olefsky JM. Inflammation and lipid signaling in the etiology of insulin resistance. Cell Metab. 2012;15:635–45.
8. Qiu Y, Nguyen KD, Odegaard JI, Cui X, Tian X, Locksley RM, Palmiter RD, Chawla A. Eosinophils and type 2 cytokine signaling in macrophages orchestrate development of functional beige fat. Cell. 2014;157:1292–308.
9. Li P, Spann NJ, Kaikkonen MU, Lu M, Oh da Y, Fox JN, Bandyopadhyay G, Talukdar S, Xu J, Lagakos WS, Patsouris D, Armando A, Quehenberger O, Dennis EA, Watkins SM, Auwerx J, Glass CK, Olefsky JM. NCoR repression of LXRs restricts macrophage biosynthesis of insulin-sensitizing omega 3 fatty acids. Cell. 2013;155:200–14.
10. Yu H, Dilbaz S, Coßmann J, Hoang AC, Diedrich V, Herwig A, Harauma A, Hoshi Y, Moriguchi T, Landgraf K, Körner A, Lucas C, Brodesser S, Balogh L, Thuróczy J, Karemore G, Kuefner MS, Park EA, Rapp C, Travers JB, Röszer T. Breast milk alkylglycerols sustain beige adipocytes through adipose tissue macrophages. J Clin Invest. 2019;129:2485–99.
11. Waqas SFH, Noble A, Hoang A, Ampem G, Popp M, Strauß S, Guille M, Röszer T. Adipose tissue macrophages develop from bone marrow-independent progenitors in *Xenopus laevis* and mouse. J Leukoc Biol. 2017;102:845–55.
12. Amano SU, Cohen JL, Vangala P, Tencerova M, Nicoloro SM, Yawe JC, Shen Y, Czech MP, Aouadi M. Local proliferation of macrophages contributes to obesity-associated adipose tissue inflammation. Cell Metab. 2014;19:162–71.
13. Odegaard JI, Ganeshan K, Chawla A. Adipose tissue macrophages: "amicus adipem?". Cell Metab. 2013;18:767–8.
14. Alcalá M, Calderon-Dominguez M, Bustos E, Ramos P, Casals N, Serra D, Viana M, Herrero L. Increased inflammation, oxidative stress and mitochondrial respiration in brown adipose tissue from obese mice. Sci Rep. 2017;7:16082.

15. Lumeng CN, Bodzin JL, Saltiel AR. Obesity induces a phenotypic switch in adipose tissue macrophage polarization. J Clin Invest. 2007;117:175–84.
16. Lumeng CN, DelProposto JB, Westcott DJ, Saltiel AR. Phenotypic switching of adipose tissue macrophages with obesity is generated by spatiotemporal differences in macrophage subtypes. Diabetes. 2008;57:3239–46.
17. Weisberg SP, McCann D, Desai M, Rosenbaum M, Leibel RL, Ferrante AW. Obesity is associated with macrophage accumulation in adipose tissue. J Clin Invest. 2003;112:1796–808.
18. Xu H, Barnes GT, Yang Q, Tan G, Yang D, Chou CJ, Sole J, Nichols A, Ross JS, Tartaglia LA, Chen H. Chronic inflammation in fat plays a crucial role in the development of obesity-related insulin resistance. J Clin Invest. 2003;112:1821–30.
19. Haase J, Weyer U, Immig K, Kloting N, Bluher M, Eilers J, Bechmann I, Gericke M. Local proliferation of macrophages in adipose tissue during obesity-induced inflammation. Diabetologia. 2014;57:562–71.
20. Fuentes L, Röszer T, Ricote M. Inflammatory mediators and insulin resistance in obesity: role of nuclear receptor signaling in macrophages. Mediat Inflamm. 2010;2010:219583.
21. Boutens L, Stienstra R. Adipose tissue macrophages: going off track during obesity. Diabetologia. 2016;59:879–94.
22. Waqas SFH, Hoang A, Lin Y, Ampem G, et al. Neuropeptide FF increases M2 activation and self-renewal of adipose tissue macrophages. J Clin Invest. 2017;127:2842–54.
23. Hruby A, Hu FB. The epidemiology of obesity: a big picture. PharmacoEconomics. 2015;33:673–89.
24. Sherwin R, Jastreboff AM. Year in diabetes 2012: the diabetes tsunami. J Clin Endocrinol Metab. 2012;97:4293–301.
25. Ng M, Fleming T, Robinson M, et al. Global, regional, and national prevalence of overweight and obesity in children and adults during 1980–2013: a systematic analysis for the Global Burden of Disease Study 2013. Lancet. 2014;384:766–81.
26. Kumar A, Bharti SK, Kumar A. Therapeutic molecules against type 2 diabetes: what we have and what are we expecting? Pharmacol Rep. 2017;69:959–70.
27. World Health Organization. Global report on diabetes. Geneva: WHO Press; 2016.
28. Griffin C, Eter L, Lanzetta N, Abrishami S, Varghese M, McKernan K, Muir L, Lane J, Lumeng CN, Singer K. TLR4, TRIF, and MyD88 are essential for myelopoiesis and $CD11c^+$ adipose tissue macrophage production in obese mice. J Biol Chem. 2018;293:8775–86.
29. Nagareddy PR, Kraakman M, Masters SL, Stirzaker RA, Gorman DJ, Grant RW, Dragoljevic D, Hong ES, Abdel-Latif A, Smyth SS, Choi SH, Korner J, Bornfeldt KE, Fisher EA, Dixit VD, Tall AR, Goldberg IJ, Murphy AJ. Adipose tissue macrophages promote myelopoiesis and monocytosis in obesity. Cell Metab. 2014;19:821–35.
30. Huang SC-C, Smith AM, Everts B, Colonna M, Pearce EL, Schilling JD, Pearce EJ. Metabolic reprogramming mediated by the mTORC2–IRF4 signaling axis is essential for macrophage alternative activation. Immunity. 2016;45:817–30.
31. Li C, Menoret A, Farragher C, Ouyang Z, Bonin C, Holvoet P, Vella AT, Zhou B. Single cell transcriptomics based-MacSpectrum reveals novel macrophage activation signatures in diseases. JCI Insight. 2019;5:e126453.
32. Blaszczak AM, Jalilvand A, Liu J, Wright VP, Suzo A, Needleman B, Noria S, Lafuse W, Hsueh WA, Bradley D. Human visceral adipose tissue macrophages are not adequately defined by standard methods of characterization. J Diabetes Res. 2019;2019:8124563.
33. Aron-Wisnewsky J, Tordjman J, Poitou C, Darakhshan F, Hugol D, Basdevant A, Aissat A, Guerre-Millo M, Clement K. Human adipose tissue macrophages: m1 and m2 cell surface markers in subcutaneous and omental depots and after weight loss. J Clin Endocrinol Metab. 2009;94:4619–23.
34. Basinska K, Marycz K, Śmieszek A, Nicpoń J. The production and distribution of IL-6 and TNF-α in subcutaneous adipose tissue and their correlation with serum concentrations in Welsh ponies with equine metabolic syndrome. J Vet Sci. 2015;16:113–20.

35. Chung S, Cuffe H, Marshall SM, McDaniel AL, Ha JH, Kavanagh K, Hong C, Tontonoz P, Temel RE, Parks JS. Dietary cholesterol promotes adipocyte hypertrophy and adipose tissue inflammation in visceral, but not in subcutaneous, fat in monkeys. Arterioscler Thromb Vasc Biol. 2014;34:1880–7.
36. Bigornia SJ, Farb MG, Mott MM, Hess DT, Carmine B, Fiscale A, Joseph L, Apovian CM, Gokce N. Relation of depot-specific adipose inflammation to insulin resistance in human obesity. Nutr Diabetes. 2012;2:e30.
37. Hotamisligil G, Shargill N, Spiegelman B. Adipose expression of tumor necrosis factor-alpha: direct role in obesity-linked insulin resistance. Science. 1993;259:87–91.
38. Rosen ED, Spiegelman BM. What we talk about when we talk about fat. Cell. 2014;156:20–44.
39. Dai L, Bhargava P, Stanya KJ, Alexander RK, Liou Y-H, Jacobi D, Knudsen NH, Hyde A, Gangl MR, Liu S, Lee C-H. Macrophage alternative activation confers protection against lipotoxicity-induced cell death. Mol Metabol. 2017;6:1186–97.
40. Kumari M, Wang X, Lantier L, Lyubetskaya A, Eguchi J, Kang S, Tenen D, Roh HC, Kong X, Kazak L, Ahmad R, Rosen ED. IRF3 promotes adipose inflammation and insulin resistance and represses browning. J Clin Invest. 2016;126:2839–54.
41. Bassaganya-Riera J, Misyak S, Guri AJ, Hontecillas R. PPAR gamma is highly expressed in F4/80(hi) adipose tissue macrophages and dampens adipose-tissue inflammation. Cell Immunol. 2009;258:138–46.
42. Borgeson E, McGillicuddy FC, Harford KA, Corrigan N, Higgins DF, Maderna P, Roche HM, Godson C. Lipoxin A4 attenuates adipose inflammation. FASEB J. 2012;26:4287–94.
43. Caesar R, Reigstad CS, Backhed HK, Reinhardt C, Ketonen M, Lunden GO, Cani PD, Backhed F. Gut-derived lipopolysaccharide augments adipose macrophage accumulation but is not essential for impaired glucose or insulin tolerance in mice. Gut. 2012;61:1701–7.
44. Filgueiras LR, Koga MM, Quaresma PG, Ishizuka EK, Montes MB, Prada PO, Saad MJ, Jancar S, Rios FJ. PAFR in adipose tissue macrophages is associated with anti-inflammatory phenotype and metabolic homoeostasis. Clin Sci. 2016;130:601–12.
45. Liu Z, Gan L, Zhang T, Ren Q, Sun C. Melatonin alleviates adipose inflammation through elevating α-ketoglutarate and diverting adipose-derived exosomes to macrophages in mice. J Pineal Res. 2018;64:e12455.
46. Pan Y, Hui X, Hoo RLC, Ye D, Chan CYC, Feng T, Wang Y, Lam KSL, Xu A. Adipocyte-secreted exosomal microRNA-34a inhibits M2 macrophage polarization to promote obesity-induced adipose inflammation. J Clin Invest. 2019;129:834–49.
47. Braune J, Weyer U, Hobusch C, Mauer J, Brüning JC, Bechmann I, Gericke M. IL-6 regulates M2 polarization and local proliferation of adipose tissue macrophages in obesity. J Immunol. 2017;198:2927–34.
48. Yi W-J, Kim TS. Melatonin protects mice against stress-induced inflammation through enhancement of M2 macrophage polarization. Int Immunopharmacol. 2017;48:146–58.
49. Odegaard JI, Ricardo-Gonzalez RR, Goforth MH, Morel CR, Subramanian V, Mukundan L, Red Eagle A, Vats D, Brombacher F, Ferrante AW, Chawla A. Macrophage-specific PPARgamma controls alternative activation and improves insulin resistance. Nature. 2007;447:1116–20.
50. Liu R, Nikolajczyk BS. Tissue immune cells fuel obesity-associated inflammation in adipose tissue and beyond. Front Immunol. 2019;10:1587.
51. Russo L, Lumeng CN. Properties and functions of adipose tissue macrophages in obesity. Immunology. 2018;155:10.
52. Spencer M, Yao-Borengasser A, Unal R, Rasouli N, Gurley CM, Zhu B, Peterson CA, Kern PA. Adipose tissue macrophages in insulin-resistant subjects are associated with collagen VI and fibrosis and demonstrate alternative activation. Am J Physiol Endocrinol Metab. 2010;299:E1016–27.
53. Chung K-J, Chatzigeorgiou A, Economopoulou M, Garcia-Martin R, Alexaki VI, Mitroulis I, Nati M, Gebler J, Ziemssen T, Goelz SE, Phieler J, Lim J-H, Karalis KP, Papayannopoulou T, Blüher M, Hajishengallis G, Chavakis T. A self-sustained loop of inflammation-driven inhibition of beige adipogenesis in obesity. Nat Immunol. 2017;18:654–64.

54. Röszer T. Transcriptional control of apoptotic cell clearance by macrophage nuclear receptors. Apoptosis. 2017;22:284–94.
55. Sidossis L, Kajimura S. Brown and beige fat in humans: thermogenic adipocytes that control energy and glucose homeostasis. J Clin Invest. 2015;125:478–86.
56. Samartsev VN, Marchik EI, Shamagulova LV. Free fatty acids as inducers and regulators of uncoupling of oxidative phosphorylation in liver mitochondria with participation of ADP/ATP- and aspartate/glutamate-antiporter. Biochemistry (Mosc). 2011;76:217–24.
57. Babaei R, Schuster M, Meln I, Lerch S, Ghandour RA, Pisani DF, Bayindir-Buchhalter I, Marx J, Wu S, Schoiswohl G, Billeter AT, Krunic D, Mauer J, Lee Y-H, Granneman JG, Fischer L, Müller-Stich BP, Amri E-Z, Kershaw EE, Heikenwälder M, Herzig S, Vegiopoulos A. Jak-TGFβ cross-talk links transient adipose tissue inflammation to beige adipogenesis. Sci Signal. 2018;11:527.
58. Tostevin P, Ellis H. The buccal pad of fat: a review. Clin Anat. 1995;8:403–6.
59. Stave U. Perinatal physiology. New York/London: Plenum; 1970.
60. Clara M. Entwicklungsgeschichte des Menschen. Leipzig: Georg Thieme Verlag; 1955.
61. Lu X, Altshuler-Keylin S, Wang Q, Chen Y, Henrique Sponton C, Ikeda K, Maretich P, Yoneshiro T, Kajimura S. Mitophagy controls beige adipocyte maintenance through a Parkin-dependent and UCP1-independent mechanism. Sci Signal. 2018;11:527.
62. Asano H, Kanamori Y, Higurashi S, Nara T, Kato K, Matsui T, Funaba M. Induction of beige-like adipocytes in 3T3-L1 cells. J Vet Med Sci. 2014;76:57–64.
63. Paulo E, Wang B. Towards a better understanding of beige adipocyte plasticity. Cell. 2019;8:1552.
64. Harms M, Seale P. Brown and beige fat: development, function and therapeutic potential. Nat Med. 2013;19:1252–63.
65. Finlin BS, Zhu B, Confides AL, Westgate PM, Harfmann BD, Dupont-Versteegden EE, Kern PA. Mast cells promote seasonal white adipose beiging in humans. Diabetes. 2017;66:1237–46.
66. Alexaki VI, Chavakis T. The role of innate immunity in the regulation of brown and beige adipogenesis. Rev Endocr Metab Disord. 2016;17:41–9.
67. Hui X, Gu P, Zhang J, Nie T, Pan Y, Wu D, Feng T, Zhong C, Wang Y, Lam KS, Xu A. Adiponectin enhances cold-induced browning of subcutaneous adipose tissue via promoting M2 macrophage proliferation. Cell Metab. 2015;22:279–90.
68. Nguyen KD, Qiu Y, Cui X, Goh YP, Mwangi J, David T, Mukundan L, Brombacher F, Locksley RM, Chawla A. Alternatively activated macrophages produce catecholamines to sustain adaptive thermogenesis. Nature. 2011;480:104–8.
69. Fischer K, Ruiz HH, Jhun K, Finan B, Oberlin DJ, van der Heide V, Kalinovich AV, Petrovic N, Wolf Y, Clemmensen C, Shin AC, Divanovic S, Brombacher F, Glasmacher E, Keipert S, Jastroch M, Nagler J, Schramm K-W, Medrikova D, Collden G, Woods SC, Herzig S, Homann D, Jung S, Nedergaard J, Cannon B, Tschöp MH, Müller TD, Buettner C. Alternatively activated macrophages do not synthesize catecholamines or contribute to adipose tissue adaptive thermogenesis. Nat Med. 2017;23:623.
70. Jun H, Yu H, Gong J, Jiang J, Qiao X, Perkey E, Kim D-I, Emont MP, Zestos AG, Cho J-S, Liu J, Kennedy RT, Maillard I, Xu XZS, Wu J. An immune-beige adipocyte communication via nicotinic acetylcholine receptor signaling. Nat Med. 2018;24:814–22.
71. Wolf Y, Boura-Halfon S, Cortese N, Haimon Z, Sar Shalom H, Kuperman Y, Kalchenko V, Brandis A, David E, Segal-Hayoun Y, Chappell-Maor L, Yaron A, Jung S. Brown-adipose-tissue macrophages control tissue innervation and homeostatic energy expenditure. Nat Immunol. 2017;18:665.
72. Bolus WR, Hasty AH. Contributions of innate type 2 inflammation to adipose function. J Lipid Res. 2019;60:1698–709.
73. Derecka M, Gornicka A, Koralov SB, Szczepanek K, Morgan M, Raje V, Sisler J, Zhang Q, Otero D, Cichy J, Rajewsky K, Shimoda K, Poli V, Strobl B, Pellegrini S, Harris TE, Seale P, Russell AP, McAinch AJ, O'Brien PE, Keller SR, Croniger CM, Kordula T, Larner

AC. Tyk2 and Stat3 regulate brown adipose tissue differentiation and obesity. Cell Metab. 2012;16:814–24.
74. Rajbhandari P, Thomas BJ, Feng A-C, Hong C, Wang J, Vergnes L, Sallam T, Wang B, Sandhu J, Seldin MM, Lusis AJ, Fong LG, Katz M, Lee R, Young SG, Reue K, Smale ST, Tontonoz P. IL-10 signaling remodels adipose chromatin architecture to limit thermogenesis and energy expenditure. Cell. 2018;172:218–33.. e217
75. Naito M, Hasegawa G, Takahashi K. Development, differentiation, and maturation of Kupffer cells. Microsc Res Tech. 1997;39:350–64.
76. Crofton R, Diesselhoff-Den Dulk M, Furth R. The origin, kinetics, and characteristics of the kupffer cells in the normal steady state. J Exp Med. 1978;148:1–17.
77. Gale R, Sparkes R, Golde D. Bone marrow origin of hepatic macrophages (Kupffer cells) in humans. Science. 1978;201:937–8.
78. Dey A, Allen J, Hankey-Giblin PA. Ontogeny and polarization of macrophages in inflammation: blood monocytes versus tissue macrophages. Front Immunol. 2015;5:683.
79. Krenkel O, Tacke F. Liver macrophages in tissue homeostasis and disease. Nat Rev Immunol. 2017;17:306–21.
80. Devisscher L, Scott CL, Lefere S, Raevens S, Bogaerts E, Paridaens A, Verhelst X, Geerts A, Guilliams M, Van Vlierberghe H. Non-alcoholic steatohepatitis induces transient changes within the liver macrophage pool. Cell Immunol. 2017;322:74–83.
81. Zigmond E, Samia-Grinberg S, Pasmanik-Chor M, Brazowski E, Shibolet O, Halpern Z, Varol C. Infiltrating monocyte-derived macrophages and resident Kupffer cells display different ontogeny and functions in acute liver injury. J Immunol. 2014;193:344–53.
82. Heymann F, Peusquens J, Ludwig-Portugall I, Kohlhepp M, Ergen C, Niemietz P, Martin C, van Rooijen N, Ochando JC, Randolph GJ, Luedde T, Ginhoux F, Kurts C, Trautwein C, Tacke F. Liver inflammation abrogates immunological tolerance induced by Kupffer cells. Hepatology. 2015;62:279–91.
83. Knoll P, Schlaak J, Uhrig A, Kempf P, zum Büschenfelde K-HM, Gerken G. Human Kupffer cells secrete IL-10 in response to lipopolysaccharide (LPS) challenge. J Hepatol. 1995;22:226–9.
84. Mandal P, Pratt BT, Barnes M, McMullen MR, Nagy LE. Molecular mechanism for adiponectin-dependent M2 macrophage polarization: link between the metabolic and innate immune activity of full-length adiponectin. J Biol Chem. 2011;286:13460–9.
85. Han Y-H, Kim H-J, Na H, Nam M-W, Kim J-Y, Kim J-S, Koo S-H, Lee M-O. RORα induces KLF4-mediated M2 polarization in the liver macrophages that protect against nonalcoholic steatohepatitis. Cell Rep. 2017;20:124–35.
86. Odegaard JI, Ricardo-Gonzalez RR, Red Eagle A, Vats D, Morel CR, Goforth MH, Subramanian V, Mukundan L, Ferrante AW, Chawla A. Alternative M2 activation of Kupffer cells by PPARdelta ameliorates obesity-induced insulin resistance. Cell Metab. 2008;7:496–507.
87. Wan J, Benkdane M, Teixeira-Clerc F, Bonnafous S, Louvet A, Lafdil F, Pecker F, Tran A, Gual P, Mallat A, Lotersztajn S, Pavoine C. M2 Kupffer cells promote M1 Kupffer cell apoptosis: a protective mechanism against alcoholic and nonalcoholic fatty liver disease. Hepatology. 2014;59:130–42.
88. de las Casas-Engel M, Dominguez-Soto A, Sierra-Filardi E, Bragado R, Nieto C, Puig-Kroger A, Samaniego R, Loza M, Corcuera MT, Gomez-Aguado F, Bustos M, Sanchez-Mateos P, Corbi AL. Serotonin skews human macrophage polarization through HTR2B and HTR7. J Immunol. 2013;190:2301–10.
89. Ju C, Tacke F. Hepatic macrophages in homeostasis and liver diseases: from pathogenesis to novel therapeutic strategies. Cell Mol Immunol. 2016;13:316–27.
90. Villalta SA, Rinaldi C, Deng B, Liu G, Fedor B, Tidball JG. Interleukin-10 reduces the pathology of mdx muscular dystrophy by deactivating M1 macrophages and modulating macrophage phenotype. Hum Mol Genet. 2011;20:790–805.

91. Tabula Muris Consortium. Single-cell transcriptomics of 20 mouse organs creates a *Tabula Muris*. Nature. 2018;562:367–72.
92. Wolf SA, Boddeke HWGM, Kettenmann H. Microglia in physiology and disease. Annu Rev Physiol. 2017;79:619–43.
93. Perdiguero EG, Klapproth K, Schulz C, Busch K, Azzoni E, Crozet L, Garner H, Trouillet C, de Bruijn MF, Geissmann F, Rodewald HR. Tissue-resident macrophages originate from yolk-sac-derived erythro-myeloid progenitors. Nature. 2015;518(7540):547–51.
94. Zöller T, Schneider A, Kleimeyer C, Masuda T, Potru PS, Pfeifer D, Blank T, Prinz M, Spittau B. Silencing of TGFβ signalling in microglia results in impaired homeostasis. Nat Commun. 2018;9:4011.
95. Taylor RA, Chang C-F, Goods BA, Hammond MD, Mac Grory B, Ai Y, Steinschneider AF, Renfroe SC, Askenase MH, McCullough LD, Kasner SE, Mullen MT, Hafler DA, Love JC, Sansing LH. TGF-β1 modulates microglial phenotype and promotes recovery after intracerebral hemorrhage. J Clin Invest. 2017;127:280–92.
96. Qin Y, Garrison BS, Ma W, Wang R, Jiang A, Li J, Mistry M, Bronson RT, Santoro D, Franco C, Robinton DA, Stevens B, Rossi DJ, Lu C, Springer TA. A milieu molecule for TGF-β required for microglia function in the nervous system. Cell. 2018;174:156–171. e116.
97. Ransohoff RM. A polarizing question: do M1 and M2 microglia exist? Nat Neurosci. 2016;19:987–91.
98. Calka J, Wolf G, Schmidt W. Induction of cytosolic NADPH-diaphorase/nitric oxide synthase in reactive microglia/macrophages after quinolinic acid lesions in the rat striatum: an electron and light microscopical study. Histochem Cell Biol. 1996;105:81–9.
99. Spitzbarth I, Bock P, Haist V, Stein VM, Tipold A, Wewetzer K, Baumgartner W, Beineke A. Prominent microglial activation in the early proinflammatory immune response in naturally occurring canine spinal cord injury. J Neuropathol Exp Neurol. 2011;70:703–14.
100. Hegyi B, Kornyei Z, Ferenczi S, Fekete R, Kudlik G, Kovacs KJ, Madarasz E, Uher F. Regulation of mouse microglia activation and effector functions by bone marrow-derived mesenchymal stem cells. Stem Cells Dev. 2014;23:2600–12.
101. Jiang MH, Chung E, Chi GF, Ahn W, Lim JE, Hong HS, Kim DW, Choi H, Kim J, Son Y. Substance P induces M2-type macrophages after spinal cord injury. Neuroreport. 2012;23:786–92.
102. Pepe G, Calderazzi G, De Maglie M, Villa A, Vegeto E. Heterogeneous induction of microglia M2a phenotype by central administration of interleukin-4. J Neuroinflammation. 2014;11:1031.
103. Stratoulias V, Venero JL, Tremblay M-È, Joseph B. Microglial subtypes: diversity within the microglial community. EMBO J. 2019;38:e101997.
104. Crain JM, Nikodemova M, Watters JJ. Microglia express distinct M1 and M2 phenotypic markers in the postnatal and adult central nervous system in male and female mice. J Neurosci Res. 2013;91:1143–51.
105. Liu W, Molnar M, Garnham C, Benav H, Rask-Andersen H. Macrophages in the human cochlea: Saviors or predators – a study using super-resolution immunohistochemistry. Front Immunol. 2018;9:223.
106. Taylor JM. Tissue resident macrophages are sufficient for demyelination during peripheral nerve myelin induced experimental autoimmune neuritis? J Neuroimmunol. 2017;313:69–76.
107. Rask-andersen H, Danckwardt-lillieström N, Friberg U, House W. Lymphocyte-macrophage activity in the human endolymphatic sac. Acta Otolaryngol. 1991;111:15–7.
108. Kämpfe Nordström C, Danckwardt-Lillieström N, Laurell G, Liu W, Rask-Andersen H. The human endolymphatic sac and inner ear immunity: macrophage interaction and molecular expression. Front Immunol. 2019;9:3181.
109. Ide C, Tohyama K. Macrophages in Pacinian corpuscles. Cells Tissues Organs. 1985;121:110–4.
110. Zelená JI. Rapid degeneration of developing rat Pacinian corpuscles after denervation. Brain Res. 1980;187:97–111.

111. Carrero JA, McCarthy DP, Ferris ST, Wan X, Hu H, Zinselmeyer BH, Vomund AN, Unanue ER. Resident macrophages of pancreatic islets have a seminal role in the initiation of autoimmune diabetes of NOD mice. Proc Natl Acad Sci USA. 2017;114:E10418–27.
112. Unanue ER. Macrophages in endocrine glands, with emphasis on pancreatic islets. Microbiol Spectr. 2016;4 https://doi.org/10.1128/microbiolspec.MCHD-0048-2016.
113. González-Hernández JA, Bornstein SR, Ehrhart-Bornstein M, Geschwend JE, Adler G, Scherbaum WA. Macrophages within the human adrenal gland. Cell Tissue Res. 1994;278:201–5.
114. Fujiwara K, Yatabe M, Tofrizal A, Jindatip D, Yashiro T, Nagai R. Identification of M2 macrophages in anterior pituitary glands of normal rats and rats with estrogen-induced prolactinoma. Cell Tissue Res. 2017;368:371–8.
115. Delgado M, Leceta J, Sun W, Gomariz RP, Ganea D. VIP and PACAP induce shift to a Th2 response by upregulating B7.2 expression. Ann N Y Acad Sci. 2000;921:68–78.
116. Lee JK, Chung J, Kannarkat GT, Tansey MG. Critical role of regulator G-protein signaling 10 (RGS10) in modulating macrophage M1/M2 activation. PLoS One. 2013;8:e81785.
117. Ganea D, Delgado M. Inhibitory neuropeptide receptors on macrophages. Microbes Infect. 2001;3:141–7.
118. Csoka B, Selmeczy Z, Koscso B, Nemeth ZH, Pacher P, Murray PJ, Kepka-Lenhart D, Morris SM Jr, Gause WC, Leibovich SJ, Hasko G. Adenosine promotes alternative macrophage activation via A2A and A2B receptors. FASEB J. 2012;26:376–86.
119. Nemethova A, Michel K, Gomez-Pinilla PJ, Boeckxstaens GE, Schemann M. Nicotine attenuates activation of tissue resident macrophages in the mouse stomach through the beta2 nicotinic acetylcholine receptor. PLoS One. 2013;8:e79264.
120. Tsuchida Y, Hatao F, Fujisawa M, Murata T, Kaminishi M, Seto Y, Hori M, Ozaki H. Neuronal stimulation with 5-hydroxytryptamine 4 receptor induces anti-inflammatory actions via alpha7nACh receptors on muscularis macrophages associated with postoperative ileus. Gut. 2011;60:638–47.
121. Bok E, Chung YC, Kim K-S, Baik HH, Shin W-H, Jin BK. Modulation of M1/M2 polarization by capsaicin contributes to the survival of dopaminergic neurons in the lipopolysaccharide-lesioned substantia nigra in vivo. Exp Mol Med. 2018;50:76.
122. Grailer JJ, Haggadone MD, Sarma JV, Zetoune FS, Ward PA. Induction of M2 regulatory macrophages through the β2-adrenergic receptor with protection during endotoxemia and acute lung injury. J Innate Immun. 2014;6:607–18.
123. Boomershine CS, Lafuse WP, Zwilling BS. Beta2-adrenergic receptor stimulation inhibits nitric oxide generation by *Mycobacterium avium* infected macrophages. J Neuroimmunol. 1999;101:68–75.
124. Pirzgalska RM, Seixas E, Seidman JS, Link VM, Sánchez NM, Mahú I, Mendes R, Gres V, Kubasova N, Morris I, Arús BA, Larabee CM, Vasques M, Tortosa F, Sousa AL, Anandan S, Tranfield E, Hahn MK, Iannacone M, Spann NJ, Glass CK, Domingos AI. Sympathetic neuron–associated macrophages contribute to obesity by importing and metabolizing norepinephrine. Nat Med. 2017;23:1309.
125. Blum A, Setiawan T, Hang L, Stoyanoff K, Weinstock JV. Interleukin-12 (IL-12) and IL-23 induction of substance P synthesis in murine T cells and macrophages is subject to IL-10 and transforming growth factor beta regulation. Infect Immun. 2008;76:3651–6.
126. Sakaguchi S, Shono J, Suzuki T, Sawano S, Anderson JE, Do MK, Ohtsubo H, Mizunoya W, Sato Y, Nakamura M, Furuse M, Yamada K, Ikeuchi Y, Tatsumi R. Implication of anti-inflammatory macrophages in regenerative moto-neuritogenesis: promotion of myoblast migration and neural chemorepellent semaphorin 3A expression in injured muscle. Int J Biochem Cell Biol. 2014;54:272–85.
127. Singer K, Morris DL, Oatmen KE, Wang T, DelProposto J, Mergian T, Cho KW, Lumeng CN. Neuropeptide Y is produced by adipose tissue macrophages and regulates obesity-induced inflammation. PLoS One. 2013;8:e57929.

Practical Approaches in M2 Macrophage Biology: Analysis, Pharmacology, and Didactical Interpretation of M2 Macrophage Functions

Keywords Macrophage biology · M2 macrophage · Th2 cytokines · Signaling · Therapy · Cancer · Diabetes · Obesity · Allergy · Tumor · Wound · Healing · Tissue repair · Pharmacology · Aging · Regeneration · Didactics · University teaching

1 Introduction

This chapter overviews the practical aspects of M2 macrophage analysis, such as the work with various types of macrophages and the choice and analysis of M2 marker mRNA and proteins. M2 macrophages have biomedical impact, and pharmacological reprogramming of macrophage activation has gained attention in the past decade. In certain diseases the induction of M2 activation is a desired therapeutic aim, most importantly in metabolic diseases, where promoting M2 activation within metabolic organs is a tool to reduce chronic inflammation. Although inducing an M2 macrophage state is desired in chronic inflammatory diseases, the restoration of anticancer immunity requires the blockage of M2 activation. The pharmacology of M2 macrophage activation, and inhibition of M2 macrophage functions, are overviewed in this chapter. Finally, some didactic approaches are shown that have practical application in interpreting M2 macrophage functions and in aiding scientific communication and university teaching of the theory of M2 macrophage biology.

2 Workflow of the Analysis of M2 Macrophages

2.1 Macrophage Models to Study M2 Activation

In vitro macrophage cultures provide valuable models for understanding how M2 macrophage activation is controlled. Macrophage cultures can be used to test the effects of certain signals, of ablation or overexpression of genes, epigenetic factors,

---

The original version of this chapter was revised: The original version of this chapter was inadvertently published with incomplete captions of Figs. 6 and 7 which has been updated now. The correction to this chapter is available at https://doi.org/10.1007/978-3-030-50480-9_11

T. Röszer, *The M2 Macrophage*, Progress in Inflammation Research 86,
https://doi.org/10.1007/978-3-030-50480-9_10

the nuclear shuttle and DNA binding of transcription factors, and many other mechanisms with potential to affect macrophage activation. The laboratory mouse is the most widely studied model of macrophage biology, genetics, and immunity–metabolism research. For this practical reason, most of the available research tools of macrophage activation are designed for use in the mouse. The J774A.1 and RAW264.7 cell lines are the mouse monocyte-macrophage cell lines often used in M2 macrophage research [1]. Both cell lines are derived from the BALB/c mouse strain. However, BALB/c and C57Black/6 are distinct in the ability to mount a Th2/M2 immune response (see chapter "What Is an M2 Macrophage? Historical Overview of the Macrophage Polarization Model. The Th1/Th2 and M1/M2 Paradigm, the Arginine Fork"). Moreover, there are strain-specific mutations in the laboratory mouse. For instance, the p200 family members, which are associated with macrophage maturation, proliferation, interferon response, *Mycobacterium* elimination, and autoimmunity, have frameshift mutations in C57Black/6 mouse [2–7]. In a "hybrid" study, where both C57Black/6 and BALB/c macrophages are used, the p200 family member Ifi204 and Ifi202b proteins should be measured in the BALB/c macrophages, whereas their equivalents, Ifi205 and Ifi203, are measured in C57Black/6 macrophages [4]. The human equivalent of Ifi204 is IFI16 [2]. At early stages of assay planning, it is hence recommended to determine whether such strain-specific mutations can be foreseen in the genes of interest.

The J774A.1 line is derived from an ascites reticulum cell tumor of a female BALB/c mouse and is a biosafety level 1 cell line. In culture it gives mostly adherent, macrophage-like cells; but some nonadherent monocytic cells are often present in the culture. They grow on cell culture-treated plastic surfaces, are less adherent on glass, form colonies in soft agar, and their adherence can be increased by poly-L-lysine coating of the surface, such as coverslips. J774A.1 cells readily phagocytose latex beads and produce interleukin (IL)-5, lysozyme, and IL-1β. They can be activated by lipopolysaccharide (LPS) and type 1 and type 2 interferons [1]. Most interferons are produced within 4 h after LPS challenge, and within 12 h when cells are infected with an ssRNA virus (Newcastle disease virus) [8]. M2 polarization can be elicited in 2–4 h with IL-4 and IL-10.

A recloned line from the J774A.1 line, called J774.2, is available. The cell line grows in both RPMI-1640 and DMEM media, supplemented with L-glutamine, in 10% fetal bovine serum under 5% CO_2 atmosphere. Interestingly, J774A.1 is sensitive to media change, increases cellular levels of diacylglycerol, dihydroceramide, and ceramide, and shows changes in protein kinase C activity and some gene expression changes following media change [9]. Mechanical stress also affects the biophysical properties of the macrophages [10]. Moreover, *Mycoplasma* contamination activates the J774A.1 macrophage, causing tumoricidal activity and elimination of intracellular *Listeria* cells [11]. The RAW264.7 cell line was established from ascites of a tumor induced in a male mouse by intraperitoneal injection of Abselon leukaemia virus (A-MuLV), and is currently classified as a biosafety level 2 cell line [12] (Fig. 1).

Genetic drift and passage number can cause considerable variance in macrophage gene expression. A recent study shows that the expression of particular genes

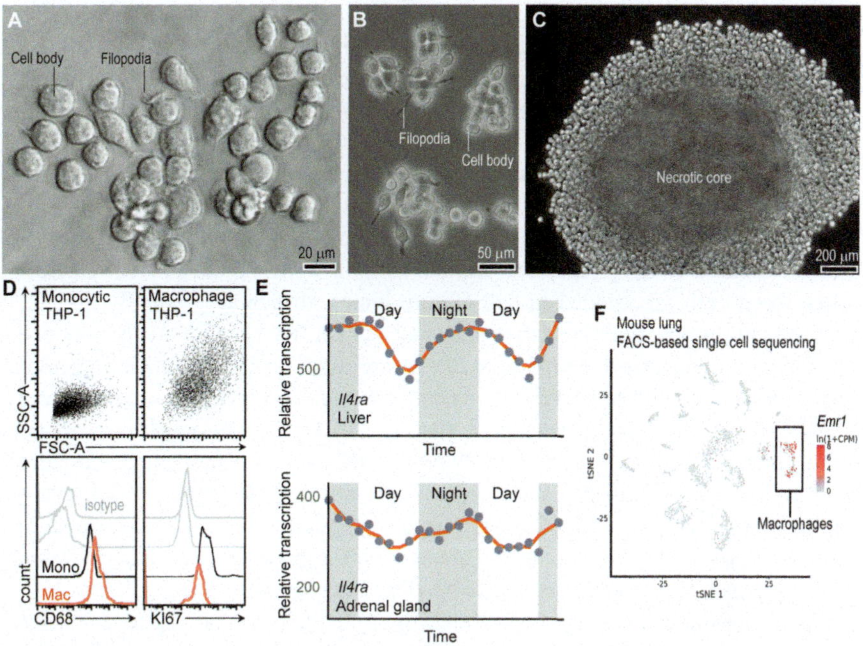

Fig. 1 Macrophage models to study M2 activation. (**a**) Mouse J774A.1 macrophages in culture. (**b**) Mouse RAW macrophages in culture. (**c**) Mouse macrophages grown in soft agar in a colony-forming assay. The founder cells have developed a large colony of daughter cells; the center of the colony dies and forms the necrotic core. Size and growth dynamics of the colonies indicate proliferation rate. (**d**) FACS analysis of human THP-1 cells. Monocytic cells grow in suspension and have a rounded, monocyte morphology. After phorbol 12-myristate 13-acetate (PMA) induction, the THP-1 cells transform into adherent, macrophage-like cells that increase in size (FSC-A) and intracellular complexity (SSC-A). These morphological changes hallmark the monocyte–macrophage transition: the cells become adherent and spread on the substrate, which increases their size; and have increased phagocytosis activity that eventually increases their intracellular content (endosomes, phagosomes, lysosomes). The monocyte–macrophage transition increases CD68 expression in macrophages (Mac) and reduces Ki-67 level. Ki-67 is expressed throughout the cell cycle, and reduced Ki-67 signal in macrophages indicates that the monocyte–macrophage transition leads to cell-cycle arrest [4]. (**e**) Expression level of M2 marker mRNA may have circadian oscillation. The example shows *Il4ra* mRNA oscillation in mouse liver and adrenal gland. (Graph retrieved from CircaDB database [13–15].) (**f**) Single-cell sequencing is valuable to collect information on macrophage activation at the single-cell level. *Emr1* (encoding murine macrophage marker F4/80 antigen) expression in macrophages of the mouse lung. (Graph retrieved from the Tabula Muris Consortium database [16])

and surface markers, phagocytosis, and NO production of the RAW 264.7 cell line remains stable from passage number 10 to passage number 30 [17]. Prolonged culture of macrophages (J774A.1, RAW264.7, and bone marrow-derived macrophages) tends to lead to an M2-like activation state with increased CD206 expression [18].

The alternative to macrophage cell lines is the in vitro culture of primary macrophages. There are established methods to harvest and culture tissue-resident macrophages such as microglia, Kupffer cells, adipose tissue macrophages, heart resident macrophages, or lung alveolar macrophages [19–23]. Isolation of these macrophages requires enzymatic digestion of brain, liver, heart or adipose tissue, followed by separation of the desired macrophages with centrifugation, then cell sorting with antibody-conjugated magnetic beads, or with preparative cell sorting. Alveolar macrophages as well as resident macrophages of the peritoneum can be isolated by lavage. These primary tissue-resident macrophages are true models of the in vivo state of the macrophages when analyzed immediately after isolation from the tissues. Hence, gene transcription profile, protein expression, and epigenetics of the freshly isolated macrophages are similar to their appearance in vivo. However, when these cells are seeded in cell culture plates, they may gradually lose their in vivo features, and their activation state may vanish. We have found changes in the lipid-metabolizing enzyme repertoire of adipose tissue macrophages during culture [24]. It is also crucial to harvest the desired macrophage population, and although this seems an easy task, in practice it may be difficult to distinguish macrophages from other myeloid cells [21, 23]. Moreover, the isolation procedure may affect macrophage activation state: mechanical stress during tissue lysis, centrifugation, and cell sorting may affect macrophage responses. Nevertheless, single-cell sequencing relies on cell sorting, and abundant information is available on antibody panels to recognize and harvest tissue-resident macrophages.

As a "universal" macrophage model, thioglycolate-elicited peritoneal macrophages and bone marrow-derived macrophages can also be used [25, 26]. Peritoneal macrophages are elicited with thioglycolate, which causes aseptic peritonitis [25]. Although the thioglycolate injection is effective in macrophage yield, it causes suffering for the mice, and it is hence obligatory to consider the ethical constraints of animal use. Moreover, the obtained macrophages are activated, have distinct metabolic and immune response features than other primary macrophages [25, 27, 28], and this may interfere with their M2 activation gene expression pattern. As an alternative, bone marrow-derived macrophages (BMDMs) can be cultured. Bone marrow hematopoietic stem cells obtained from the medullary cavity of the femur and tibia of mice are differentiated into macrophages in vitro. The main differentiation stimulus is conditioned media supernatant from L-929 fibroblast cell cultures that contains MCSF, a macrophage differentiation and M2-polarizing signal. This model is not to be confused with the use of primary bone marrow macrophages: these are primary macrophages that reside within the bone marrow and function in the hematopoietic niche. Use of BMDMs is a good choice when bone marrow is isolated from wild-type and genetically engineered (knockout, transgenic) mice, and the macrophage effect of the specific gene engineering can be easily traced.

However, BMDMs are artificial macrophages that are not present in tissues, so they are distinct from the tissue-resident macrophages. For instance, response of BMDMs to *Mycobacterium* is faster and more intense than that of J774A.1 cells in terms of number of differentially expressed genes and magnitude of induction or repression [29]. BMDMs release IL-6 during the first few days of culture, but not in J774A.1 and RAW264.7 cell lines. In contrast, tumor necrosis factor-α (TNFα) release is lacking in BMDM culture compared to J774A.1 and RAW264.7 cultures [18]. The BMDM lipidome is also distinct from that of tissue-resident macrophages.

Similarly, tissue-resident macrophages and macrophage cell lines have a distinct lipidomic profile [28].

Further circumstances that can affect the outcome and interpretation of an in vitro macrophage assay are the circadian clock of the macrophages, epigenetic factors, and the mitochondrial health of the macrophages. It is recommended to screen prior experiments in databases whether the target gene expression of the study shows a circadian oscillation. Epigenetic factors influence macrophage differentiation and activation, and it is not known what degree of epigenetic modification is retained in primary macrophages in culture. Mitochondrial oxidative phosphorylation affects antibacterial response of macrophages, and M1 macrophages are predominantly glycolytic, whereas M2 macrophages rely on oxidative phosphorylation and fatty acid oxidation [30]. However, high glucose conditions, although widely used in standard macrophage cultures, can also elicit M1 activation [31, 32]. Mitochondrial function is affected by the culture conditions, and the carbon source available may change macrophage activation. The commonly used cell culture media are DMEM and RPMI-1640 supplemented with antibiotics and fetal calf serum; additional optional supplements include EM nonessential amino acids. Metabolic precursors generated by mitochondrial intermediary metabolism are necessary for the proper ATP synthesis and growth of mammalian cells in culture. For instance, pyruvate supplementation delays cell-cycle arrest [33]. Extracellular pH also affects macrophage activation: the standard 5% CO_2 atmosphere, which may reduce extracellular pH, suppresses IL-6 release in some cell lines [34]. Atmospheric culture conditions with the appropriate pH buffering in the culture medium do not affect IL-6 or NO release [34]. Atmospheric culture is an alternative to 5% CO_2 atmosphere and in some instances is close to the physiological setting. For instance, macrophages of the carp (*Cyprinus carpio*) can be cultured in Leibovitz's L-15 medium at 26 °C without need of CO_2 to buffer the pH [35].

The THP-1 and the U937 monocytic cell lines are human monocyte-macrophage cell lines that are often used to study macrophage activation. The U937 monocytic cell line was derived from a human myeloid leukemia in 1976. The THP-1 cell line was derived from an acute monocytic leukemia in 1980, and the cells show monocyte features and grow in suspension culture [36]. Chemical induction of THP-1 with phorbol 12-myristate 13-acetate (PMA) elicits a differentiation into adherent, nonproliferating, macrophage-like cells (Fig. 1). The applied PMA dose and treatment time affect the responsiveness of the THP-1-derived macrophages [37]. Both cell lines are widely used in laboratories; they are stable up to 25 passages and can be cryopreserved and stored for a prolonged time. However, as these are malignant tumor cell lines, there is a possibility of experimental bias. To overcome this, primary human macrophages are a valuable model. Peripheral blood monocytes (PBMCs) can be differentiated into macrophages in vitro. Recently it has been shown that human monocyte-derived macrophages (hMDM) are heterogeneous with subsets that express distinct cell-surface receptors and have distinct viral susceptibility [38]. These hMDMs yield a model with utility similar to the mouse BMDM culture, which means that hMDMs are artificial macrophages, distinct from the primary tissue-resident macrophages. Primary macrophage isolation and in vitro culture is possible in the human, although it is more challenging than the techniques already discussed [4].

2.2 Analysis of M2 Activation

The many possible tools to study M2 activation include measurement of the mRNA level of M2 activation genes. In this regard one may use quantitative polymerase chain reaction (qPCR), DNA chip, and next-generation sequencing (NGS). One consideration to note here is that the conventional qPCR, DNA chip, and NGS yield information about a macrophage population and do not provide information on the single-cell transcript level of M2 genes (Fig. 2a). Why is this important? Let us suppose our study is about the antiinflammatory effect of a newly defined chemical compound, and our hypothesis is that this drug reverts M1 activation of

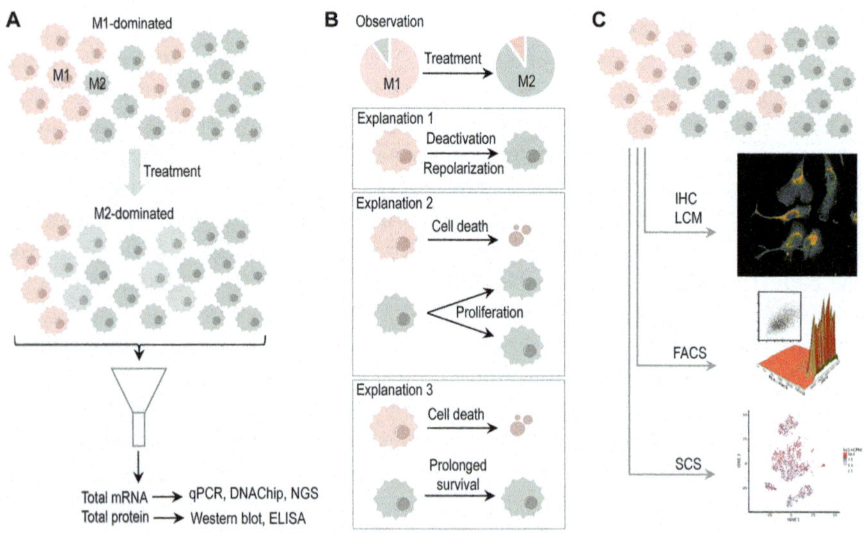

Fig. 2 Workflow of macrophage analysis. (**a**) Conventional macrophage analysis means the analysis of macrophage populations, rather than individual macrophages. The in vitro cultured macrophages, or tissue-resident macrophages isolated from organs, are lysed for mRNA or protein purification. The resulting mRNA and protein samples represent the macrophage population as a whole and do not reflect the activation state of individual macrophages. (**b**) Example of possible interpretation of observations obtained by the conventional macrophage analysis. The observed increase in M2-associated mRNA or protein level may be interpreted in at least three different ways: the macrophages have been deactivated and repolarized toward the M2 state; the M1 macrophages have died and naïve macrophages have generated new macrophages activated toward the M2 state; or the already-existing M2 macrophages have a prolonged survival whereas the M1 macrophages have died. (**c**) Macrophage analysis options at the single-cell level. *IHC* immunohistochemistry, *LCM* laser capture microdissection, *FACS* fluorescence-assisted cell sorting, *SCS* single-cell sequencing

macrophages and leads to M2 activation. We treat the in vitro cultured macrophages with vehicle and our drug of interest, lyse the cells with a phenolic extraction protocol, then proceed with RNA extraction and subsequent qPCR analysis. We find that the treated cells have a reduced *Nos2* expression and an increased *Arg1* expression level. We conclude that our hypothesis is proven: the drug reverts macrophage M1 activation and leads to M2 activation (Fig. 2b). Albeit it seems a logical assumption, the conclusion is unfortunately not correct. The macrophages divide and undergo apoptosis in the culture plates while we treat them with vehicle or the test material. We collect all the cells present in the cell culture plates at the time point of sample collection. Hence, either the cells have changed their activation state, or (more likely), the M1-activated cells did their job and died of apoptosis or programmed necrosis, and we are actually studying a population of bystander macrophages, newly generated naïve or M2-activated macrophages (Fig. 2b). The drug hence may increase the death of M1 macrophages, can trigger inflammasome activation or mitophagy, can change lipid droplet composition or the prevalence of citric acid cycle intermediates, and can induce cell-cycle entry of macrophages: quite a different conclusion, indeed.

Fortunately, live cell imaging, FACS analysis, laser capture microdissection, and the recently available single-cell sequencing provide information at the single macrophage level. We indeed can measure and take into account cell death and proliferation of macrophages with these techniques (Fig. 2c).

After establishing the transcriptional changes associated with macrophage activation, it is necessary to continue with the analysis of M2 activation-associated proteins. The available techniques include FACS, immunohistochemistry, live cell imaging, in-cell ELISA, and Western blotting. The signal mechanisms of macrophage activation need Western blotting, in-cell ELISA, and FACS (e.g., to measure STAT6 and pSTAT6, phosphorylated JAK, STAT1, or STAT3 levels). DNA binding of transcriptional activators and repressors requires chromatin immunoprecipitation and sequencing. Moreover, miR and other noncoding RNA species of activated macrophages require special RNA analysis that allows the recognition and quantification of short RNA species. Metabolites and metabolic changes associated with macrophages are the subject of a specific research field, called immunometabolism. Specific lipid species, mitochondrial dynamics, enzymes, and intermediates of the citric acid cycle are associated with M2 activation, and metabolomics of macrophage activation is currently a rapidly growing area of research.

Increased mRNA expression of a particular gene is not necessarily mirrored by changes in the level of the corresponding protein. It does result from the presence of miR species that can impede the successful translation of proteins from mRNA, and the effects of miRs on the control of macrophage activation are prevalent (chapter "Signal Mechanisms of M2 Macrophage Activation"). Proteasomal degradation may also cause discrepancy between measured mRNA and protein levels: mRNA may be abundant, but the protein itself may be rapidly degraded by the proteasome. In the context of macrophages, one important consideration is the high amount of hydrolytic enzymes in macrophages, which can easily degrade mRNA and protein during sample processing. Moreover, macrophages are phagocytic cells, and they take up M2 proteins produced by other cells. A unique phenomenon is thus possible: the presence of M2 marker proteins without the expression of the corresponding mRNA. Arginase-1, for instance, is released by extracellular vesicles, which may be taken up by macrophages [39]. The presence of Ym1 protein in wound-associated macrophages results from the uptake of Ym1 released by wound neutrophil granulocytes without upregulation of its mRNA in the macrophages [40].

3 Typical M2 Marker Proteins[1]

An important step is the choice of the M2 activation markers. Extensive reviews of M2 macrophage activation markers and effectors are available, including tissue-specific considerations (e.g., specific M2 macrophage types in atherosclerosis, or in the skin) (Fig. 3). How to choose the right M2 marker? It depends on circumstances such as the type of treatment of macrophages, type of macrophage, immune context,

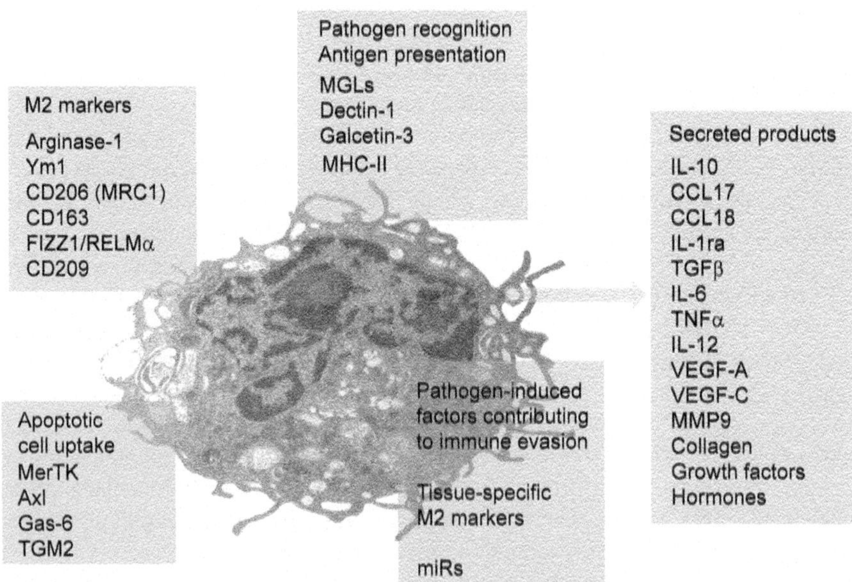

Fig. 3 Hallmarks of M2 macrophages. M2 markers are typically expressed in response to IL-4/STAT6 and IL-10 or TGFβ. Arginase-1 is a prototypical M2 marker; however, many tissue-resident M2 macrophages lack arginase-1. Effector functions of M2 macrophages also have their own characteristic protein and mRNA repertoire. M2 macrophages secrete pro-resolving and immunosuppressing mediators, hormones, and growth factors. Intracellular pathogens elicit specific M2-associated gene expression patterns that aid immune evasion

[1]This section is available in extended form in [41] Rőszer, T. (2015) Understanding the Mysterious M2 Macrophage Through Activation Markers and Effector Mechanisms. *Mediators of Inflammation* 2015, 16.

and mouse strain. For instance, if we treat macrophages with IL-10, we expect that *Cd163* gene transcription is increased in macrophages. When we isolate primary macrophages and would like to classify them as M2 macrophages, we should consider that circulating monocytes and most of the tissue-resident macrophages are $CD163^+$ cells, whereas inflammatory macrophages lack CD163. Hence, it is plausible that a $CD163^+$ tissue macrophage is M2 or, at least, not an M1 macrophage. However, some tissue-resident macrophage types lack CD163 expression (e.g., mantle zone and germinal center macrophages), and this should be considered when interpreting the observations.

3.1 Arginase-1

Arginase-1 (EC 3.5.3.1) is considered a prototypic M2 marker in the mouse because it is a key enzyme of the arginine fork in the M1/M2 dichotomy. Its expression is increased in response to IL-4, IL-13, and TGFβ, *Arg1* has a STAT6-binding site in its promoter region, and arginase-1 is expressed in the Th2 immune context such as in helminth infection and airway inflammation [42–46]. Arginase may have been primarily a wound-healing protein, transcriptionally induced by proteins of the TGFβ family [43] (also see chapters "Evolutionary Origin of M2 Macrophage Activation: Invertebrates" and "Immune Functions of the M2 Macrophages: Host Defense, Self-tolerance, and Autoimmunity"). Arginase-1 is constitutively expressed by tissue-resident macrophages [43, 47]. The tissue niche provides M2-polarizing stimuli, including Th2 cytokines: in microglia, arginase-1 expression is sustained by endogenously produced TGFβ [48], and by IL-10 in the intestinal mucosa [49–51]. Several tissue-specific Th2 cytokine-independent cues have also been identified as inducing arginase-1 expression in tissue-resident macrophages: netrin [52], adenosine [53], neuropeptides [4, 54, 55], retinal pigments [56], and many other signals, such as extracellular vesicles released by surrounding cells [57, 58].

3.2 Chitinase-3-Like Protein 3 or Ym1

Chitinase-3-like protein 3 (Chi3l3, Chil3, also called Ym1) is another M2 marker frequently used in the mouse [59], because IL-4 and IL-13 upregulate its expression through IL-4Rα and STAT6 [42, 60], and its steady-state expression is sustained by TGFβ [48]. Expression of Ym1 is inhibited by Th1 cytokines, and its lack leads to inflammatory cytokine production [60, 61]. Macrophages synthesize Ym1 during Th2 immune response: parasitic or fungal infection [62, 63], allergy [62], and eosinophilic meningitis and meningoencephalitis [64], making Ym1 a bona fide M2 marker. The Ym1-immunoreactive protein is associated with the rough endoplasmic reticulum and with needle-shaped crystalline bodies in the cytoplasm [65–67]. The precise mechanism of Ym1 actions in macrophages is uncertain: it binds chitin, but lacks chitinase activity. It also binds heparin and heparan sulfate, and it might lyse glycosaminoglycans because of its weak β-*N*-acetylglucosaminidase (EC 3.2.2.11) activity [65, 67, 68]. Ym1 is thought to control macrophage heparan sulfate levels, which impact macrophage activation [41]. It also may compete for extracellular matrix binding with leukocytes and eventually inhibit their extravasation [62]. Tissue-resident macrophages expressing Ym1 include alveolar macrophages, splenic macrophages, bone marrow macrophages, and microglia in the mouse [64]. Peritoneal macrophages lack Ym1 expression; however, they become Ym1$^+$ in response to parasite infection [60, 69]. Humans lack Ym1: its closest homologue is the eosinophil chemotactic cytokine [59].

3.3 C-type Mannose Receptor CD206

CD206, also termed MRC1 (C-type mannose receptor 1) is an M2 macrophage marker in both mouse and human [70, 71]: it is a 175-kDa type I transmembrane glycoprotein that binds and internalizes glycoproteins and collagen ligands. Its expression is induced by IL-4, helminth infection, TGFβ, GM-CSF, and STAT6-independent tissue-specific cues [72]. Cardiac-resident macrophages, peritoneal macrophages, adipose tissue macrophages, dermal macrophages, and placental macrophages (Hofbauer cells) express CD206 [72–77]. The CD206-expressing macrophages have unfavorable profibrotic effects because they promote fibroblast growth through TGFβ and chemokine (C-C) motif ligand 18 (CCL18) secretion [78]. They may also undergo a fibrocyte-like phenotype switch and produce collagen [79].

3.4 Haptoglobin–Hemoglobin Scavenger Receptor CD163

CD163 is a haptoglobin–hemoglobin scavenger receptor [73, 74, 80, 81]. It is referred to as an M2 marker protein, principally because of its upregulated expression in response to IL-4 [71]. Its expression is amplified also by M-CSF, IL-6, IL-10, and glucocorticoids, but TNFα, TGFβ, IFNγ, and LPS reduce its expression [70, 82]. In human monocytes CD163 has a high basal expression, and surprisingly IL-4 represses its expression [70]. Macrophages co-expressing CD206 and CD163 are high IL-10, IL-1 receptor antagonist (IL-1ra), and CCL18 producers [77]; they also have high capacity for apoptotic cell uptake [83]. CD206-expressing tissue-resident macrophages, such as mouse and human adipose tissue macrophages and placental macrophages, also express CD163. CD163 expression is increased in tumor-associated macrophages (TAMs) [84]. The expression of CD163 is however not restricted to M2 macrophages, so it is advisable to detect CD163 in combination with other M2 markers [85].

3.5 Resistin-Like Molecule-Alpha and DC-SIGN

Found in inflammatory zone 1 (FIZZ1), and also known as hypoxia-induced mitogenic factor (HIMF), resistin-like molecule-α (RELMα) is a 9.4-kDa cysteine-rich secreted protein [60]. Its expression is upregulated by helminth infection, IL-4, and IL-13 via the STAT6 pathway and suppressed by IFNγ [60, 86, 87]. In helminth infection, FIZZ1 diminishes inflammation [88, 89]. Another molecule in which macrophage expression is increased by IL-4, IL-10, and MCSF is the dendritic cell-specific ICAM-3 grabbing nonintegrin (DC-SIGN), also known as CD209 [82, 90]. It is a marker of dendritic cells; however, certain tissue-resident macrophages express it, such as colorectal mucosal macrophages, placental macrophages, alveolar macrophages, and adipose tissue macrophages in the mouse [82, 91–94] and in human microglia cultured in vitro [95].

3.6 Galactose-Type C-Type Lectins and Dectin-1

Further possible M2 markers are galactose-type C-type lectins (MGL1 and -2). In the mouse, MGL1 and MGL2 are expressed in peritoneal macrophages elicited during parasite infection and in alveolar macrophages in allergic asthma. IL-4 and IL-13 upregulate both MGL1 and MGL2 expression [96]. The human homologue of MGL1 is MGL, which recognizes antigens and increases IL-10 and TNFα expression [96]; it is also expressed by TAMs [97]. Dectin-1 (CLEC7A) is a lectin-like innate immune receptor that binds beta-glucans [98]. The fungal cell wall is rich in beta-glucans and thus has a key function in the recognition and phagocytosis of pathogenic fungi by macrophages [99]. Tumor cell surfaces can express N-glycans, which are also recognized by dectin-1, allowing the uptake of tumor cells [100]. Absence of dectin-1 impairs the phagocytic and fungicidal abilities of macrophages and alters NO and cytokines production. Increasing dectin-1 by IL-13 improves antifungal defense [101]. It is suggested that M2a macrophages have high dectin-1 expression while M2b macrophages express low levels of dectin-1 [99]. However, contrary to expectations, the lack of dectin-1 amplifies the expression of other M2 markers such as Ym1, arginase-1, and FIZZ1 [101]. Of note, dectin-1 is involved in M1 macrophage activation as well and increases pathogen killing [101]. In the mouse, various resident macrophage types express dectin-1: alveolar macrophages, Kupffer cells, intestinal macrophages, and splenic macrophages, with a possible role in pathogen recognition [102].

3.7 Mediators Associated with M2 Macrophages

M2 macrophages secrete antiinflammatory cytokines IL-10, IL-1 receptor antagonist (IL-1ra), and moreover TGFβ, chemokine C-C motif ligand 17 (CCL17), vascular endothelial growth factor A (VEGF-A), growth factors, and chemotactic factors that promote tumor cell motility [103, 104]. TGFβ suppresses Th1 response, promotes Th2 lymphocyte functions, and is required for natural killer cell, B cell, and dendritic cell functions [105]; it is responsible for the pro-fibrogenic effect of M2 macrophages. Similar to TGFβ, IL-10 and CCL17 are cytokines with pleiotropic effects. In the context of M2 macrophages, they are responsible for pro-resolving and antiinflammatory effects. However, cytokines associated with M1 activation may also be produced by M2 macrophages, such as IL-6, TNFα, and IL-12 [71, 106]. M2 macrophages also secrete metastasis-related mediators such as VEGF-A, VEGF-C, MMP9 [107], and collagen. M2 macrophages also release hormones, growth factors, and neurotransmitters, and hence they are constituents of the neuroendocrine system (see in chapters "Signal Mechanisms of M2 Macrophage Activation" and "M2 Macrophages in the Circulatory, Respiratory, and Excretory Organs"). Further molecules associated with M2 macrophages are necessary for phagocytosis of apoptotic cells: galectin-3 [108], Mer tyrosine kinase, Axl receptor tyrosine kinase, growth arrest-specific 6 (Gas-6) [83, 109], signaling lymphocyte-activation molecule [83, 109], and transglutaminase 2 (TGM2) [109–111]. Antigen-presenting M2 macrophages also express major histocompatibility complex-II [71]. The pattern of M2 marker expression is largely dependent of the tissue environment in which the macrophages reside (Graphic Abstract 2).

3.8 M2 Activation as Seen by NGS

NGS and single-cell sequencing provide a genome-wide landscape of gene transcription, and if completed with lipidomic and proteomic information, generates abundant information about macrophage physiology. With the use of appropriate databases and proper skills in massive data analysis, bioinformatics, and statistics, it is possible to obtain information about the gene ontology enrichment of the gene set, transcription factors, diseases, cellular compartments, knockout mouse phenotypes, metabolites, and pathways associated with the transcripts. It is also possible to check if the transcripts are known to have circadian oscillations, are associated with specific cell types, or are known targets of miRs. However, this abundance of information may also render the interpretation difficult. Gene ontology (GO) terms and associated pathways may be misleading in certain settings. For instance, the combination of *Cd206*, *Cd163*, *Il10*, and *Il4ra* leads to the following set of gene ontology terms: growth factor activity (GO:0008083), growth factor receptor binding (GO:0070851), cytokine receptor binding (GO:0005126), cytokine activity (GO:0005125); regulation of response to wounding (GO:1903034), regulation of immunoglobulin-mediated immune response (GO:0002889), and cytoplasmic sequestering of NF-kappaB (GO:0007253). The same set of transcripts are associated with inflammatory bowel disease (IBD), asthma, and African trypanosomiasis Kegg pathway terms.

M2 activation hence may lead to the enrichment of GO terms associated with growth factors, cytokines, antigen presentation, arginine metabolism, and inhibition of cytokine signaling. Decreased expression of interferon-responsive genes also hallmarks M2 macrophages.

4 M2 Activation in Nonmurine, Nonhuman Macrophages

M2 macrophage biology is centered on mouse and human macrophages. However, M2 activation can be elicited and studied in other experimental models as well. M2 activation in invertebrates and nonmammalian vertebrates is detailed in chapters "Evolutionary Origin of M2 Macrophage Activation: Invertebrates" and "Evolution of M2 Macrophage Functions in Chordates". In mammals, in addition to the conventional mouse or human models, M2 activation has been shown in rat, porcine, canine, feline, ruminant, equine, and nonhuman primate macrophages [112–115]. For instance, porcine M2 macrophages express IL4-Rα, CCL11, CCL17, CCL22, CCL26, IL-1ra, TGM2, adenosine receptor A2B, and VDR, and have reduced pathogenicity toward *Staphylococcus* [116]. Canine macrophages express arginase-1, and in the setting of allergy, canine macrophages express CD206, high-affinity IgE receptor (FcεRI) [117], and membrane-spanning 4-domains, subfamily A, member 2 (MS4A2) [118]. Canine mammary tumor is associated with M2-like macrophages [119]. Equine peritoneal macrophages and lung macrophages express certain M2 markers, such as *VEGFA*, *CD163*, *CCL22*, *IL4R*, and *IL10* [113].

5 Pharmacology and Medical Use of M2 Macrophages

5.1 Promoting M2 Activation

M2 macrophages have biomedical impact, and pharmacological reprogramming of macrophage activation has gained attention. In certain diseases the induction of M2 activation is a desired therapeutic aim, most importantly in metabolic diseases, where promoting M2 activation within metabolic organs is a tool to reduce chronic inflammation (Fig. 4). Moreover, M2 macrophages are necessary for proper wound healing, tissue regeneration (e.g., regeneration of bone fractures, muscle and tendon injuries, and recovery after myocardial infarction; see chapters "M2 Macrophages in the Integument and in the Musculoskeletal System" and "M2 Macrophages in the Circulatory, Respiratory, and Excretory Organs"). Drug candidates that induce M2

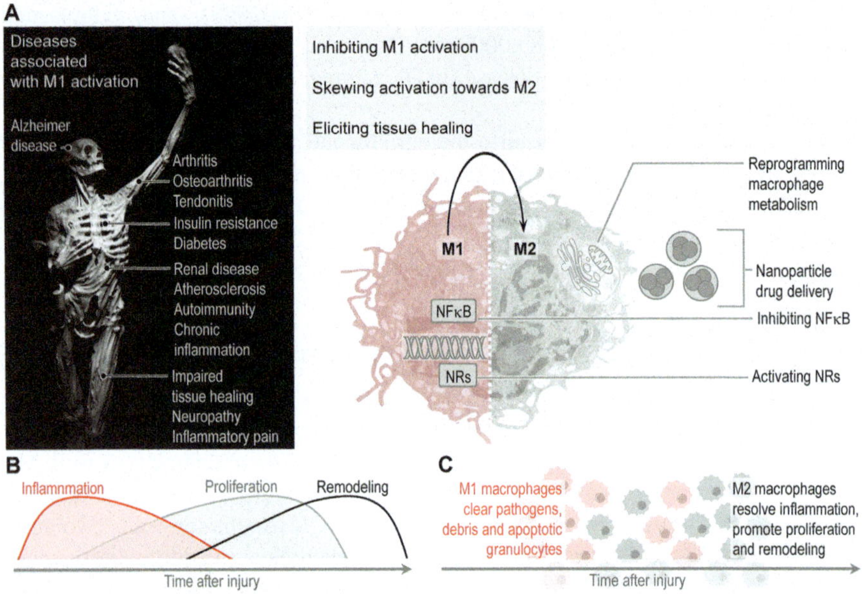

Fig. 4 Increasing M2 macrophage activation to combat chronic inflammation and improve tissue repair. (**a**) Chronic inflammatory diseases are associated with the prevalence of M1 macrophages in the affected tissues. Skewing macrophage polarization toward the M2 state can alleviate disease severity and can slow disease progression. However, pleiotropy of macrophage functions limits the utility of systemically acting M1-inhibiting–M2-activating drugs. "Allegory of disease" is an artwork by the sculptor Ligier Richer (c. 1500–1567) (image with courtesy). (**b**) Schematic representation of major phases of tissue healing [144]. (**c**) Macrophage functions in the course of tissue healing. Macrophages are recruited to the wounded tissue and also proliferate locally. In the inflammatory phase, they clear debris and pathogens and degrade extracellular matrix directly or via stimulating fibroblasts. In the proliferative and remodeling phase, macrophages promote fibroblast proliferation, extracellular matrix synthesis, and angiogenesis

activation in metabolic diseases are mostly molecules which target NR signaling. NRs such as PPARs, LXRs, or RXRs are associated with apoptotic cell clearance, and they can promote the transcription of M2-associated genes and repress Th1 cytokines and M1 activation [109, 120–123] (also see chapter "Signal Mechanisms of M2 Macrophage Activation"). The insulin sensitizer thiazolidinediones (TZDs) activate PPARγ signaling, which ameliorates obesity-associated inflammation and insulin resistance. This effect involves the reduction of M1 activation of adipose tissue macrophages [124]. Safety issues of TZDs, however, led to the withdrawal of the TZDs troglitazone and rosiglitazone from the market by 2010 [125]. Pioglitazone, another TZD, is still in clinical use, and it exerts similar effects on macrophages as other TZDs [81, 126, 127]. Endogenous PPARγ ligands such as alkyl phospholipids [128], and plant-derived PPARγ activator molecules such as procyanidins, induce M2 gene expression in macrophages [129], and such molecules may be alternatives for TZDs. However, pleiotropic effects of PPARγ activation impede the clinical utility of systemically acting PPARγ ligands. For instance, PPARγ is expressed in kidney nephron tubules, and PPARγ activation causes water retention [130]. Also, PPARγ is a master regulator of adipogenesis, and TZDs can induce adipogenesis in a disease setting in which reducing obesity would be a desired outcome [131]. Moreover, PPARγ expression is increased in certain cancers [132], and PPARγ in tumor-associated macrophages impedes their M2-like traits. Of note, pioglitazone also decreases CD163 expression in macrophages [127]. Another NR, PPARδ/β, has a similar role as PPARγ and mitigates proinflammatory Kupffer cell activation in the liver [133]. More recently, dual PPAR agonists are being tested in the treatment of metabolic diseases. For instance, the PPARα/PPARγ dual agonist MD001 decreases the expression of TNFα, MCP-1, CD11b, and CD11c in adipose tissue in mice, most likely by abrogating macrophage infiltration [134]. Further NR ligands, such as retinoic acid [116, 135], LXR ligands, glucocorticoids, and FXR ligands, have antiinflammatory effects [136–138]. However, the pleiotropy of NR ligand effects also impedes their use in clinical settings [139]. Glucocorticoids are applied today as antiinflammatory drugs, but they have side effects such as the induction of insulin resistance [140], which underlines the importance of selective delivery of the NR ligands into macrophages. The delivery of drugs into the desired macrophage population is challenging. First, specific cell-surface proteins are needed to distinguish the distinct macrophage populations. Second, an effective career vehicle is necessary to deliver the drug to the macrophages. For instance, liposomal targeting of glucocorticoids to the synovial lining may curb inflammation in arthritis without metabolic side effects [141]. The most recently studied M2-promoting mechanisms target macrophage metabolic processes. For instance, vitamin D elicits M2 activation of dermal macrophages by increasing autophagy [142]. Inhibition of the sodium-glucose cotransporter 2 by ipragliflozin increases M2 macrophage number in adipose tissue [143]. Collectively, pharmacological interventions that promote M2 macrophage polarization, or increase the self-renewal of M2 macrophage niches in metabolic organs, have therapeutic utility in treating metabolic "superinflammation," can increase insulin sensitivity, and resolve obesity-associated diseases.

An evolutionarily conserved role of M2 macrophages is related to wound healing (also see chapter "Evolutionary Origin of M2 Macrophage Activation: Invertebrates"). Macrophages are enriched in the wounded tissues, either by extravasation from the blood vessels or by a local redistribution or proliferation of resident macrophages. Ablation of macrophages impairs tissue regeneration [145, 146]. The wounded tissue releases danger signals, the so-called endogenous damage-associated pattern molecules that attract monocytes to the site of injury, and the extravasating (or infiltrating) monocytes differentiate into macrophages [147]. Tissue-resident macrophages may also be first responders to tissue damage, for instance, the dermal macrophages in skin injury [148].

The wounds are exposed to microbial pathogens, which makes tissues vulnerable for infections. To overcome this, macrophages adopt a pathogen-killing phenotype in response to pathogen-associated patterns and clear the wounded tissue from pathogens. Inflammatory cytokines such as IL-6, TNFα, and MCP1 attract further monocytes and macrophages to the site of injury. This initial stage of wound healing is the inflammatory stage, in which inflammatory granulocytes and macrophages have the major role. In the initial stage of wound healing, macrophages hence display M1-like traits, have increased phagocytosis activity, and clear damaged cell debris, the potentially harmful cellular contents of the dead cells, eliminate pathogens, and phagocytose apoptotic granulocytes [147]. These processes ensure debridement and sanitation of the damaged tissue. The inflammatory phase is essential for the progression of the healing process [149, 150]. For instance, inflammatory cytokine signaling initiates metalloproteinase secretion from fibroblasts, hence promoting extracellular matrix remodeling [151, 152].

The M2 traits of wound macrophages function in the later repair phase, including the proliferative phase and the remodeling phase. M2 macrophages are hence often termed as healing or repair macrophages. Their major tasks are the resolution of the initial inflammatory phase (e.g., by secreting IL-10), promoting fibroblast proliferation, and secreting growth factors and tissue repair molecules [153]. These molecules include extracellular matrix components and angiogenic factors (e.g., VEGFs), which aid capillarization and proper blood perfusion of the tissue [43, 153, 154]. M2 macrophage traits are hence associated with the proliferative and remodeling stages of tissue repair and promote the proliferation and collagen secretion of fibroblasts [152]. Such effects of M2 macrophages can be seen in distinct tissue repair processes: wounded skin regeneration, nerve regeneration, skeletal muscle and tendon healing, bone fracture repair, and extracellular matrix remodeling following a myocardial infarct [152, 155–158].

Tissue healing is impaired in certain diseases. For example, diabetes and insulin resistance cause chronic wounds and ulcers and impair bone fracture healing [147, 159]. The inflammatory phase is essential for wound healing [149]; however, when there is a lack of transition toward resolution of inflammation and remodeling, the wound remains chronic and does not heal [158]. Antiinflammatory therapy has been shown to accelerate diabetic wound healing [147]. Pressure ulcers and diabetic ulcers may be mitigated by interventions that mimic M2 macrophage activation (e.g., exogenous administration of M2-like monocytes, increasing the level of M2-polarizing mediators, increasing IL-4Rα signaling) [147, 160, 161]. Neuropeptide FF (NPFF), which is known to increase M2 macrophage replenishment and sustains the M2 state by impeding the inactivation of STAT6 signaling [4], also accelerates wound healing and nerve regeneration in diabetic mice [162]. ATP through purinergic receptors also accelerates skin wound healing in mice [163]. Exosomes released by mesenchymal stem cells also induce M2 polarization, which in turn promotes wound healing in mice [164]. Endogenous M2-polarizing stimuli hence have the potential to promote wound healing. However, excess arginase-1 expression impairs wound healing in diabetic mice, albeit arginase-1 expression is normally increased in the healing skin wound [165]. The timing and proportion of M2 traits are hence affecting the healing process. We are still lacking pharmacological utilities that would accelerate wound healing through M2 macrophage activation [147]. Moreover, unbalanced wound repair can lead to tissue fibrosis and sclerosis [166]. One underlying mechanism is the chronic secretion of extracellular matrix components by fibroblasts, which process may also be a consequence of excess M2 activation [167].

5.2 Inhibiting M2 Activation

Diseases associated with M2 macrophages include allergic airway inflammation, asthma, chronic obstructive pulmonary disease, ectopic dermatitis, and various fibrotic diseases [167]. Tumor-associated macrophages can also adopt an M2 phenotype, eventually suppressing antitumor immunity, and increasing tumor growth and metastasis. M2 activation of subcapsular macrophages of the lymph node also compromises antitumor immunity. In the context of these diseases, the inhibition of M2 traits and increase of M1 macrophage functions disrupt the Th2 immune cytokine environment and eosinophil granulocyte infiltration of the tissues, improve the recognition and elimination of tumor cells, and impede angiogenesis and fibrosis of the diseased tissue. IL-4, IL-13, IL-10, and TGFβ are primarily responsible for M2 macrophage activation in these disease settings [168]. One possible approach is thus the neutralization of Th2 cytokines and their receptors and the inhibition of STAT6 signaling. In allergic airway inflammation and asthma, Th2 immune response is inhibited by blockage of IL-4 and IL-13 signaling. Variants of IL-4 can bind with high affinity to IL-4Rα and act as IL-4 antagonists [169], inhibiting IL-4Rα signaling in allergic airway inflammation in the mouse [170]. Another strategy is targeting IL-4Rα with a monoclonal antibody, such as AMG-317, pitrakinra (AER-001, BAY-16-9996), pascolizumab, and dupilumab. Animal studies suggested that such could reduce the severity of Th2/M2-dominated diseases such as eczema and asthma [171, 172]. Presently, only dupilumab has been developed to clinical use, and it reduces the severity of certain allergic diseases, such as atopic dermatitis and chronic sinusitis; some studies suggest that it may reduce asthma [172, 173]. Lebrikizumab and tralokinumab are humanized monoclonal antibodies that bind to IL-13 and block its action [174], with potential for use in the treatment of atopic dermatitis [175].

The hispanolone derivative 8,9-dehydrohispanolone-15,16-lactol inhibits IL-4-induced STAT6 and JAK1 phosphorylation, eventually blocking M2 activation in response to IL-4 or IL-13 [176]. This factor has potential in TAMs. RNF128 is a ubiquitin ligase that is involved in the proteasomal degradation of STAT6 and pSTAT6 and hence inactivates M2 macrophages [4, 177]. RNF128 also promotes antiviral immunity [178]. Recently it has been shown that reduced RNF128 level is correlated with severity, epithelial to mesenchyme transition, and poor prognosis of melanoma and urothelial cancer [179, 180]. This role of RNF128 calls for attention, because it may evoke an antitumor immunity. However, RNF128 causes T cell anergy [181], has a potentially wide substrate preference [182], and it has an increased level in squamous cell carcinoma [183].

IL-10/STAT3 signaling is another possible target of pharmacological intervention when Th2/M2 immune response should be curbed [184]. IL-10 in the tumor environment has a dual role, however, and in addition to its immune suppressor role it increases natural killer cell activity, reduces tumor metastasis, and is necessary for antitumor immunity [185, 186].

Inhibition of TGFβ signaling may also limit M2 functions. M2 macrophages produce TGFβ, which augments the M2 state in an autocrine loop, increases Th2 cell response through IL-10, inhibits Th1 response and M1 activation, and suppresses natural killer cell and dendritic cell functions. TGFβ also has a pro-fibrogenic effect. Anti-TGFβ therapy is hence a possible way to inhibit Th2/M2 immune response [187]. Monoclonal antibodies against TGFβ, small molecule inhibitors of TGFβ receptor, and antisense oligonucleotides that interfere with TGFβ-induced signal pathways were in preclinical phase or in phase II clinical trials in 2016 [188]. TGFβ has pleiotropic effects and a complex signaling network, which renders its effects context dependent and cell type specific. Albeit TGFβ is considered as an M2 activating stimulus [105, 189], TGFβ signaling suppresses early tumor growth, and cancer cells have a reduced responsiveness to TGFβ [190]. Delivery of drugs specifically to tumor-associated macrophages may reduce unwanted effects that develop from the pleiotropic effects of a skewed Th2/Th1 cytokine signaling. Today, nanoparticle delivery is a potential route to reach macrophages and reprogram their activation state [191]. Antitumor immunity may also be elicited by provoking a Th1 immune response: for instance, BCG installation therapy of bladder cancer serves this aim [192, 193].

The most recent drug candidates target noncanonical M2-activating mechanisms, specifically metabolic functions of M2 macrophages. For instance, TGFβ induces lipid droplet formation in macrophages, which can be inhibited with the fatty acid synthase inhibitor C75. Inhibition of lipid droplet synthesis switches macrophage polarization from M2 to M1 [194]. A specific cyclooxygenase-2, etodolac, inhibits human M2 macrophage differentiation and increases TNFα production. Etodolac also inhibits metastasis-related macrophage mediators VEGF-A and MMP9 [107]. M1 macrophage synthesize a metabolite, called itaconate, which is taken up by M2 macrophages. The itaconate-synthesizing enzyme is aconitase decarboxylase (encoded by *Irg1* or *Acod1*), upregulated in M1 macrophages. Overexpression of *Irg1* or a cell-permeable derivative of itaconate inhibits M2 activation in mouse [195]. Lipid metabolism of macrophages also has the potential to evoke antitumor immunity. For instance, nanoliposome-loaded C6-ceremide reduces macrophage number and immunosuppression [196].

6 Didactics of M2 Macrophage Biology

This book also aims to serve in university classes as a lecture text. Some didactical aids are shown next that which can help to explain and comprehend concepts of M2 macrophage biology in university-level teaching. Moreover, visualization is an integral element in effective scientific communication, and these examples also serve for better transmission of the framework of the M2 macrophage activation model.

6.1 Impact of M2 Macrophages in Health and Disease

The first learning objective is to understand that M2 macrophages have context-dependent functions. The same M2 function can be physiologically crucial, or can cause disease. In a teamwork approach, participating students should list the major functions of M2 macrophages. Chapters "What Is an M2 Macrophage? Historical Overview of the Macrophage Polarization Model. The Th1/Th2 and M1/M2 Paradigm, the Arginine Fork" and "Signal Mechanisms of M2 Macrophage Activation" provides resources for listing M2 functions. As a next task, these functions should be grouped as beneficial or as disease-inducing functions. Students should discuss whether the physiological outcome of M2 macrophage functions depends on their context. The discussion is facilitated by reading and discussing the quote: "*There's nothing on earth that's so evil that it does not also provide the earth with some kind of good, nor is there anything so good that it can't be turned bad if it's abused.*"[2] As visual aids for the discussion, Figs. 4 and 5 in chapter "What Is an M2 Macrophage? Historical Overview of the Macrophage Polarization Model. The Th1/Th2 and M1/M2 Paradigm, the Arginine Fork", 4, and 5 can be used.

The discussion can be continued with identifying context-specific M2 markers. For instance, consider which organs provide hypoxic conditions that can induce the hypoxia-responsive expression of VEGF-A. The peritoneum is a hypoxic niche, and resident macrophages tend to express VEGF-A in the peritoneum. In a disease context, the necrotic core of tumors is a hypoxic environment (Fig. 3 in chapter "Evolutionary Origin of the M2 Macrophage Activation: Invertebrates" and 1).

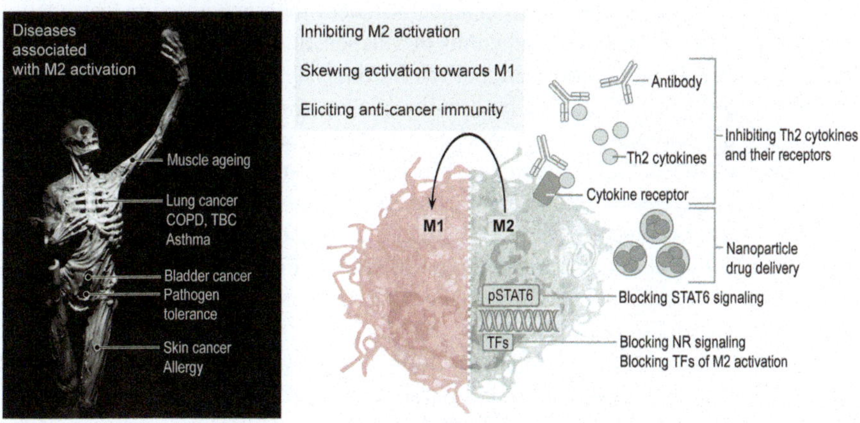

Fig. 5 Inhibiting M2 macrophage activation to combat disease. The potential of sustaining Th2 inflammation and causing the loss of anticancer immunity are the main adverse effects of M2 macrophages. Skewing macrophage polarization toward the M1 state can alleviate disease severity and can slow disease progression. "Allegory of disease" is an artwork by the sculptor Ligier Richer (c. 1500–1567). (Image with courtesy)

[2] William Shakespeare: Romeo and Juliet, Act 2, Scene 3, monologue of Friar Lawrence.

What can be the consequence of VEGF-A synthesis in the hypoxic tumor environment? Consider the effect of VEGF-A on angiogenesis.

6.2 Recognition and Classification of M2 Macrophages

A macrophage may not present all the hallmarks of an M1 or an M2 macrophage, just as nothing in everyday life can fall into clear categories. For instance, consider how one would assign historical buildings to categories of architectural styles. In some instances it is easy to recognize hallmarks of a historical era in architecture, and sometimes these are less clear. For example, the representative elements of Gothic architecture are stone masonry, rose windows, rib vaults, pointed arches and façade, flying buttresses, and gargoyles (Fig. 6). The Gothic style, which flourished in Europe between the twelfth and sixteenth centuries, represented a technological revolution, with mathematical and geometric perfection in architecture [197], and these hallmarks can be easily recognized in the cathedrals of Valencia and Barcelona in Spain (Fig. 6a–d). A contrasting artistic movement was the Baroque style between the seventeenth and eighteenth centuries, which expressed divine perfection by a different geometric concept with oval- and shell-shaped architectural and ornamental elements (Fig. 6e, f). Baroque buildings have a richly decorated façade with dizzying arrays of statues and twisting elements, dynamic and emotional paintings and statuary, and extravagant use of textures and materials, giving a dramatic and illusory effect [198]. Examples are the main façade of the Cathedral of Valencia (Fig. 6e), and the Melk Abbey in Austria (Fig. 6f). Medieval cathedrals were often built over a span of centuries, and in some instances their construction was continued in the Baroque style. The cathedrals of Valencia and Toledo have cores built in the Gothic era, but large parts were built in Baroque style (Fig. 6g, h). Consider the Almudena Cathedral of Madrid, built in the nineteenth–twentieth century. Its exterior shows the hallmarks of Baroque, but inside a very different image appears, resembling medieval Gothic churches (Fig. 6i, j). The artistic concept was to tribute two artistic languages of church architecture within one modern building. And last, the Ulm Minster (Ulmer Münster), which has today the tallest church tower in the world, is a Gothic church without Baroque additions (Fig. 6k). However, it also has some features unusual for Gothic architecture, such as the excessive use of brick during its construction.

Does the mixture of the two architectural styles invalidate the existence of the two principal categories of Gothic and Baroque styles? Similarly, does the existence of mixed M1 and M2 traits in certain macrophages invalidate the existence of the two principal categories of M1 and M2 macrophage activation?

Fig. 6 An example of the feature-based classification system: features of Gothic and Baroque architecture. (**a**) Gothic gate with rose window of the Metropolitan Cathedral, Basilica of the Assumption of Our Lady of Valencia, Spain, consecrated in 1238. (**b, c**) Spire and stone masonry from the Cathedral of the Holy Cross and Saint Eulalia, Valencia, Spain. The church was built in the thirteenth to fifteenth centuries. This picture was taken during renovation work in 2010. (**d**) Mythical stone figure from a Gothic façade, Valencia, Spain. Gothic sculpture favored featuring grotesque, mythical creatures. (**e**) Baroque façade of the Metropolitan Cathedral–Basilica of the Assumption of Our Lady of Valencia, Spain. The Baroque gate (also called the "Gate of Irons") was built in the eighteenth century. Baroque interior (**f**) of Gothic façade (**g**) of the Primate Cathedral of Saint Mary of Toledo, built in 1226. (**h**) Baroque elements over the Gothic core of the same cathedral. (**i, j**) Neo-Baroque exterior and neo-Gothic interior of the Almudena Cathedral (Santa María la Real de La Almudena) in Madrid, Spain, consecrated in 1993. (**k**) The Ulm Minster (Ulmer Münster) in Ulm, Germany. As of 2020, it is the tallest church tower in the world. (**a–e**, **g–i**, **k**, images by author; **f**, **j**, images with courtesy: see Acknowledgments)

6.3 How to Interpret the Presence of Absence of M2 Markers: Production of Meaning from Observations

The so-called M2 marker genes or proteins are attributes of biological functions (Fig. 4 in chapter "What Is an M2 Macrophage? Historical Overview of The Macrophage Polarization Model. The Th1/Th2 and M1/M2 Paradigm, the Arginine Fork" and 3). However, the interpretation of the presence orf absence of these attributes is largely dependent on biological context (Graphic Abstract 2). As an everyday analogy, one may consider examples of semiotics. In everyday visual communication, we use symbols that express often complex messages with simple visual representations. When we see signs (e.g., pictograms, symbols, icons) we process the primary visual information and produce a meaning for the signs. Often the same symbol can have multiple meanings, and the consequent meaning depends on factors such as context and the knowledge base of the person who perceives the sign. For instance, the pictorial representation of a serpent can have context-dependent meanings (Fig. 7): it can label a pharmacy or the science of pharmacology. This interpretation has roots in Greek antiquity, which associated snakes with healing and rebirth because of the ability to shed their skin, representing letting go of the past and rebirth. Snake venoms were also used as remedies, and the serpent was an attribute of Asclepius, the healing god in antique Greek mythology. If the person has this knowledge base, it is very likely that the serpent as a sign will be understood as the symbol of pharmacy, pharmacology, and healing. The snake that swallows its tail has a different meaning: it is a symbol of eternity. Producing this meaning requires a different knowledge base. Ouroboros, the snake that eats its own tail, is a symbol of eternity with roots in ancient Egypt, and in Europe it was known as a symbol used by alchemists in the Middle Ages. Because snakes live close to the ground and hide by means of their camouflage, in some cultures they are associated with the underground. For instance, the Biblical meaning of a snake is eternal sin and betrayal. In these different settings the serpent represents opposing qualities: healing, eternity and rebirth, destruction, and sin and betrayal (Fig. 7).

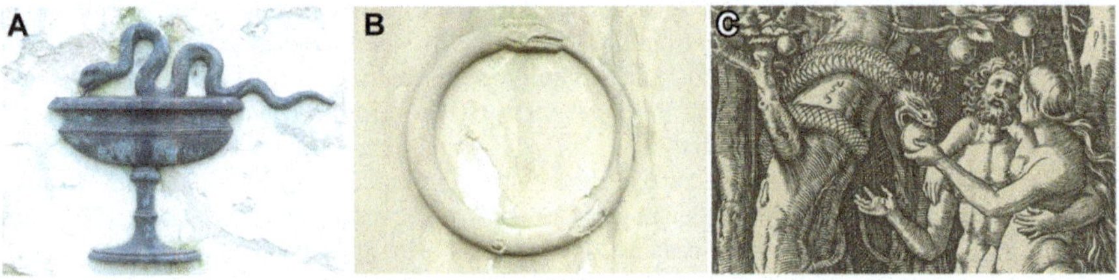

Fig. 7 Example of context-dependent interpretation of observations. This semiotic example illustrates how many ways the same sign – a serpent in this example – can have diverse and context-dependent meanings. (**a**) The serpent represents pharmacy, pharmacology, and healing. (**b**) The serpent represents eternity. (**c**) The serpent represents sin and shame. Images **a** and **b** show ornaments of nineteenth-century tombstones in Ulm, Germany. (**a** and **b** are images by author; **c** is with courtesy: see Acknowledgments)

This example illustrate how the same piece of information may be represented in different ways, and the same is true for data sets one may obtain during research work. For instance, IL-6 expression is traditionally a hallmark of M1 macrophages. However, IL-6 induces M2 activation and in certain settings augments STAT6 signaling. The observation that a macrophage produces IL-6 may indicate both M1 and M2 responses. More information is needed for a proper interpretation: for instance, the amount of IL-6, the coexpression of other molecules (e.g., NOS2), and the context (e.g., the tissue from which the macrophages were obtained). Arginase-1 is an M2 marker protein, but there are macrophages with M2 functions that are tolerogenic and antiinflammatory but which do not express arginase-1 (for instance, see the single-cell sequencing data in Fig. 1 in chapter "M2 Macrophages in the Circulatory, Respiratory and Excretory Organs"). The lack of arginase-1 expression alone does not indicate the lack of M2 functions.

6.4 *Evolutionary Outlook of M2 Macrophages*

Key functions of macrophages are evolutionarily conserved (Graphic Abstract 1). As an interesting phenomenon, some M2 traits of mammalian macrophages can be recognized in M1-like macrophages of invertebrates. Figure 8 displays a functional analogy between invertebrate macrophages and mammalian M2 macrophages. In mollusks, a foreign body is isolated from the surrounding tissues by forming $CaCO_3$ layers around the foreign body. This process is called pearl formation and is an ancient form of innate immune response that is aided by macrophages. In this instance, extracellular matrix is formed around the foreign body with the aid of macrophage-equivalent immune cells. It is plausible that the function of the cells is

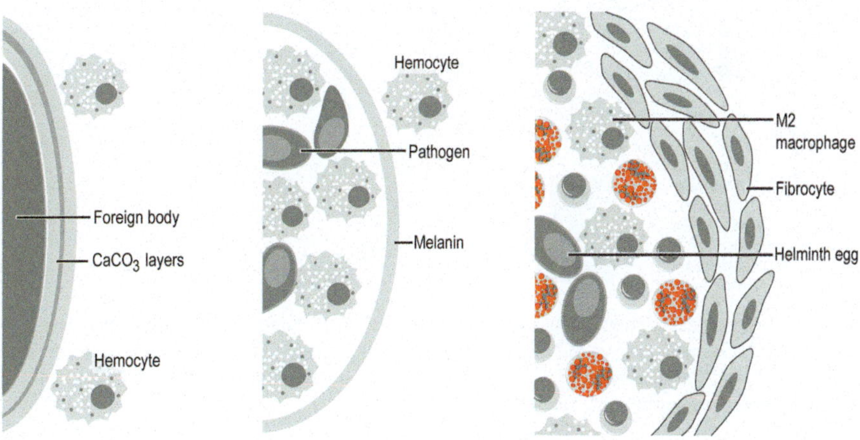

Fig. 8 Functional analogies of macrophage functions across species. *Left:* scheme of pearl formation in bivalves; *middle:* scheme of melanization in arthropods; *right:* scheme of a Th2 granuloma in mammals

a forerunner of the extracellular matrix synthesizing M2 trait in mammals. In arthropods, a melanization process serves the same function as the pearl formation, isolating the pathogen from the tissues. This process requires macrophage-equivalent cells and phenoloxidase activity: an M1-like macrophage function is associated with shielding of the pathogens. Last, M2 macrophages promote the development of a fibrous layer around pathogens, which is called granuloma formation. (For discussion, see chapters "Evolutionary Origin of M2 Macrophage Activation: Invertebrates" and "Immune Functions of the M2 Macrophages: Host Defense, Self-tolerance, and Autoimmunity").

Finally, to illustrate how physiological context dictates immune functions, and eventually macrophage functions, consider three examples (Fig. 9). First, amphibian metamorphosis generates neoantigens, and larval and adult-type antigens are present in overlap. This transition period requires M2-like tolerogenic traits (Fig. 9a). Metamorphosis also demands apoptotic cell clearance. This demand is present in all tissues that have a high turnover. However, in some instance, both M1 and M2 macrophages traits are necessary in the same tissue, such as in the tonsils (Fig. 9b). Tonsils serve pathogen recognition and elimination, and this function requires M1-like traits of macrophages. However, the safe disposal of apoptotic cells require simultaneously the presence of an M2 macrophage population. Last, there are organs that require nonspecific immune mechanisms, which secondarily can also affect macrophage functions. For instance, the skin of the human auditory tube is rich in sebaceous glands which secrete antibacterial substances. At the same time, sebaceous glands produce lipids that can inhibit M1-like macrophage activation (Fig. 9c). Altogether these didactic approaches illustrate the diversity and the impact of M2 macrophage biology.

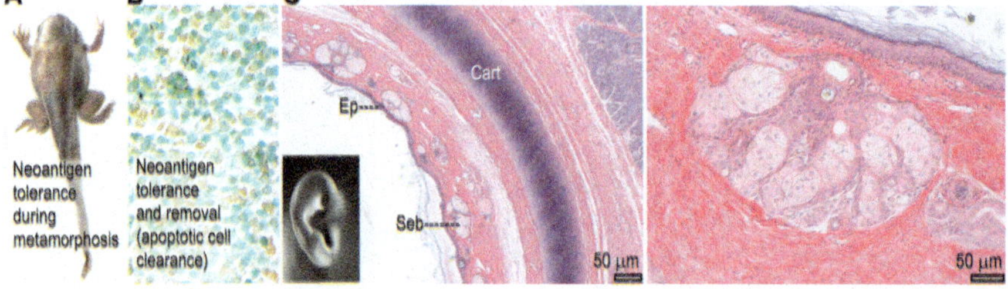

Fig. 9 Physiological context dictates immune functions. Amphibian metamorphosis (**a**) and apoptosis (**b**) generate neoantigens. M2 macrophage functions are associated with the tolerance of these neoantigens and the clearance of apoptotic debris. (**c**) Human auditory tube. Hematoxylin and eosin. *Ep* cornified epithelium, *Seb* sebaceous gland

7 Concluding Remarks

This monograph intended to summarize the manifold biological functions vested in M2 macrophages, key members of the innate immune system. Beyond its involvement in host defense against infection, the innate immune system also participates in the normal physiological function of organs, maintains tissue integrity, tolerates self-antigens, safely disposes of unwanted cells, and eventually shapes the metabolic performance of the body [199–201]. M2 macrophage functions are hence among the principal traits of all living organisms, and are crucial in sustaining growth and metabolic homeostasis. Diseases can occur when the M2 macrophage system is compromised, overactivated, or unbalanced. The pharmacology of M2 macrophage activation is hence a key to the therapy of many diseases. Some aspects of macrophage activation, such as antiviral activities of macrophages, are yet to be further explored for better understanding of the ways pathogens can avoid immune response [202]. Research areas deserving attention for the M2 macrophage system include immunometabolism, metabolic biochemistry, cancer immunity, molecular endocrinology, ecophysiology, tissue engineering, biomaterial development, disease immunity, tissue development, and regeneration [199–201, 203].

I hope that this monograph will be a useful reference for teaching and research, and that it will open new avenues of research in the understanding of M2 macrophage biology.

References

1. Russell SW, Pace JL. Both the kind and magnitude of stimulus are important in overcoming the negative regulation of macrophage activation by PGE2. J Leukoc Biol. 1984;35:291–301.
2. Bourette RP, Mouchiroud G. The biological role of interferon-inducible P204 protein in the development of the mononuclear phagocyte system. Front Biosci. 2008;13:879–86.
3. Luan Y, Lengyel P, Liu C-J. p204, a p200 family protein, as a multifunctional regulator of cell proliferation and differentiation. Cytokine Growth Factor Rev. 2008;19:357–69.
4. Waqas SFH, Hoang A, Lin Y, Ampem G, Röszer T, et al. Neuropeptide FF increases M2 activation and self-renewal of adipose tissue macrophages. J Clin Invest. 2017;127:2842–54.
5. Choubey D, Panchanathan R. Interferon-inducible Ifi200-family genes in systemic lupus erythematosus. Immunol Lett. 2008;119:32–41.
6. Chunfa L, Xin S, Qiang L, Sreevatsan S, Yang L, Zhao D, Zhou X. The central role of IFI204 in IFN-β release and autophagy activation during Mycobacterium bovis infection. Front Cell Infect Microbiol. 2017;7
7. Dauffy J, Mouchiroud G, Bourette RP. The interferon-inducible gene, Ifi204, is transcriptionally activated in response to M-CSF, and its expression favors macrophage differentiation in myeloid progenitor cells. J Leukoc Biol. 2006;79:173–83.
8. Salo RJ, Bleam DK, Greer VL, Ortega AP. Interferon production in murine macrophage-like cell lines. J Leukoc Biol. 1985;37:395–406.
9. Smith ER, Jones PL, Boss JM, Merrill AH. Changing J774A.1 cells to new medium perturbs multiple signaling pathways, including the modulation of protein kinase C by endogenous sphingoid bases. J Biol Chem. 1997;272:5640–6.

10. Lam J, Herant M, Dembo M, Heinrich V. Baseline mechanical characterization of J774 macrophages. Biophys J. 2009;96:248–54.
11. Dietz JN, Cole BC. Direct activation of the J774.1 Murine macrophage cell line by mycoplasma arthritidis. Infect Immun. 1982;37:811–9.
12. Lindl T, Steubing R. Atlas of living cell cultures. Hoboken: Wiley-Blackwell; 2013.
13. Circadian Expression Profiles Database (CircaDB). http://circadb.hogeneschlab.org/.
14. Pizarro A, Hayer K, Lahens NF, Hogenesch JB. CircaDB: a database of mammalian circadian gene expression profiles. Nucleic Acids Res. 2013;41:D1009–13.
15. Hughes ME, DiTacchio L, Hayes KR, Vollmers C, Pulivarthy S, Baggs JE, Panda S, Hogenesch JB. Harmonics of circadian gene transcription in mammals. PLoS Genet. 2009;5:e1000442–e1000442.
16. Tabula Muris Consortium; Overall coordination; Logistical coordination; Organ collection and processing; Library preparation and sequencing; Computational data analysis; Cell type annotation; Writing group; Supplemental text writing group; Principal investigators. Single-cell transcriptomics of 20 mouse organs creates a Tabula Muris. Nature. 2018;562:367–72.
17. Taciak B, Białasek M, Braniewska A, Sas Z, Sawicka P, Kiraga Ł, Rygiel T, Król M. Evaluation of phenotypic and functional stability of RAW 264.7 cell line through serial passages. PLoS One. 2018;13:e0198943.
18. Chamberlain LM, Holt-Casper D, Gonzalez-Juarrero M, Grainger DW. Extended culture of macrophages from different sources and maturation results in a common M2 phenotype. J Biomed Mater Res A. 2015;103:2864–74.
19. Lee JK, Tansey MG. Microglia isolation from adult mouse brain. Methods Mol Biol. 2013;1041:17–23.
20. Li PZ, Li JZ, Li M, Gong JP, He K. An efficient method to isolate and culture mouse Kupffer cells. Immunol Lett. 2014;158:52–6.
21. Cho KW, Morris DL, Lumeng CN. Flow cytometry analyses of adipose tissue macrophages. Methods Enzymol. 2014;537:297–314.
22. Engwall KS, Li AP. Isolation and culturing of rat pulmonary alveolar macrophages. J Tissue Cult Methods. 1983;8:91–4.
23. Orr JS, Kennedy AJ, Hasty AH. Isolation of adipose tissue immune cells. J Vis Exp. 2013:e50707–e50707.
24. Yu H, Dilbaz S, Coßmann J, Hoang AC, Diedrich V, Herwig A, Harauma A, Hoshi Y, Moriguchi T, Landgraf K, Körner A, Lucas C, Brodesser S, Balogh L, Thuróczy J, Karemore G, Kuefner MS, Park EA, Rapp C, Travers JB, Röszer T. Breast milk alkylglycerols sustain beige adipocytes through adipose tissue macrophages. J Clin Invest. 2019;129:2485–99.
25. Shaw DR, Griffin FM. Thioglycollate-elicited mouse peritoneal macrophages are less efficient than resident macrophages in antibody-dependent cell-mediated cytolysis. J Immunol. 1982;128:433–40.
26. Weischenfeldt J, Porse B. Bone marrow-derived macrophages (BMM): isolation and applications. CSH Protocols 2008, pdb.prot5080. 2008.
27. Pavlou S, Wang L, Xu H, Chen M. Higher phagocytic activity of thioglycollate-elicited peritoneal macrophages is related to metabolic status of the cells. J Inflamm. 2017;14:4.
28. Norris PC, Dennis EA. A lipidomic perspective on inflammatory macrophage eicosanoid signaling. Adv Biol Regul. 2014;54:99–110.
29. Andreu N, Phelan J, de Sessions PF, Cliff JM, Clark TG, Hibberd ML. Primary macrophages and J774 cells respond differently to infection with *Mycobacterium tuberculosis*. Sci Rep. 2017;7:42225.
30. Ramond E, Jamet A, Coureuil M, Charbit A. Pivotal role of mitochondria in macrophage response to bacterial pathogens. Front Immunol. 2019;10:2461–2461.
31. Grosick R, Alvarado-Vazquez PA, Messersmith AR, Romero-Sandoval EA. High glucose induces a priming effect in macrophages and exacerbates the production of pro-inflammatory cytokines after a challenge. J Pain Res. 2018;11:1769–78.

32. Torres-Castro I, Arroyo-Camarena Ú, Martínez-Reyes C, Gómez-Arauz A, Dueñas-Andrade Y, Ruiz J, Béjar Y, Zaga-Clavellina V, Morales-Montor J, Terrazas L, Kzhyshkowska J, Escobedo G. Human monocytes and macrophages undergo M1-type inflammatory polarization in response to high levels of glucose. Immunol Lett. 2016;176:181–9.
33. Van Den Bogert C, Spelbrink JN, Dekker HL. Relationship between culture conditions and the dependency on mitochondrial function of mammalian cell proliferation. J Cell Physiol. 1992;152:632–8.
34. Kos M, Kuebler JF, Jesch NK, Vieten G, Bax NM, van der Zee DC, Busche R, Ure BM. Carbon dioxide differentially affects the cytokine release of macrophage subpopulations exclusively via alteration of extracellular pH. Surg Endosc Other Interv Tech. 2006;20:570–6.
35. Qiu W, Liu S, Chen J, Hu L, Wu M, Yang M. The primary culture of carp (*Cyprinus carpio*) macrophages and the verification of its phagocytosis activity. In Vitro Cell Dev Biol Anim. 2016;52:10–9.
36. Chanput W, Peters V, Wichers H. THP-1 and U937 cells. In: Verhoeckx K, Cotter P, López-Expósito I, Kleiveland C, Lea T, Mackie A, Requena T, Swiatecka D, Wichers H, editors. The impact of food bioactives on health: in vitro and ex vivo models. Cham: Springer; 2015. p. 147–59.
37. Lund ME, To, J, O'Brien BA, Donnelly S. The choice of phorbol 12-myristate 13-acetate differentiation protocol influences the response of THP-1 macrophages to a pro-inflammatory stimulus. J Immunol Methods. 2016;430:64–70.
38. Jobe O, Kim J, Tycksen E, Onkar S, Michael NL, Alving CR, Rao M. Human primary macrophages derived in vitro from circulating monocytes comprise adherent and non-adherent subsets with differential expression of Siglec-1 and CD4 and permissiveness to HIV-1 infection. Front Immunol. 2017;8:1352.
39. Czystowska-Kuzmicz M, Sosnowska A, Nowis D, Ramji K, Szajnik M, Chlebowska-Tuz J, Wolinska E, Gaj P, Grazul M, Pilch Z, Zerrouqi A, Graczyk-Jarzynka A, Soroczynska K, Cierniak S, Koktysz R, Elishaev E, Gruca S, Stefanowicz A, Blaszczyk R, Borek B, Gzik A, Whiteside T, Golab J. Small extracellular vesicles containing arginase-1 suppress T-cell responses and promote tumor growth in ovarian carcinoma. Nat Commun. 2019;10:3000.
40. Goren I, Pfeilschifter J, Frank S. Uptake of neutrophil-derived ym1 protein distinguishes wound macrophages in the absence of interleukin-4 signaling in murine wound healing. Am J Pathol. 2014;184:3249–61.
41. Röszer T. Understanding the mysterious M2 macrophage through activation markers and effector mechanisms. Mediat Inflamm. 2015;2015:16.
42. Welch JS, Escoubet-Lozach L, Sykes DB, Liddiard K, Greaves DR, Glass CK. TH2 cytokines and allergic challenge induce Ym1 expression in macrophages by a STAT6-dependent mechanism. J Biol Chem. 2002;277:42821–9.
43. Dzik JM. Evolutionary roots of arginase expression and regulation. Front Immunol. 2014;5:544.
44. Munder M. Arginase: an emerging key player in the mammalian immune system. Br J Pharmacol. 2009;158:638–51.
45. Muller U, Stenzel W, Kohler G, Werner C, Polte T, Hansen G, Schutze N, Straubinger RK, Blessing M, McKenzie AN, Brombacher F, Alber G. IL-13 induces disease-promoting type 2 cytokines, alternatively activated macrophages and allergic inflammation during pulmonary infection of mice with *Cryptococcus neoformans*. J Immunol. 2007;179:5367–77.
46. Stempin CC, Dulgerian LR, Garrido VV, Cerban FM. Arginase in parasitic infections: macrophage activation, immunosuppression, and intracellular signals. J Biomed Biotechnol. 2010;2010:683485.
47. Ahn M, Yang W, Kim H, Jin JK, Moon C, Shin T. Immunohistochemical study of arginase-1 in the spinal cords of Lewis rats with experimental autoimmune encephalomyelitis. Brain Res. 2012;1453:77–86.
48. Spittau B, Wullkopf L, Zhou X, Rilka J, Pfeifer D, Krieglstein K. Endogenous transforming growth factor-beta promotes quiescence of primary microglia in vitro. Glia. 2013;61:287–300.

49. Li B, Alli R, Vogel P, Geiger TL. IL-10 modulates DSS-induced colitis through a macrophage-ROS-NO axis. Mucosal Immunol. 2014;7:869–78.
50. Singh V, Kumar M, San Yeoh B, Xiao X, Saha P, Kennett MJ, Vijay-Kumar M. Inhibition of interleukin-10 signaling induces microbiota-dependent chronic colitis in apolipoprotein E deficient mice. Inflamm Bowel Dis. 2016;22:841–52.
51. Shouval DS, Biswas A, Goettel JA, McCann K, Conaway E, Redhu NS, Mascanfroni ID, Al Adham Z, Lavoie S, Ibourk M, Nguyen DD, Samsom JN, Escher JC, Somech R, Weiss B, Beier R, Conklin LS, Ebens CL, Santos FGMS, Ferreira AR, Sherlock M, Bhan AK, Müller W, Mora JR, Quintana FJ, Klein C, Muise AM, Horwitz BH, Snapper SB. Interleukin-10 receptor signaling in innate immune cells regulates mucosal immune tolerance and anti-inflammatory macrophage function. Immunity. 2014;40:706–19.
52. Ranganathan PV, Jayakumar C, Ramesh G. Netrin-1-treated macrophages protect the kidney against ischemia-reperfusion injury and suppress inflammation by inducing M2 polarization. Am J Physiol Renal Physiol. 2013;304:F948–57.
53. Csoka B, Selmeczy Z, Koscso B, Nemeth ZH, Pacher P, Murray PJ, Kepka-Lenhart D, Morris SM Jr, Gause WC, Leibovich SJ, Hasko G. Adenosine promotes alternative macrophage activation via A2A and A2B receptors. FASEB J. 2012;26:376–86.
54. Kawanaka N, Taylor AW. Localized retinal neuropeptide regulation of macrophage and microglial cell functionality. J Neuroimmunol. 2011;232:17–25.
55. Chen W, Liu J, Meng J, Lu C, Li X, Wang E, Shan F. Macrophage polarization induced by neuropeptide methionine enkephalin (MENK) promotes tumoricidal responses. Cancer Immunol Immunother. 2012;61:1755–68.
56. Liu J, Copland DA, Horie S, Wu WK, Chen M, Xu Y, Paul Morgan B, Mack M, Xu H, Nicholson LB, Dick AD. Myeloid cells expressing VEGF and arginase-1 following uptake of damaged retinal pigment epithelium suggests potential mechanism that drives the onset of choroidal angiogenesis in mice. PLoS One. 2013;8:e72935.
57. Hegyi B, Kornyei Z, Ferenczi S, Fekete R, Kudlik G, Kovacs KJ, Madarasz E, Uher F. Regulation of mouse microglia activation and effector functions by bone marrow-derived mesenchymal stem cells. Stem Cells Dev. 2014;23:2600–12.
58. Jung SH, Saxena A, Kaur K, Fletcher E, Ponemone V, Nottingham JM, Sheppe JA, Petroni M, Greene J, Graves K, Baliga MS, Fayad R. The role of adipose tissue-associated macrophages and T lymphocytes in the pathogenesis of inflammatory bowel disease. Cytokine. 2013;61:459–68.
59. Raes G, Van den Bergh R, De Baetselier P, Ghassabeh GH. Arginase-1 and Ym1 are markers for murine, but not human, alternatively activated myeloid cells. J Immunol. 2005;174:6561–2.
60. Raes G, De Baetselier P, Noel W, Beschin A, Brombacher F, Hassanzadeh GG. Differential expression of FIZZ1 and Ym1 in alternatively versus classically activated macrophages. J Leukoc Biol. 2002;71:597–602.
61. Arora S, Hernandez Y, Erb-Downward JR, McDonald RA, Toews GB, Huffnagle GB. Role of IFN-gamma in regulating T2 immunity and the development of alternatively activated macrophages during allergic bronchopulmonary mycosis. J Immunol. 2005;174:6346–56.
62. Chang NC, Hung SI, Hwa KY, Kato I, Chen JE, Liu CH, Chang AC. A macrophage protein, Ym1, transiently expressed during inflammation is a novel mammalian lectin. J Biol Chem. 2001;276:17497–506.
63. Guo L, Johnson RS, Schuh JC. Biochemical characterization of endogenously formed eosinophilic crystals in the lungs of mice. J Biol Chem. 2000;275:8032–7.
64. Zhao J, Lv Z, Wang F, Wei J, Zhang Q, Li S, Yang F, Zeng X, Wu X, Wu Z. Ym1, an eosinophilic chemotactic factor, participates in the brain inflammation induced by *Angiostrongylus cantonensis* in mice. Parasitol Res. 2013;112:2689–95.
65. Harbord M, Novelli M, Canas B, Power D, Davis C, Godovac-Zimmermann J, Roes J, Segal AW. Ym1 is a neutrophil granule protein that crystallizes in p47phox-deficient mice. J Biol Chem. 2002;277:5468–75.

66. Waern I, Jia J, Pejler G, Zcharia E, Vlodavsky I, Li JP, Wernersson S. Accumulation of Ym1 and formation of intracellular crystalline bodies in alveolar macrophages lacking heparanase. Mol Immunol. 2010;47:1467–75.
67. Nio J, Fujimoto W, Konno A, Kon Y, Owhashi M, Iwanaga T. Cellular expression of murine Ym1 and Ym2, chitinase family proteins, as revealed by in situ hybridization and immunohistochemistry. Histochem Cell Biol. 2004;121:473–82.
68. Ward JM, Yoon M, Anver MR, Haines DC, Kudo G, Gonzalez FJ, Kimura S. Hyalinosis and Ym1/Ym2 gene expression in the stomach and respiratory tract of 129S4/SvJae and wild-type and CYP1A2-null B6, 129 mice. Am J Pathol. 2001;158:323–32.
69. Hung SI, Chang AC, Kato I, Chang NC. Transient expression of Ym1, a heparin-binding lectin, during developmental hematopoiesis and inflammation. J Leukoc Biol. 2002;72:72–82.
70. Porcheray F, Viaud S, Rimaniol AC, Leone C, Samah B, Dereuddre-Bosquet N, Dormont D, Gras G. Macrophage activation switching: an asset for the resolution of inflammation. Clin Exp Immunol. 2005;142:481–9.
71. Murray PJ, Allen JE, Biswas SK, Fisher EA, Gilroy DW, Goerdt S, Gordon S, Hamilton JA, Ivashkiv LB, Lawrence T, Locati M, Mantovani A, Martinez FO, Mege JL, Mosser DM, Natoli G, Saeij JP, Schultze JL, Shirey KA, Sica A, Suttles J, Udalova I, van Ginderachter JA, Vogel SN, Wynn TA. Macrophage activation and polarization: nomenclature and experimental guidelines. Immunity. 2014;41:14–20.
72. Dupasquier M, Stoitzner P, Wan H, Cerqueira D, van Oudenaren A, Voerman JS, Denda-Nagai K, Irimura T, Raes G, Romani N, Leenen PJ. The dermal microenvironment induces the expression of the alternative activation marker CD301/mMGL in mononuclear phagocytes, independent of IL-4/IL-13 signaling. J Leukoc Biol. 2006;80:838–49.
73. Zeyda M, Farmer D, Todoric J, Aszmann O, Speiser M, Gyori G, Zlabinger GJ, Stulnig TM. Human adipose tissue macrophages are of an anti-inflammatory phenotype but capable of excessive pro-inflammatory mediator production. Int J Obes. 2007;31:1420–8.
74. Aron-Wisnewsky J, Tordjman J, Poitou C, Darakhshan F, Hugol D, Basdevant A, Aissat A, Guerre-Millo M, Clement K. Human adipose tissue macrophages: m1 and m2 cell surface markers in subcutaneous and omental depots and after weight loss. J Clin Endocrinol Metab. 2009;94:4619–23.
75. Haase J, Weyer U, Immig K, Kloting N, Bluher M, Eilers J, Bechmann I, Gericke M. Local proliferation of macrophages in adipose tissue during obesity-induced inflammation. Diabetologia. 2014;57:562–71.
76. Titos E, Rius B, Gonzalez-Periz A, Lopez-Vicario C, Moran-Salvador E, Martinez-Clemente M, Arroyo V, Claria J. Resolvin D1 and its precursor docosahexaenoic acid promote resolution of adipose tissue inflammation by eliciting macrophage polarization toward an M2-like phenotype. J Immunol. 2011;187:5408–18.
77. Svensson-Arvelund J, Mehta RB, Lindau R, Mirrasekhian E, Rodriguez-Martinez H, Berg G, Lash GE, Jenmalm MC, Ernerudh J. The human fetal placenta promotes tolerance against the semiallogeneic fetus by inducing regulatory T cells and homeostatic M2 macrophages. J Immunol. 2015;194:1534–44.
78. Bellon T, Martinez V, Lucendo B, del Peso G, Castro MJ, Aroeira LS, Rodriguez-Sanz A, Ossorio M, Sanchez-Villanueva R, Selgas R, Bajo MA. Alternative activation of macrophages in human peritoneum: implications for peritoneal fibrosis. Nephrol Dial Transplant. 2011;26:2995–3005.
79. Medbury HJ, James V, Ngo J, Hitos K, Wang Y, Harris DC, Fletcher JP. Differing association of macrophage subsets with atherosclerotic plaque stability. Int Angiol. 2013;32:74–84.
80. Tang Z, Niven-Fairchild T, Tadesse S, Norwitz ER, Buhimschi CS, Buhimschi IA, Guller S. Glucocorticoids enhance CD163 expression in placental Hofbauer cells. Endocrinology. 2013;154:471–82.
81. Shaul ME, Bennett G, Strissel KJ, Greenberg AS, Obin MS. Dynamic, M2-like remodeling phenotypes of CD11c^{+} adipose tissue macrophages during high-fat diet-induced obesity in mice. Diabetes. 2010;59:1171–81.

82. Svensson J, Jenmalm MC, Matussek A, Geffers R, Berg G, Ernerudh J. Macrophages at the fetal-maternal interface express markers of alternative activation and are induced by M-CSF and IL-10. J Immunol. 2011;187:3671–82.
83. Zizzo G, Hilliard BA, Monestier M, Cohen PL. Efficient clearance of early apoptotic cells by human macrophages requires M2c polarization and MerTK induction. J Immunol. 2012;189:3508–20.
84. Komohara Y, Ohnishi K, Kuratsu J, Takeya M. Possible involvement of the M2 anti-inflammatory macrophage phenotype in growth of human gliomas. J Pathol. 2008;216:15–24.
85. Barros MH, Hauck F, Dreyer JH, Kempkes B, Niedobitek G. Macrophage polarisation: an immunohistochemical approach for identifying M1 and M2 macrophages. PLoS One. 2013;8:e80908.
86. Stutz AM, Pickart LA, Trifilieff A, Baumruker T, Prieschl-Strassmayr E, Woisetschlager M. The Th2 cell cytokines IL-4 and IL-13 regulate found in inflammatory zone 1/resistin-like molecule alpha gene expression by a STAT6 and CCAAT/enhancer-binding protein-dependent mechanism. J Immunol. 2003;170:1789–96.
87. Pepe G, Calderazzi G, De Maglie M, Villa A, Vegeto E. Heterogeneous induction of microglia M2a phenotype by central administration of interleukin-4. J Neuroinflammation. 2014;11:1031.
88. Pesce JT, Ramalingam TR, Wilson MS, Mentink-Kane MM, Thompson RW, Cheever AW, Urban JF Jr, Wynn TA. Retnla (relmalpha/fizz1) suppresses helminth-induced Th2-type immunity. PLoS Pathog. 2009;5:e1000393.
89. Nair MG, Du Y, Perrigoue JG, Zaph C, Taylor JJ, Goldschmidt M, Swain GP, Yancopoulos GD, Valenzuela DM, Murphy A, Karow M, Stevens S, Pearce EJ, Artis D. Alternatively activated macrophage-derived RELM-α is a negative regulator of type 2 inflammation in the lung. J Exp Med. 2009;206:937–52.
90. Relloso M, Puig-Kroger A, Pello OM, Rodriguez-Fernandez JL, de la Rosa G, Longo N, Navarro J, Munoz-Fernandez MA, Sanchez-Mateos P, Corbi AL. DC-SIGN (CD209) expression is IL-4 dependent and is negatively regulated by IFN, TGF-beta, and anti-inflammatory agents. J Immunol. 2002;168:2634–43.
91. Gustafsson C, Mjosberg J, Matussek A, Geffers R, Matthiesen L, Berg G, Sharma S, Buer J, Ernerudh J. Gene expression profiling of human decidual macrophages: evidence for immunosuppressive phenotype. PLoS One. 2008;3:e2078.
92. Kammerer U, Eggert AO, Kapp M, McLellan AD, Geijtenbeek TB, Dietl J, van Kooyk Y, Kampgen E. Unique appearance of proliferating antigen-presenting cells expressing DC-SIGN (CD209) in the decidua of early human pregnancy. Am J Pathol. 2003;162:887–96.
93. Preza GC, Tanner K, Elliott J, Yang OO, Anton PA, Ochoa MT. Antigen-presenting cell candidates for HIV-1 transmission in human distal colonic mucosa defined by CD207 dendritic cells and CD209 macrophages. AIDS Res Hum Retrovir. 2014;30:241–9.
94. Fujisaka S, Usui I, Kanatani Y, Ikutani M, Takasaki I, Tsuneyama K, Tabuchi Y, Bukhari A, Yamazaki Y, Suzuki H, Senda S, Aminuddin A, Nagai Y, Takatsu K, Kobayashi M, Tobe K. Telmisartan improves insulin resistance and modulates adipose tissue macrophage polarization in high-fat-fed mice. Endocrinology. 2011;152:1789–99.
95. Durafourt BA, Moore CS, Zammit DA, Johnson TA, Zaguia F, Guiot MC, Bar-Or A, Antel JP. Comparison of polarization properties of human adult microglia and blood-derived macrophages. Glia. 2012;60:717–27.
96. van Kooyk Y, Ilarregui JM, van Vliet SJ. Novel insights into the immunomodulatory role of the dendritic cell and macrophage-expressed C-type lectin MGL. Immunobiology. 2015;220:185–92.
97. Allavena P, Chieppa M, Bianchi G, Solinas G, Fabbri M, Laskarin G, Mantovani A. Engagement of the mannose receptor by tumoral mucins activates an immune suppressive phenotype in human tumor-associated macrophages. Clin Dev Immunol. 2010;2010:547179.
98. Dambuza IM, Brown GD. C-type lectins in immunity: recent developments. Curr Opin Immunol. 2014;32C:21–7.

99. Lefevre L, Gales A, Olagnier D, Bernad J, Perez L, Burcelin R, Valentin A, Auwerx J, Pipy B, Coste A. PPARgamma ligands switched high fat diet-induced macrophage M2b polarization toward M2a thereby improving intestinal *Candida* elimination. PLoS One. 2010;5:e12828.
100. Okabe Y, Medzhitov R. How the immune system spots tumors. eLife. 2014;3:e04476.
101. Loures FV, Araujo EF, Feriotti C, Bazan SB, Costa TA, Brown GD, Calich VL. Dectin-1 induces M1 macrophages and prominent expansion of CD8⁺IL-17⁺ cells in pulmonary Paracoccidioidomycosis. J Infect Dis. 2014;210:762–73.
102. Reid DM, Montoya M, Taylor PR, Borrow P, Gordon S, Brown GD, Wong SY. Expression of the beta-glucan receptor, Dectin-1, on murine leukocytes in situ correlates with its function in pathogen recognition and reveals potential roles in leukocyte interactions. J Leukoc Biol. 2004;76:86–94.
103. Staples KJ, Hinks TS, Ward JA, Gunn V, Smith C, Djukanovic R. Phenotypic characterization of lung macrophages in asthmatic patients: overexpression of CCL17. J Allergy Clin Immunol. 2012;130(1404–1412):e1407.
104. Solinas G, Schiarea S, Liguori M, Fabbri M, Pesce S, Zammataro L, Pasqualini F, Nebuloni M, Chiabrando C, Mantovani A, Allavena P. Tumor-conditioned macrophages secrete migration-stimulating factor: a new marker for M2-polarization, influencing tumor cell motility. J Immunol. 2010;185:642–52.
105. Gong D, Shi W, Yi S-J, Chen H, Groffen J, Heisterkamp N. TGFβ signaling plays a critical role in promoting alternative macrophage activation. BMC Immunol. 2012;13:31–31.
106. Sica A, Mantovani A. Macrophage plasticity and polarization: in vivo veritas. J Clin Invest. 2012;122:787–95.
107. Na Y-R, Yoon Y-N, Son D-I, Seok S-H. Cyclooxygenase-2 inhibition blocks M2 macrophage differentiation and suppresses metastasis in murine breast cancer model. PLoS One. 2013;8:e63451.
108. MacKinnon AC, Farnworth SL, Hodkinson PS, Henderson NC, Atkinson KM, Leffler H, Nilsson UJ, Haslett C, Forbes SJ, Sethi T. Regulation of alternative macrophage activation by galectin-3. J Immunol. 2008;180:2650–8.
109. Röszer T, Menendez-Gutierrez MP, Lefterova MI, Alameda D, Nunez V, Lazar MA, Fischer T, Ricote M. Autoimmune kidney disease and impaired engulfment of apoptotic cells in mice with macrophage peroxisome proliferator-activated receptor gamma or retinoid X receptor alpha deficiency. J Immunol. 2011;186:621–31.
110. Nadella V, Wang Z, Johnson TS, Griffin M, Devitt A. Transglutaminase 2 interacts with syndecan-4 and CD44 at the surface of human macrophages to promote removal of apoptotic cells. Biochim Biophys Acta. 2015;1853:201–12.
111. Martinez FO, Gordon S. The M1 and M2 paradigm of macrophage activation: time for reassessment. F1000Prime Rep. 2014;6:13.
112. Ampem G, Azegrouz H, Bacsadi A, Balogh L, Schmidt S, Thuroczy J, Roszer T. Adipose tissue macrophages in non-rodent mammals: a comparative study. Cell Tissue Res. 2016;363:461–78.
113. Karagianni AE, Kapetanovic R, Summers KM, McGorum BC, Hume DA, Pirie RS. Comparative transcriptome analysis of equine alveolar macrophages. Equine Vet J. 2017;49:375–82.
114. Hirosawa N, Uchida K, Kuniyoshi K, Murakami K, Inoue G, Miyagi M, Matsuura Y, Orita S, Inage K, Suzuki T, Takaso M, Ohtori S. Vein wrapping promotes M2 macrophage polarization in a rat chronic constriction injury model. J Orthop Res. 2018;36:2210–7.
115. Smith KA, Pearson CB, Hachey AM, Xia DL, Wachtman LM. Alternative activation of macrophages in rhesus macaques (*Macaca mulatta*) with endometriosis. Comp Med. 2012;62:303–10.
116. Chen C, Perry TL, Chitko-McKown CG, Smith AD, Cheung L, Beshah E, Urban JF, Dawson HD. The regulatory actions of retinoic acid on M2 polarization of porcine macrophages. Dev Comp Immunol. 2019;98:20–33.

117. Herrmann I, Gotovina J, Fazekas-Singer J, Fischer MB, Hufnagl K, Bianchini R, Jensen-Jarolim E. Canine macrophages can like human macrophages be in vitro activated toward the M2a subtype relevant in allergy. Dev Comp Immunol. 2018;82:118–27.
118. Heinrich F, Lehmbecker A, Raddatz B, Kegler K, Tipold A, Stein V, Kalkuhl A, Deschl U, Baumgärtner W, Ulrich R, Spitzbarth I. Morphologic, phenotypic, and transcriptomic characterization of classically and alternatively activated canine blood-derived macrophages in vitro. PLoS One. 2017;12:e0183572.
119. Beirão BCB, Raposo T, Pang LY, Argyle DJ. Canine mammary cancer cells direct macrophages toward an intermediate activation state between M1/M2. BMC Vet Res. 2015;11:151.
120. Röszer T, Menendez-Gutierrez MP, Cedenilla M, Ricote M. Retinoid X receptors in macrophage biology. Trends Endocrinol Metab. 2013;24(9):460–8.
121. Gonzalez AN, Hidalgo A. Nuclear receptors and clearance of apoptotic cells: stimulating the macrophage's appetite. Front Immunol. 2014;5:211.
122. Gonzalez AN, Bensinger SJ, Hong C, Beceiro S, Bradley MN, Zelcer N, Deniz J, Ramirez C, Diaz M, Gallardo G, de Galarreta CR, Salazar J, Lopez F, Edwards P, Parks J, Andujar M, Tontonoz P, Castrillo A. Apoptotic cells promote their own clearance and immune tolerance through activation of the nuclear receptor LXR. Immunity. 2009;31:245–58.
123. Sarang Z, Garabuczi E, Joos G, Kiss B, Toth K, Ruhl R, Szondy Z. Macrophages engulfing apoptotic thymocytes produce retinoids to promote selection, differentiation, removal and replacement of double positive thymocytes. Immunobiology. 2013;218:1354–60.
124. Hevener AL, Olefsky JM, Reichart D, Nguyen MT, Bandyopadyhay G, Leung HY, Watt MJ, Benner C, Febbraio MA, Nguyen AK, Folian B, Subramaniam S, Gonzalez FJ, Glass CK, Ricote M. Macrophage PPAR gamma is required for normal skeletal muscle and hepatic insulin sensitivity and full antidiabetic effects of thiazolidinediones. J Clin Invest. 2007;117:1658–69.
125. Menendez-Gutierrez MP, Röszer T, Ricote M. Biology and therapeutic applications of peroxisome proliferator-activated receptors. Curr Top Med Chem. 2012;12:548–84.
126. Fujisaka S, Usui I, Bukhari A, Ikutani M, Oya T, Kanatani Y, Tsuneyama K, Nagai Y, Takatsu K, Urakaze M, Kobayashi M, Tobe K. Regulatory mechanisms for adipose tissue M1 and M2 macrophages in diet-induced obese mice. Diabetes. 2009;58:2574–82.
127. Matsutani T, Tamura K, Kutsukake M, Matsuda A, Tachikawa E, Uchida E. Impact of pioglitazone on macrophage dynamics in adipose tissues of cecal ligation and puncture-treated mice. Biol Pharm Bull. 2017;40:638–44.
128. Davies SS, Pontsler AV, Marathe GK, Harrison KA, Murphy RC, Hinshaw JC, Prestwich GD, Hilaire AS, Prescott SM, Zimmerman GA, McIntyre TM. Oxidized alkyl phospholipids are specific, high affinity peroxisome proliferator-activated receptor gamma ligands and agonists. J Biol Chem. 2001;276:16015–23.
129. Tian Y, Yang C, Yao Q, Qian L, Liu J, Xie X, Ma W, Nie X, Lai B, Xiao L, Wang N. Procyanidin B2 activates PPARγ to induce M2 polarization in mouse macrophages. Front Immunol. 2019;10:1895–1895.
130. Kiss-Toth E, Röszer T. PPARgamma in kidney physiology and pathophysiology. PPAR Res. 2008;2008:183108.
131. Fuentes L, Röszer T, Ricote M. Inflammatory mediators and insulin resistance in obesity: role of nuclear receptor signaling in macrophages. Mediat Inflamm. 2010;2010:219583.
132. Goldstein JT, Berger AC, Shih J, Duke FF, Furst L, Kwiatkowski DJ, Cherniack AD, Meyerson M, Strathdee CA. Genomic activation of PPARG reveals a candidate therapeutic axis in bladder cancer. Cancer Res. 2017;77:6987–98.
133. Odegaard JI, Ricardo-Gonzalez RR, Red Eagle A, Vats D, Morel CR, Goforth MH, Subramanian V, Mukundan L, Ferrante AW, Chawla A. Alternative M2 activation of Kupffer cells by PPARdelta ameliorates obesity-induced insulin resistance. Cell Metab. 2008;7:496–507.

134. Kim S-H, Hong SH, Park Y-J, Sung J-H, Suh W, Lee KW, Jung K, Lim C, Kim J-H, Kim H, Park KS, Park SG. MD001, a novel peroxisome proliferator-activated receptor α/γ agonist, improves glucose and lipid metabolism. Sci Rep. 2019;9:1656.
135. Rebe C, Raveneau M, Chevriaux A, Lakomy D, Sberna AL, Costa A, Bessede G, Athias A, Steinmetz E, Lobaccaro JM, Alves G, Menicacci A, Vachenc S, Solary E, Gambert P, Masson D. Induction of transglutaminase 2 by a liver X receptor/retinoic acid receptor alpha pathway increases the clearance of apoptotic cells by human macrophages. Circ Res. 2009;105:393–401.
136. Calkin AC, Tontonoz P. Transcriptional integration of metabolism by the nuclear sterol-activated receptors LXR and FXR. Nat Rev Mol Cell Biol. 2012;13:213.
137. Bijsmans ITGW, Guercini C, Ramos Pittol JM, Omta W, Milona A, Lelieveld D, Egan DA, Pellicciari R, Gioiello A, van Mil SWC. The glucocorticoid mometasone furoate is a novel FXR ligand that decreases inflammatory but not metabolic gene expression. Sci Rep. 2015;5:14086.
138. Fiorucci S, Biagioli M, Zampella A, Distrutti E. Bile acids activated receptors regulate innate immunity. Front Immunol. 2018;9:1853.
139. Leussink S, Aranda-Pardos I, A-Gonzalez N. Lipid metabolism as a mechanism of immunomodulation in macrophages: the role of liver X receptors. Curr Opin Pharmacol. 2020;53:18–26.
140. Liu D, Ahmet A, Ward L, Krishnamoorthy P, Mandelcorn ED, Leigh R, Brown JP, Cohen A, Kim H. A practical guide to the monitoring and management of the complications of systemic corticosteroid therapy. Allergy Asthma Clin Immunol. 2013;9:30.
141. Hofkens W, Storm G, van den Berg W, van Lent P. Inhibition of M1 macrophage activation in favour of M2 differentiation by liposomal targeting of glucocorticoids to the synovial lining during experimental arthritis. Ann Rheum Dis. 2011;70:A40.
142. Das LM, Binko AM, Traylor ZP, Peng H, Lu KQ. Vitamin D improves sunburns by increasing autophagy in M2 macrophages. Autophagy. 2019;15:813–26.
143. Miyachi Y, Tsuchiya K, Shiba K, Mori K, Komiya C, Ogasawara N, Ogawa Y. A reduced M1-like/M2-like ratio of macrophages in healthy adipose tissue expansion during SGLT2 inhibition. Sci Rep. 2018;8:16113.
144. Shaw TJ, Martin P. Wound repair at a glance. J Cell Sci. 2009;122:3209–13.
145. Mirza R, DiPietro LA, Koh TJ. Selective and specific macrophage ablation is detrimental to wound healing in mice. Am J Pathol. 2009;175:2454–62.
146. Simkin J, Sammarco MC, Marrero L, Dawson LA, Yan M, Tucker C, Cammack A, Muneoka K. Macrophages are required to coordinate mouse digit tip regeneration. Development (Cambridge). 2017;144:3907–16.
147. Krzyszczyk P, Schloss R, Palmer A, Berthiaume F. The role of macrophages in acute and chronic wound healing and interventions to promote pro-wound healing phenotypes. Front Physiol. 2018;9:419–419.
148. Malissen B, Tamoutounour S, Henri S. The origins and functions of dendritic cells and macrophages in the skin. Nat Rev Immunol. 2014;14:417–28.
149. Röszer T. Inflammation as death or life signal in diabetic fracture healing. Inflamm Res. 2011;60:3–10.
150. Forbes SJ, Rosenthal N. Preparing the ground for tissue regeneration: from mechanism to therapy. Nat Med. 2014;20:857–69.
151. Kaikita K, Hayasaki T, Okuma T, Kuziel WA, Ogawa H, Takeya M. Targeted deletion of CC chemokine receptor 2 attenuates left ventricular remodeling after experimental myocardial infarction. Am J Pathol. 2004;165:439–47.
152. Ploeger DTA, Hosper NA, Schipper M, Koerts JA, de Rond S, Bank RA. Cell plasticity in wound healing: paracrine factors of M1/M2 polarized macrophages influence the phenotypical state of dermal fibroblasts. Cell Commun Signal. 2013;11:29.
153. Ley K. M1 means kill; M2 means heal. J Immunol. 2017;199:2191–3.

154. Minutti CM, Knipper JA, Allen JE, Zaiss DMW. Tissue-specific contribution of macrophages to wound healing. Semin Cell Dev Biol. 2017;61:3–11.
155. O'Rourke SA, Dunne A, Monaghan MG. The role of macrophages in the infarcted myocardium: orchestrators of ECM remodeling. Front Cardiovasc Med. 2019;6:101–101.
156. Chamberlain CS, Clements AEB, Kink JA, Choi U, Baer GS, Halanski MA, Hematti P, Vanderby R. Extracellular vesicle-educated macrophages promote early Achilles tendon healing. Stem Cells. 2019;37:652–62.
157. Duffield JS, Forbes SJ, Constandinou CM, Clay S, Partolina M, Vuthoori S, Wu S, Lang R, Iredale JP. Selective depletion of macrophages reveals distinct, opposing roles during liver injury and repair. J Clin Invest. 2005;115:56–65.
158. Ferrante CJ, Leibovich SJ. Regulation of macrophage polarization and wound healing. Adv Wound Care. 2012;1:10–6.
159. Röszer T, Jozsa T, Kiss-Toth ED, De Clerck N, Balogh L. Leptin receptor deficient diabetic (db/db) mice are compromised in postnatal bone regeneration. Cell Tissue Res. 2014;356:195–206.
160. Minutti CM, Jackson-Jones LH, García-Fojeda B, Knipper JA, Sutherland TE, Logan N, Ringqvist E, Guillamat-Prats R, Ferenbach DA, Artigas A, Stamme C, Chroneos ZC, Zaiss DM, Casals C, Allen JE. Local amplifiers of IL-4Rα-mediated macrophage activation promote repair in lung and liver. Science. 2017;356:1076–80.
161. Kotwal GJ, Chien S. Macrophage differentiation in normal and accelerated wound healing. Results Probl Cell Differ. 2017;62:353–64.
162. Dai Y, Zhao X, Chen P, Yu Y, Wang Y, Xie L. Neuropeptide FF promotes recovery of corneal nerve injury associated with hyperglycemia. Invest Ophthalmol Vis Sci. 2015;56:7754–65.
163. Chiang B, Essick E, Ehringer W, Murphree S, Hauck MA, Li M, Chien S. Enhancing skin wound healing by direct delivery of intracellular adenosine triphosphate. Am J Surg. 2007;193:213–8.
164. He X, Dong Z, Cao Y, Wang H, Liu S, Liao L, Jin Y, Yuan L, Li B. MSC-derived exosome promotes M2 polarization and enhances cutaneous wound healing. Stem Cells Int. 2019;2019:7132708.
165. Kämpfer H, Pfeilschifter J, Frank S. Expression and activity of arginase isoenzymes during normal and diabetes-impaired skin repair. J Invest Dermatol. 2003;121:1544–51.
166. Ho YY, Lagares D, Tager AM, Kapoor M. Fibrosis—a lethal component of systemic sclerosis. Nat Rev Rheumatol. 2014;10:390–402.
167. Laurent P, Sisirak V, Lazaro E, Richez C, Duffau P, Blanco P, Truchetet M-E, Contin-Bordes C. Innate immunity in systemic sclerosis fibrosis: recent advances. Front Immunol. 2018;9:1702–1702.
168. He S, Xie L, Lu J, Sun S. Characteristics and potential role of M2 macrophages in COPD. Int J Chron Obstruct Pulmon Dis. 2017;12:3029–39.
169. Mueller TD, Zhang J-L, Sebald W, Duschl A. Structure, binding, and antagonists in the IL-4/IL-13 receptor system. Biochim Biophys Acta Mol Cell Res. 2002;1592:237–50.
170. Tomkinson A, Duez C, Cieslewicz G, Pratt JC, Joetham A, Shanafelt M-C, Gundel R, Gelfand EW. A murine IL-4 receptor antagonist that inhibits IL-4- and IL-13-induced responses prevents antigen-induced airway eosinophilia and airway hyperresponsiveness. J Immunol. 2001;166:5792–800.
171. Antoniu SA. Pitrakinra, a dual IL-4/IL-13 antagonist for the potential treatment of asthma and eczema. Curr Opin Investig Drugs. 2010;11:1286–94.
172. Bagnasco D, Ferrando M, Varricchi G, Passalacqua G, Canonica GW. A critical evaluation of anti-IL-13 and anti-IL-4 strategies in severe asthma. Int Arch Allergy Immunol. 2016;170:122–31.
173. Pennington EJ, Wechsler ME, Ortega VE. Chapter 9. Pharmacogenomics and applications to asthma management. In: Szefler SJ, Holguin F, Wechsler ME, editors. Personalizing asthma management for the clinician. St. Louis: Elsevier; 2018. p. 97–112.

174. Kau AL, Korenblat PE. Anti-interleukin 4 and 13 for asthma treatment in the era of endotypes. Curr Opin Allergy Clin Immunol. 2014;14:570–5.
175. Simpson EL, Flohr C, Eichenfield LF, Bieber T, Sofen H, Taïeb A, Owen R, Putnam W, Castro M, DeBusk K, Lin C-Y, Voulgari A, Yen K, Omachi TA. Efficacy and safety of lebrikizumab (an anti-IL-13 monoclonal antibody) in adults with moderate-to-severe atopic dermatitis inadequately controlled by topical corticosteroids: a randomized, placebo-controlled phase II trial (TREBLE). J Am Acad Dermatol. 2018;78:863–871.e811.
176. Jiménez-García L, Higueras MÁ, Herranz S, Hernández-López M, Luque A, de las Heras B, Hortelano S. A hispanolone-derived diterpenoid inhibits M2-macrophage polarization in vitro via JAK/STAT and attenuates chitin induced inflammation in vivo. Biochem Pharmacol. 2018;154:373–83.
177. Sahoo A, Alekseev A, Obertas L, Nurieva R. Grail controls Th2 cell development by targeting STAT6 for degradation. Nat Commun. 2014;5:4732.
178. Song G, Liu B, Li Z, Wu H, Wang P, Zhao K, Jiang G, Zhang L, Gao C. E3 ubiquitin ligase RNF128 promotes innate antiviral immunity through K63-linked ubiquitination of TBK1. Nat Immunol. 2016;17:1342–51.
179. Wei C-Y, Zhu M-X, Yang Y-W, Zhang P-F, Yang X, Peng R, Gao C, Lu J-C, Wang L, Deng X-Y, Lu N-H, Qi F-Z, Gu J-Y. Downregulation of RNF128 activates Wnt/β-catenin signaling to induce cellular EMT and stemness via CD44 and CTTN ubiquitination in melanoma. J Hematol Oncol. 2019;12:21.
180. Lee Y-Y, Wang C-T, Huang SK-H, Wu W-J, Huang C-N, Li C-C, Chan T-C, Liang P-I, Hsing C-H, Li C-F. Downregulation of RNF128 predicts progression and poor prognosis in patients with urothelial carcinoma of the upper tract and urinary bladder. J Cancer. 2016;7:2187–96.
181. Seroogy CM, Soares L, Ranheim EA, Su L, Holness C, Bloom D, Fathman CG. The gene related to anergy in lymphocytes, an E3 ubiquitin ligase, is necessary for anergy induction in CD4 T cells. J Immunol. 2004;173:79–85.
182. Su LL, Iwai H, Lin JT, Fathman CG. The transmembrane E3 ligase GRAIL ubiquitinates and degrades CD83 on CD4 T cells. J Immunol. 2009;183:438–44.
183. Gao J, Wang Y, Yang J, Zhang W, Meng K, Sun Y, Li Y, He Q-Y. RNF128 promotes invasion and metastasis via the EGFR/MAPK/MMP-2 pathway in esophageal squamous cell carcinoma. Cancers (Basel). 2019;11(6):840.
184. Ruffell B, Chang-Strachan D, Chan V, Rosenbusch A, Ho CMT, Pryer N, Daniel D, Hwang ES, Rugo HS, Coussens LM. Macrophage IL-10 blocks $CD8^+$ T cell-dependent responses to chemotherapy by suppressing IL-12 expression in intratumoral dendritic cells. Cancer Cell. 2014;26:623–37.
185. Mocellin S, Panelli MC, Wang E, Nagorsen D, Marincola FM. The dual role of IL-10. Trends Immunol. 2003;24:36–43.
186. Kundu N, Beaty TL, Jackson MJ, Fulton AM. Antimetastatic and antitumor activities of interleukin 10 in a murine model of breast cancer. JNCI J Natl Cancer Inst. 1996;88:536–41.
187. Elise FS, Rosemary JA. TGF beta inhibition for cancer therapy. Curr Cancer Drug Targets. 2006;6:565–78.
188. de Gramont A, Faivre S, Raymond E. Novel TGF-β inhibitors ready for prime time in onco-immunology. Onco Targets Ther. 2016;6:e1257453–e1257453.
189. Mills CD, Kincaid K, Alt JM, Heilman MJ, Hill AM. M-1/M-2 macrophages and the Th1/Th2 paradigm. J Immunol. 2000;164:6166–73.
190. Bierie B, Moses HL. Transforming growth factor beta (TGF-beta) and inflammation in cancer. Cytokine Growth Factor Rev. 2010;21:49–59.
191. Hu G, Guo M, Xu J, Wu F, Fan J, Huang Q, Yang G, Lv Z, Wang X, Jin Y. Nanoparticles targeting macrophages as potential clinical therapeutic agents against cancer and inflammation. Front Immunol. 2019;10:1998–1998.
192. Suriano F, Santini D, Perrone G, Amato M, Vincenzi B, Tonini G, Muda A, Boggia S, Buscarini M, Pantano F. Tumor associated macrophages polarization dictates the efficacy

of BCG instillation in non-muscle invasive urothelial bladder cancer. J Exp Clin Cancer Res CR. 2013;32:87.
193. Lacerda Mariano L, Ingersoll MA. Bladder resident macrophages: mucosal sentinels. Cell Immunol. 2018;330:136–41.
194. Bose D, Banerjee S, Chatterjee N, Das S, Saha M, Saha KD. Inhibition of TGF-β induced lipid droplets switches M2 macrophages to M1 phenotype. Toxicol In Vitro. 2019;58:207–14.
195. Runtsch MC, O'Neill L. Itaconate inhibits alternative activation of macrophages by targeting Janus kinase 1. J Immunol. 2019;202:–58.11.
196. Li G, Liu D, Kimchi ET, Kaifi JT, Qi X, Manjunath Y, Liu X, Deering T, Avella DM, Fox T, Rockey DC, Schell TD, Kester M, Staveley-O'Carroll KF. Nanoliposome C6-ceramide increases the anti-tumor immune response and slows growth of liver tumors in mice. Gastroenterology. 2018;154:1024–1036. e1029.
197. Samper A, Herrera B. A study of the roughness of Gothic rose windows. Nexus Netw J. 2016;18:397–417.
198. Duvernoy S. Baroque oval churches: innovative geometrical patterns in early modern sacred architecture. Nexus Netw J. 2015;17:425–56.
199. Gasteiger G, D'Osualdo A, Schubert DA, Weber A, Bruscia EM, Hartl D. Cellular innate immunity: an old game with new players. J Innate Immun. 2017;9:111–25.
200. DeNardo DG, Ruffell B. Macrophages as regulators of tumour immunity and immunotherapy. Nat Rev Immunol. 2019;19:369–82.
201. Lackey DE, Olefsky JM. Regulation of metabolism by the innate immune system. Nat Rev Endocrinol. 2016;12:15–28.
202. Sang Y, Miller LC, Blecha F. Macrophage polarization in virus–host interactions. J Clin Cell Immunol. 2015;6(2):311.
203. Muraille E, Leo O, Moser M. TH1/TH2 paradigm extended: macrophage polarization as an unappreciated pathogen-driven escape mechanism? Front Immunol. 2014;5:603.

Correction to: Practical Approaches in M2 Macrophage Biology: Analysis, Pharmacology, and Didactical Interpretation of M2 Macrophage Functions

Correction to:
Chapter 10 in: T. Röszer, *The M2 Macrophage*, Progress in Inflammation Research 86, https://doi.org/10.1007/978-3-030-50480-9_10

The original version of chapter 10 in this book was inadvertently published with incomplete captions of Figs. 6 and 7. The figure captions have now been corrected – image courtesy details are added.

The updated online version of the chapter can be found at https://doi.org/10.1007/978-3-030-50480-9_10

T. Röszer, *The M2 Macrophage*, Progress in Inflammation Research 86,
https://doi.org/10.1007/978-3-030-50480-9_11

Correction to: The M2 Macrophage

Correction to:
T. Röszer, *The M2 Macrophage*, Progress in Inflammation Research 86, https://doi.org/10.1007/978-3-030-50480-9

This book published under the series "Progress in Inflammation Research" was inadvertently published with the incorrect series ISSN number. This has now been amended in the book to the correct ISSN:

Print ISSN: 1422-7746

Electronic ISSN: 2296-4525

The updated version of the book can be found at https://doi.org/10.1007/978-3-030-50480-9

T. Röszer, *The M2 Macrophage*, Progress in Inflammation Research 86,
https://doi.org/10.1007/978-3-030-50480-9_12

The manufacturer's authorised representative in the EU is Springer Nature Customer Service Centre GmbH, Europaplatz 3, 69115 Heidelberg, Germany. If you have any concerns regarding our products, please contact ProductSafety@springernature.com

Printed and bound by CPI Group (UK) Ltd, Croydon, CR0 4YY
30/07/2026
02181734-0001